MW00737555

Estado da implementação da gestão integrada dos recursos hídricos

Dieudonné Zerbo

Estado da implementação da gestão integrada dos recursos hídricos

na sub-bacia hidrográfica do Alto Como no Burkina Faso

ScienciaScripts

Imprint
Any brand names and product names mentioned in this book are subject to trademark, brand or patent protection and are trademarks or registered trademarks of their respective holders. The use of brand names, product names, common names, trade names, product descriptions etc. even without a particular marking in this work is in no way to be construed to mean that such names may be regarded as unrestricted in respect of trademark and brand protection legislation and could thus be used by anyone.

Cover image: www.ingimage.com

This book is a translation from the original published under ISBN 978-620-2-27500-2.

Publisher:
Sciencia Scripts
is a trademark of
Dodo Books Indian Ocean Ltd. and OmniScriptum S.R.L publishing group

120 High Road, East Finchley, London, N2 9ED, United Kingdom
Str. Armeneasca 28/1, office 1, Chisinau MD-2012, Republic of Moldova, Europe

ISBN: 978-620-5-81772-8

Índice

Dedicação

A Jesus Cristo, meu Senhor e Salvador, em quem encontro recusa, força e perseverança;

Em memória do meu pai, o falecido Emile ZERBO, que nos deixou enquanto eu estava a dar os meus primeiros passos na escola;

À minha mãe, Fakié ZERBO/HEMA, que nunca deixou de lutar por mim;

Ao meu irmão Mathias ZERBO que estava muito empenhado nos meus estudos;

À minha amada esposa Rachidatou e ao meu filho Anael Jonathan pelo seu apoio, paciência e ajuda incalculável;

Aos meus tios, tias, irmãos e irmãs e a todos aqueles que não pouparam esforços para me apoiarem durante esta luta.

Agradecimentos

A realização deste trabalho de fim de estudo foi possível graças ao apoio de várias pessoas e actores a quem gostaríamos de agradecer.

Agradecimentos especiais vão para :

- **Bernard TYCHON e Charles BIELDERS,** respectivos coordenadores do mestrado em Ciências Ambientais e Gestão nos Países em Desenvolvimento na Universidade de Liège (ULg) e na Universidade Católica de Lovaina (UCL), por nos terem aceite para esta formação;
- **Fundação Rei Baudouin** por nos permitir regressar ao Burkina Faso para recolher dados através do financiamento da viagem;
- **O Dr Joost WELLENS, o** nosso promotor, que aceitou supervisionar-nos, apesar das suas numerosas actividades de investigação, interessou-se muito pela nossa formação e pelo mundo científico. A sua simplicidade e os seus preciosos conselhos foram de grande interesse para o sucesso do nosso trabalho;
- **Sié PALE**, doutorando da Universidade de Liège e leitor do nosso trabalho pela confiança que depositou em nós, confiando-nos este trabalho e a sua orientação ao longo das nossas actividades de campo e durante a reacção da tese;
- **Dr Farid TRAORE,** investigador do Centro Nacional de Investigação Científica e Tecnológica (CNRST/Burkina) a quem chamo "koro Farid" pela vossa simpatia, orientação e orientações. Apesar da distância, o vosso apoio multiforme tem sido um grande contributo para o sucesso do nosso trabalho;
- **A todo o pessoal da DRAAH, DPAAH, AEC, SN SOSUCO, ONEA, CLE-HC, aos valentes produtores** e particularmente a **Ferdinand SAWADOGO,** Secretário-Geral da CLE Haute-Comoé, **Victorien SAWADOGO,** Director de Água e Ambiente da AEC e **Fousseni SANOU**, PADI DRAAH-Cascades ponto focal da sua disponibilidade, do seu encorajamento e da sua simpatia;
- **Os meus amigos Arnaud, Orelien, Hamidou, todos os meus colegas e estagiários (Delphine e Souzan)** pelo agradável ambiente de trabalho que criaram à minha volta e pela sua franca colaboração durante a formação;
- Finalmente, gostaria de agradecer a toda a família ZERBO que me apoiou durante este período difícil e laborioso,

A todos aqueles cujos nomes não puderam ser mencionados pelo seu apoio multifacetado, que encontrem aqui a expressão da nossa profunda gratidão.

Resumo

A água é um recurso finito, precioso e que sustenta a vida. No entanto, está no centro de muitas exigências e conflitos, tornando-a frágil a factores externos e à degradação dos ecossistemas. Uma boa gestão da água que seja integrada e concertada é, portanto, uma necessidade, especialmente a nível da bacia hidrográfica.

A sub-bacia de Comoé Superior, localizada no sudoeste do Burkina Faso, tem um elevado potencial para recursos hídricos que estão no centro de várias utilizações (agrícola, doméstica e industrial). A complexidade de tal gestão levou à adopção da abordagem de gestão integrada dos recursos hídricos na região.

O objectivo do nosso trabalho era fazer um balanço da implementação da política de GIRH na sub-bacia.

Foi adoptada uma metodologia baseada principalmente em entrevistas e observações para fornecer informações sobre a GIRH, o funcionamento dos sistemas de gestão da água e a implementação da Contribuição Financeira da Água (WFC) na área de estudo.

Os resultados do estudo mostram que a operacionalização da política de GIRH na sub-bacia foi alcançada através da criação de instituições e organismos de gestão da água a diferentes escalas (bacia, regional e local) e através do reforço da capacidade das administrações públicas envolvidas no sector da água. Assim, a gestão dos recursos hídricos é realizada ao nível mais baixo possível e é quase inteiramente apoiada pelos utilizadores, indicando um certo desempenho do sistema.

O estudo revelou também a presença de um sistema de tarifação da água baseado no princípio do "utilizador paga", que se baseia essencialmente na existência de textos legais que estabelecem os direitos e impostos sobre a utilização da água. Os pontos fortes do sistema de financiamento residem nos regulamentos e no nível de organização dos actores, mas a taxa de cobrança do CFE permanece baixa (cerca de 24%) desde a sua implementação.

Por conseguinte, há ainda enormes desafios a enfrentar na sub-bacia para melhorar a gestão dos recursos hídricos.

Palavras-chave: gestão da água, contribuição financeira para a água, sub-captação Haute Comoé

Siglas e abreviaturas

Siglas e abreviaturas	Significado
AEC	Agência de Água de Cascatas
AEDE	Associação para o Desenvolvimento da Água e Ambiente
AEPS	Abastecimento simplificado de água potável
ASDI	Agência Sueca de Cooperação Internacional para o Desenvolvimento
BAD	Banco Africano de Desenvolvimento
BTP	Construção Civil e Obras Públicas
CFE	Contribuição financeira para a água
CLE-HC	Comité Locau de l'Eau de la Haute Comoé
DANIDA	Agência Dinamarquesa de Desenvolvimento Internacional
DPAAH	Direcção Provincial de Agricultura e Desenvolvimento Hidráulico
DRAAH	Direcção Regional de Agricultura e Desenvolvimento Hidráulico
IWRM	Gestão integrada dos recursos hídricos
M	milhões
Biliões	biliões
ONEA	Conselho Nacional da Água e Saneamento
PADI	Programa de apoio para o desenvolvimento da agricultura irrigada
PAGIRE	Programa de apoio para a implementação da gestão integrada dos recursos hídricos
PEM	Ponto de Água Moderno
PIB	Produto Interno Bruto
PLB	Produto Local Bruto
Qrest	fluxo sanitário do curso de água
RGPH	Censo Geral da População e Habitação
SAGE	Planos de desenvolvimento e gestão da água
SDAGE	Plano director para o desenvolvimento e gestão da água
SN-SOSUCO	Société Nouvelle- Société Sucrière de la Comoé
SOFITEX	Société des Fibres Textiles du Burkina
UCL	Universidade Católica de Leuven
ULg	Universidade de Liège

Introdução geral

Sendo um país Saheliano, a economia do Burkina Faso é altamente dependente da agricultura, que por sua vez depende da pluviosidade. De facto, o sector primário (agricultura, pecuária, silvicultura, vida selvagem e pesca) contribui entre 35% e 40% do PIB real e emprega mais de 84% da população activa (INSD, 2007). Quanto à agricultura, é o sector que emprega mais pessoas e é a principal fonte de rendimento para 88-90% da população activa no Burkina Faso (Câmara de Comércio, 2006). Além disso, o contexto no Burkina Faso é marcado por uma demografia cada vez maior, mesmo que as taxas de crescimento, com uma média de 2,3 durante a última década, não sejam as mais alarmantes na sub-região. Contudo, a este ritmo, a população atingirá 20 milhões até 2026 (MEF-SNAT, 2007), o que poderá levar a uma maior insegurança alimentar e pressão sobre os recursos naturais, incluindo os recursos hídricos. O problema dos recursos hídricos no Burkina Faso é marcado por uma tendência decrescente da precipitação, pressão sobre os recursos disponíveis por parte da população em constante crescimento, e alterações climáticas resultantes das actividades humanas, o que levou ao agravamento de certos fenómenos como a destruição de habitats aquáticos, erosão eólica e hídrica, eliminação de resíduos e várias formas de poluição (MAHRH, 2009). Segundo a mesma fonte, esta situação compromete a sustentabilidade do recurso e arrisca-se a pôr em risco as opções de desenvolvimento do país. Consciente deste problema, uma das prerrogativas do governo Burkinabe é tornar a agricultura menos dependente da chuva e controlar parcial ou totalmente as inundações a fim de se espalhar, diversificar, intensificar e assegurar a produção agrícola no tempo e no espaço. Para este fim, o governo Burkinabe está a concentrar-se na agricultura irrigada. Assim, em várias regiões do Burkina, iniciativas para promover a irrigação em pequena escala estão em expansão (Traoré et *al.*, 2013; Wellens, 2014), tais como na sub-bacia hidrográfica de Haute Comoé, na região de Cascades. Esta sub-bacia contém recursos hídricos importantes que estão divididos entre várias utilizações, nomeadamente para necessidades domésticas (abastecimento de água potável), necessidades agrícolas (irrigação) e necessidades industriais (produção industrial de cana-de-açúcar, extracção mineral) (WAIPRO-CILS-IWMI, 2010). Como as necessidades de água são diversas, existem carências crónicas de água para irrigação e conflitos entre utilizadores de água, tornando delicada a gestão do recurso a nível local. Consciente deste estado de coisas, o Burkina Faso tem estado empenhado desde a Cimeira do Milénio do Rio, em 1992, no processo de Gestão Integrada dos Recursos Hídricos (GIRH) através da adopção em 2003 de um Plano de Acção para a Gestão Integrada dos Recursos Hídricos

(PAGIRE) (MAHRH, 2003). Neste processo inovador, o Estado pretende tornar a GIRH operacional em todo o país. Se ao nível central, os instrumentos dedicados à implementação da GIRH funcionam relativamente bem, ao nível descentralizado, a abordagem da GIRH enfrenta muitos desafios. Este é o contexto do nosso estudo, que visa fornecer uma visão geral da implementação da **abordagem de GIRH e do seu funcionamento, particularmente na sub-bacia hidrográfica do Alto Comoé (Burkina Faso).**

> **Objectivo geral :**

O objectivo geral do nosso trabalho é contribuir para uma melhor abordagem local da gestão integrada dos recursos hídricos.

> **Objectivos específicos :**

Especificamente, isto envolverá :

- Determinar e avaliar o nível de operacionalização da GIRH a nível local;
- avaliar as práticas locais de gestão dos recursos hídricos ;
- e analisar a implementação das "Contribuições Financeiras da Água" a nível local e avaliar a eficácia/sustentabilidade deste sistema de financiamento.

- **Pressupostos do estudo :**

Presumimos que :

- A operacionalização da GIRH é eficaz na sub-bacia hidrográfica;
- As práticas locais de gestão de recursos hídricos são eficazes;
- a implementação das "Contribuições Financeiras da Água" a nível local é eficaz, eficiente e sustentável.

Este relatório final está dividido em quatro capítulos: o primeiro é dedicado a uma revisão da literatura, o segundo à metodologia utilizada para realizar o estudo, o terceiro aos resultados obtidos, e o quarto à interpretação dos resultados e à formulação de recomendações.

CAPÍTULO 1: REVISÃO DE LITERATURA

1.1. Informações gerais sobre o sítio do estudo

1.1.1. País

O Burkina Faso é um país sem litoral situado no coração da África Ocidental, cobrindo uma área de 274.000 km^2 . É limitado a noroeste pelo Mali, a nordeste pelo Níger e a sul pela Costa do Marfim, Gana, Togo e Benim.

De acordo com a divisão administrativa de 2001, o país tem 13 regiões e 45 províncias. O nosso estudo teve lugar na província de Comoé, a região de Cascades na parte sudoeste do país, que contém a sub-bacia hidrográfica da parte superior de Comoé.

Figura 1: Localização da região de Cascade (Fonte: INSD, 2011 http://bref24.com/burkina-faso-d-1960-nos-jours/)

1.1.2. Contexto sócio-económico da província de Comoé

- **População**

A população da província de Comoé é essencialmente rural. De facto, entre 1996 e 2006, a população urbana representava apenas 26% da província (INSD, 2007). De acordo com a mesma fonte, entre este período, a área registou uma taxa média de crescimento anual de 3,1%. De acordo com os resultados do RGPH, 2006, o sexo feminino está fortemente representado com aproximadamente 51,67% da população, devido principalmente aos movimentos migratórios de homens para países estrangeiros (Costa do Marfim, Gana, etc.).

- **Sector de produção**

A actividade económica da região baseia-se essencialmente no sector primário, mais particularmente na agricultura e pecuária, que emprega cerca de 91,7% da população activa (INSD, 2005) e contribui com mais de 48% para a criação de riqueza local, ou seja, o Produto Local Bruto (BPL), ou seja, cerca de 26 mil milhões de francos CFA (40 milhões de euros) em 2003 (UN HABITAT_ECOLOC/Banfora, 2003).

A região administrativa das Cascatas, em particular a província de Comoé, é uma zona muito bem irrigada, adequada para a agricultura alimentada pela chuva, mas também para a agricultura fora da estação. A principal produção na bacia de Comoé é constituída por culturas comerciais (algodão, cana-de-açúcar), seguida por culturas cerealíferas (arroz, milho, sorgo), outras culturas alimentares (inhame rainfed yams, feijão-frade, voandzou) e a horticultura de mercado. Existem cerca de 4.980 hectares de perímetro irrigado e muitas terras baixas desenvolvidas para a produção de arroz de sequeiro (MEE, 2001). Segundo a mesma fonte, a região tem ainda um elevado potencial de desenvolvimento de terras baixas (cerca de 3.157 ha em irrigação). Existe agora uma forte expansão do cultivo irrigado, tendo em conta o seu valor económico e o crescimento da população da província, o que não está isento de consequências para os recursos hídricos disponíveis.

As actividades pecuárias contribuem consideravelmente para a economia da zona, que beneficia de vantagens tais como a existência de uma grande biomassa e de rotas de transumância. É também favorecida pela posição fronteiriça da zona. Em termos de criação de riqueza regional, a pecuária gera 7,4 mil milhões de francos CFA, ou seja, quase 1,13 milhões de euros (INSD, 2007). Existem sistemas pecuários tradicionais e modernos. Isto leva a numerosos conflitos entre agricultores e pastores, e mesmo entre pastores sedentários e transumantes (MAHRH, 2003).

As actividades industriais e comerciais estão organizadas principalmente em torno da cidade de Banfora, a capital da região. Existem algumas unidades de transformação, nomeadamente a Nouvelle Société Sucrière de la Comoé (SN-SOSUCO), SOFITEX, Grands Moulins du Burkina (GMB) e a Société Industrielle de Transformation Industrielle de l'Anacarde du Burkina (SOTRIA-B). A SN-SOSUCO, a maior unidade industrial do Burkina Faso, explora cerca de 4.000 hectares de terra para a produção irrigada de cana-de-açúcar e representa por si só 63% do PIB regional do sector secundário (UN HABITAT_ECOLOC_Banfora, 2006). Com cerca de 1.600 empregados permanentes e mais de 2.000 trabalhadores agrícolas diários, a SN-SOSUCO é o 2º maior empregador nacional depois do estado de Burkinabe (MCA, 20012).

1.1.3. Contexto ambiental da área

A bacia internacional de Comoé é partilhada entre quatro países, principalmente o Burkina Faso e a Costa do Marfim e, em segundo lugar, com o Mali e o Gana (Figura 2). No Burkina Faso, a bacia de Comoé tem uma superfície de quase 17.620 km^2 localizada 86% na região de Cascades, 7% na região de Hauts Bassins e quase 7% na região do Sudoeste (MCA, 2012). A nossa área de interesse está localizada na bacia hidrográfica do Alto Comoé, que é uma sub-bacia do Burkinabe na parte do rio Comoé que forma a fronteira entre o Burkina Faso e a Costa do Marfim (Figura 3).

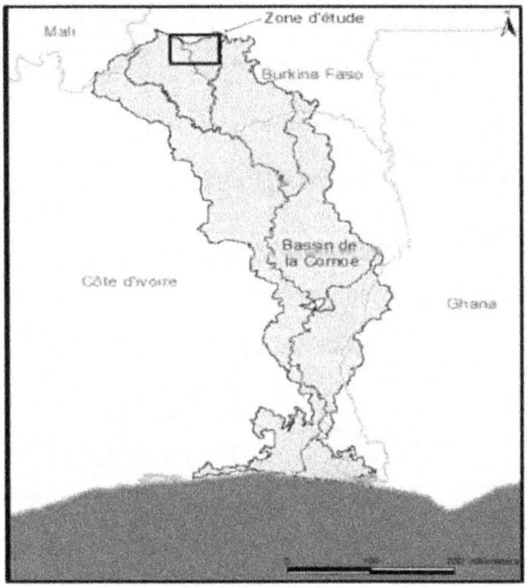

Figura 2: Localização da bacia hidrográfica de Comoé (Fonte: MCA, 2012)

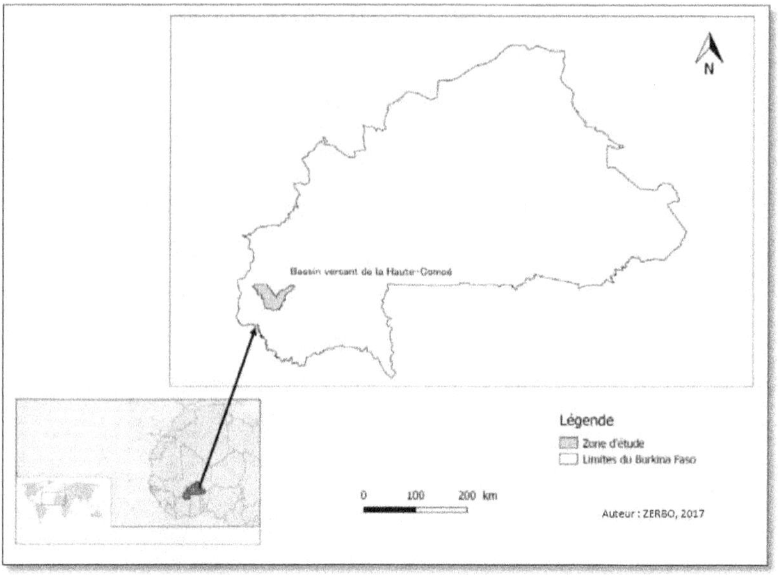

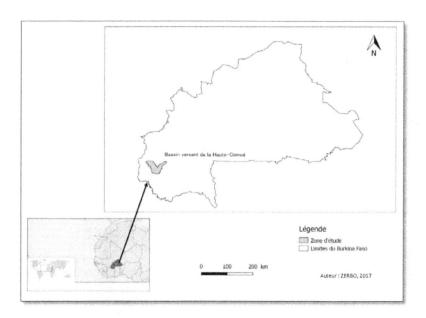

Figura 3: Localização da bacia hidrográfica superior de Comoé

- Clima

A sub-bacia de Comoé Superior tem um clima do Sul do Sudão caracterizado por duas estações principais: uma estação chuvosa (Abril a Outubro) com precipitações que podem exceder 1400 mm e uma estação seca (Novembro a Março) (Figura 4).

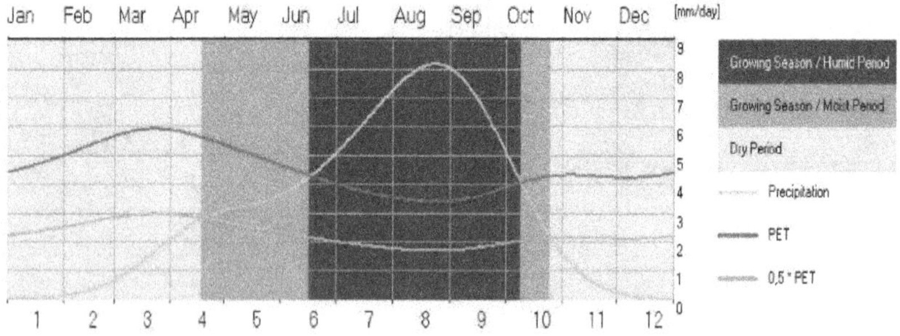

Figura 4: Ilustração das estações do ano através dos períodos favoráveis à vegetação (Fonte: FAO NewLocClim)

A Bacia de Comoé é a parte mais chuvosa do Burkina Faso. De acordo com os dados de precipitação para o período 1971 a 2011, a precipitação média anual é de 1032 mm, com decénios secos de menos de 800 mm e decénios húmidos de mais de 1400 mm registados na estação Banfora (VREO, 2010). No entanto, há uma grande variabilidade na precipitação ao longo do tempo (Figura 5), com uma tendência decrescente de tempos a tempos. No entanto, estudos realizados na bacia hidrográfica de Comoé mostraram que, apesar da tendência negativa, ainda existem oportunidades reais para mobilizar recursos hídricos e satisfazer as necessidades de água até 2025 (VREO, 2010).

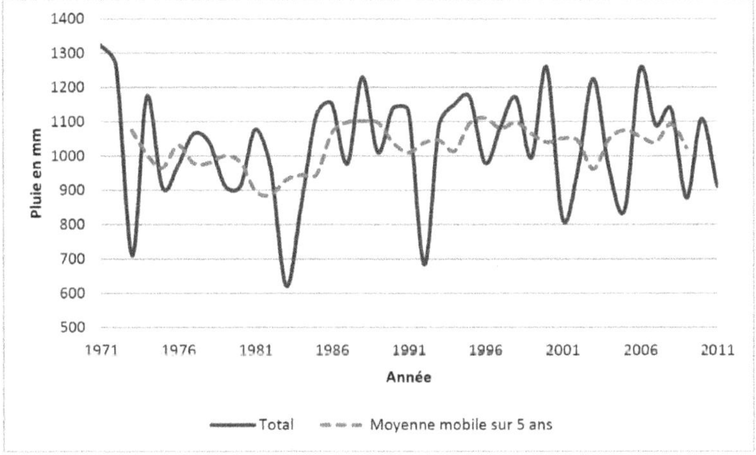

Figura 5: Evolução da precipitação média anual na estação de Berégadougou (Fonte, VREO, 2010)

As temperaturas médias anuais no Comoé variam entre os 17 e 36°C.

- **Hidrografia**

A província é drenada pelo rio Comoé e o seu afluente, o rio Yanon, numa bacia que cobre cerca de 16.810 km^2 . O Comoé é um rio perene muito importante, com cerca de 750 km de comprimento. Tem origem na parte norte da província de Comoé, nos departamentos de Bérégadougou, Banfora, Tiéfora e corre para sul, onde encontra a Léraba com a qual forma uma fronteira natural entre a província e a Costa do Marfim (Lankoandé e Sébégo, 2008). O Comoé e o seu afluente são regulados pelas barragens de Moussodougou, Lobi e Toussiana e constituem um imenso recurso para actividades agrícolas (irrigação), usos domésticos da água, criação de gado, pesca e turismo.

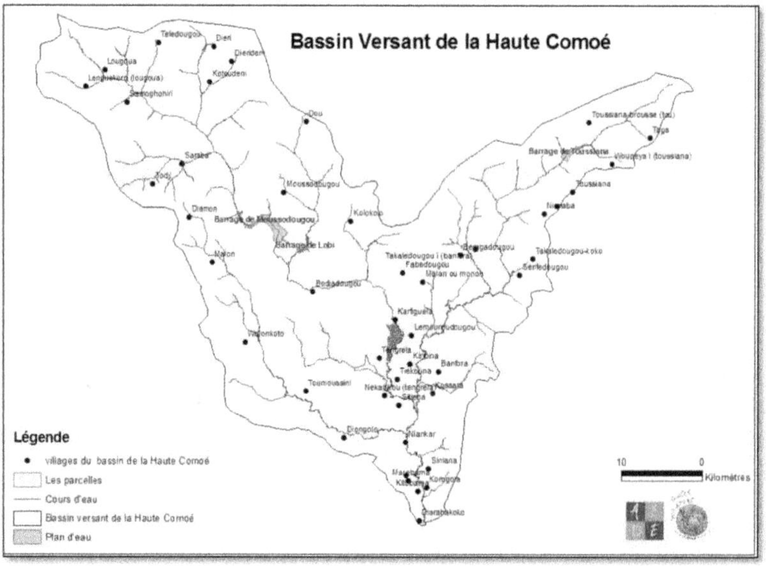

Figura 6: Rede hidrográfica da bacia do Alto Comoé (Fonte: AEDE, 2012)

- **Geologia e Hidrogeologia**

A bacia do Alto Comoé é estruturada por duas grandes zonas geológicas:

- A zona de cave no sul consiste em formações cristalinas, incluindo rochas graníticas e xistos sedimentares vulcânicos com a presença de aquíferos descontínuos (MCA, 2012);
- e a zona sedimentar a norte, constituída principalmente por formações de arenito em que são alojadas veias dolerite (MCA, 2012).

- **Topografia**

A bacia de Comoé é marcada por duas unidades topográficas: planícies e planaltos. A altitude varia entre 300 m e 700 m e todos os planaltos se inclinam para o sul (CRPA, 1994). As planícies, com uma altitude média entre 200 m e 300 m, são vastas e são atravessadas por grandes rios.

- **Pedologia**

A bacia é amplamente ocupada por solos arenosos a argilo-arenosos, ou solos mais profundos, ferruginosos, lixiviados, de cascalho profundo ou de média profundidade. Estes solos são geralmente de baixa fertilidade (Lankoande e Sebego, 2005).

No entanto, na parte noroeste da bacia, existem solos argilo-argilosos à superfície, e solos argilosos em profundidade, que são menos bem drenados e de baixa fertilidade. Os solos são bastante arenosos a argilo-arenosos com boa drenagem e de alta fertilidade para a parte sul (Lankoande e Sebego, 2005).

Nas zonas baixas dos principais rios, os solos sedosos aluviais são encontrados à superfície, com argila nas profundidades. Estes são solos pouco drenados com fertilidade dependente da petrografia da bacia.

Para áreas de afloramentos de armaduras e áreas de afloramentos de granito, são encontrados solos adequados para culturas de sequeiro e silvicultura.

A diversidade destes solos proporciona assim um enorme potencial para as actividades agrícolas na área.

1.2. Conceito de Gestão Integrada dos Recursos Hídricos (GIRH)

1.2.1. Definição de GIRH

A gestão integrada dos recursos hídricos é um conceito desenvolvido a nível internacional para dar respostas à crise da água e à complexidade da sua gestão, particularmente a sua governação. Este conceito, criado em 1992 como uma parceria entre nações e na sequência das conferências internacionais de Dublin e do Rio de Janeiro, foi aprovado e melhorado através de numerosas conferências internacionais (Marraquexe em 1997, Haia em 2000, Bona em 2001, Joanesburgo em 2002, Quioto em 2003, Cidade do México em 2006 e Istambul em 2009). De acordo com a Global Water Partnership (2000), a *GIRH é um processo que promove o desenvolvimento e a gestão coordenada da água, da terra e dos recursos associados, a fim de maximizar, de forma equitativa, o bem-estar económico e social resultante sem comprometer a sustentabilidade de ecossistemas vitais.*

A GIRH baseia-se, portanto, no facto de haver uma interacção entre os muitos usos diferentes da água. É uma abordagem holística e, portanto, implica que todos os níveis de utilização da água sejam reunidos e que as decisões relacionadas com a água tenham em conta os impactos de um sobre o outro a uma escala mínima de gestão da bacia hidrográfica.

A gestão integrada dos recursos hídricos é, portanto, um processo sistemático para o desenvolvimento sustentável, atribuição e monitorização da utilização dos recursos hídricos no contexto dos objectivos sociais, económicos e ambientais (Global Water Partnership, 2005). Por conseguinte, é diferente da gestão sectorial da água e deve ter em conta todos os elementos relevantes e associados, todos os sectores associados à partilha equitativa dos recursos hídricos e à sua utilização equilibrada, ecológica e sustentável.

1.2.2. Os princípios da GIRH

A GIRH baseia-se em quatro princípios fundamentais (Conferência de Dublin, 1992).

- ***Princípio 1: Abordagem integrada***

A água doce é um recurso escasso e vulnerável, essencial para a vida, o desenvolvimento e os ecossistemas. A água está, portanto, no centro de uma variedade de empregos e serviços. Em contraste com a gestão sectorial da água, a abordagem integrada é uma abordagem de gestão coordenada que tem em conta a variabilidade das necessidades, usos e todos os sectores de actividade afectados pela atribuição de água.

- ***Princípio 2: Abordagem participativa***

O desenvolvimento e a gestão da água devem basear-se numa abordagem participativa envolvendo utilizadores, gestores e decisores a todos os níveis.

- ***Princípio 3: O papel das mulheres***

As mulheres estão no centro dos mecanismos de abastecimento, gestão e conservação da água.

- ***Princípio 4: Valor económico***

Considerando as diferentes utilizações da água, a maioria das quais são competitivas, a água tem um carácter económico e deve, portanto, ser reconhecida como um bem económico.

Pelo seu valor económico, social e ambiental, os países precisam de instituir reformas políticas que promovam a gestão sustentável dos recursos hídricos, neste caso através da adopção de um "ambiente propício" para a GIRH (regulamentação, quadro jurídico, institucional e financeiro).

1.2.3. A introdução da GIRH no Burkina Faso

Há quase 15 anos que o Burkina Faso está empenhado na implementação de programas de gestão integrada dos recursos hídricos adaptados ao contexto nacional e de acordo com as recomendações internacionais sobre gestão sustentável e ecológica dos recursos hídricos.

As principais etapas desta reforma foram (PAGIRE, 2009):

- a elaboração do Documento sobre Política e Estratégia da Água em 1998;
- a elaboração da Lei de Orientação da Gestão da Água, adoptada pelo poder legislativo na sua sessão de 8 de Fevereiro de 2001;
- o desenvolvimento do Plano de Acção para a Gestão Integrada dos Recursos Hídricos (PAGIRE), adoptado pelo Estado Burkinabe em Março de 2003, que abrange duas fases (2003-2008 prorrogado até 2009, e 2010-2015);

O Burkina fez muitas realizações graças aos esforços para melhorar a gestão da água, nomeadamente através da criação de organismos de GIRH a diferentes níveis (Conselho Nacional da Água, Comité Técnico da Água, Agências de Água, Comités Locais da Água) e através da concepção e implementação da Contribuição Financeira para a Água (CFE).

Quanto ao CFE, nasceu da vontade política do Estado burquinês de fazer participar as populações beneficiárias nos custos de gestão do sistema de mobilização de recursos hídricos (MEE, 2001). Este sistema, no entanto, desenvolveu uma política fiscal para a água que se baseia principalmente em impostos e taxas sobre a água.

Todas estas medidas colocaram o Burkina Faso à frente de muitos países da África Ocidental em termos de política de implementação da GIRH.

Até 2030, o país criou um programa operacional sobre a água através do Programa Nacional de GIRH 2016 - 2030, cujos principais objectivos são apoiar e promover as realizações das fases anteriores, assegurar a protecção efectiva dos recursos hídricos e reforçar o envolvimento e a apropriação da GIRH pelos utilizadores do recurso e para todos os sectores de actividade relevantes (MEA, 2016).

CAPÍTULO 2: METODOLOGIA DE ESTUDO

2.1. Antecedentes do estudo

A bacia hidrográfica do rio Comoé está localizada na parte ocidental de África entre as longitudes 2,45° e 5,58° oeste e as latitudes 5,10° e 10,29° norte. A escolha do Comoé para realizar o presente estudo justifica-se pela importância dos recursos hídricos na área e pela complexidade da sua gestão devido à forte mobilização das partes interessadas para diferentes utilizações da água. De facto, Comoé é uma das áreas do Burkina Faso marcada pela intensidade da produção agrícola fora de época através do desenvolvimento da irrigação. Com a sua população crescente e multicultural, a Bacia Alta de Comoé encontra-se assim na encruzilhada de todas as utilizações para o desenvolvimento. A bacia de Comoé, incluindo a sub-bacia, contém cerca de 4.980 ha de perímetros irrigados e cerca de 1.880 ha de terras baixas desenvolvidas (MEE, 2001).

Os principais actores presentes são a Nouvelle Société Sucrière de la Comoé (SN SOSUCO) com cerca de 4.000 ha de plantações de cana de açúcar, o Office National de l'Eau et de l'Assainissement (ONEA) que abastece a cidade de Banfora e Bérégadougou com água doméstica, o perímetro irrigado de Karfiguela (350 ha) e os horticultores instalados ao longo dos cursos de água. Estes numerosos usos de água conduzem a uma grande pressão sobre as reservas de água na sub-bacia, a qual tem visto numerosos conflitos - por vezes violentos - entre os diferentes utilizadores dos pontos de água durante muitos anos. Os fluxos migratórios para a área aumentam a pressão sobre os recursos hídricos locais. Por exemplo, em 1995, a tensão entre os grupos envolvidos tornou-se extrema e levou à perda de vidas (YEO, 2008).

O nosso estudo visa uma melhor gestão dos recursos hídricos para um desenvolvimento sócio-económico harmonioso da zona.

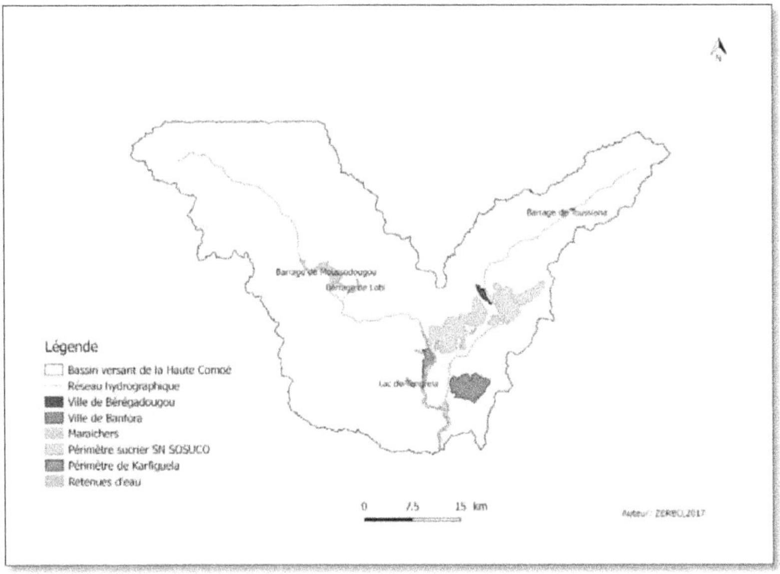

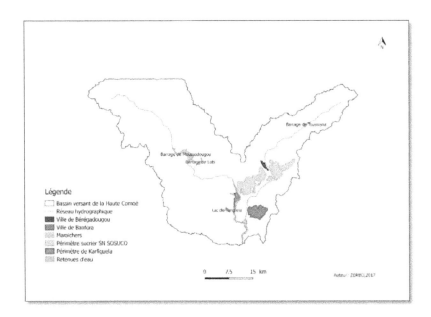

Figura 7: Principais utilizadores na sub-bacia hidrográfica de Comoé Superior

2.2. Métodos

Sendo este um processo de investigação participativa, a abordagem metodológica centrar-se-á em duas actividades principais: uma fase de revisão bibliográfica e uma fase de campo.

2.2.1. Pesquisa bibliográfica

A revisão bibliográfica consistiu principalmente na pesquisa e exploração de documentos. As actividades centraram-se principalmente nos dados disponíveis (relatórios de actividade, vários estudos, planos ou esboços da área) sobre o acesso e gestão dos recursos hídricos no Burkina Faso, particularmente na região de Cascades, dos Serviços Regionais e Nacionais de Agricultura e Gestão dos Recursos Hídricos, ONGs, SN-SOSUCO, ULg, UCL, bem como o recurso à Webografia

Foram utilizados documentos oficiais (PAGIRE, monografia da comuna, plano de desenvolvimento comunitário, textos legislativos e regulamentares, etc.) descrevendo políticas, estratégias, métodos e planos de acção para melhorar o conhecimento do sítio.

Esta fase permitiu-nos preparar melhor a segunda fase do estudo, que foi levada a cabo no terreno.

2.2.2. Levantamento e recolha de dados no terreno

Nesta fase, realizámos inquéritos, visitas e entrevistas com os intervenientes envolvidos a todos os níveis da gestão da água na sub-bacia.

Os inquéritos foram realizados com base num questionário semi-estruturado com os principais utilizadores de água e também com as instituições envolvidas na gestão da água (ver ficha de inquérito em anexo).

A realização de uma sessão de "grupo focal" com os chefes das cooperativas agrícolas dos perímetros e os comités de irrigadores dos produtores de hortas de mercado permitiu recolher dados qualitativos sobre o funcionamento destas estruturas e a sua consideração nas decisões relativas à gestão da água na sub-bacia.

Os levantamentos foram complementados por observações, fotografia e dados quantitativos sobre os recursos hídricos (necessidades de água, caudais, precipitações, etc.) no terreno.

O painel de inquiridos está resumido no quadro seguinte:

Quadro 1: Número de pessoas inquiridas

Categoria	Estruturas/Instituições	Número de inquiridos
Estruturas de apoio	DRAAH AEC	4 3
Organismo de gestão	CLE	1
Actores da indústria	SN SOSUCO ONEA	1 2
Agricultores	Karfiguela cooperativas de planícies e comité de jardineiros de mercado	17
Total		28

2.2.3 Processamento e análise dos dados dos inquéritos

Todos os dados recolhidos foram registados numa base de dados e depois analisados utilizando ferramentas estatísticas em Excel. O software QGIS versão 2.14.6 foi utilizado para produzir mapas da estrutura de gestão da água na sub-bacia.

2.2.4. Método de monitorização dos fluxos na sub-bacia superior de Comoé

Tendo em conta a multiplicidade de utilizadores de água e o aumento da sua procura, é necessário realizar estudos mais detalhados para uma estimativa óptima das necessidades actuais de água por categoria de utilização. A importância deste estudo é especificamente a de compreender o potencial de irrigação da área, a fim de melhorar a gestão da água.

Neste contexto, o PADI monitorizou os caudais através de medições em quatro estações hidrométricas, nomeadamente na entrada de Bodadiougou (ponto situado a montante das abstracções SN-SOSUCO e ONEA), a entrada de Karfiguela (abstracção da planície desenvolvida e jardins de mercado ao longo do rio Comoé), a ponte Tengrela e a ponte Diarabakoko na saída da sub-bacia (ver figura 8). Centrar-nos-emos em estudos desde Novembro de 2014 (fim da estação chuvosa) até Novembro de 2015, a fim de monitorizar a evolução dos caudais no fim da estação chuvosa e durante o ano.

A estimativa dos caudais nos locais foi baseada em dados de aferição, mas também em dados de medição do nível da água a partir de sondas.

O método de medição do campo de velocidade, vulgarmente conhecido como medição de bobinas, é o mais comummente utilizado. Esta aferição é a determinação do caudal de um rio através da medição, numa secção recta, da velocidade da corrente num determinado número de pontos cuja profundidade também é medida, e fazendo a soma dos produtos das velocidades médias pelas áreas elementares a que se aplicam.

As sondas estão equipadas com um sistema de medição automático baseado em sensores e são fixadas em diferentes locais do rio. Cada sonda mede regularmente alterações de pressão que correspondem a alterações na altura da água acima da sonda.

Ao combinar os dados das sondas e das curvas de medição, a análise da informação recolhida permitiu-nos estimar os caudais e compreender a variabilidade dos recursos hídricos no tempo e no espaço ao nível da sub-bacia.

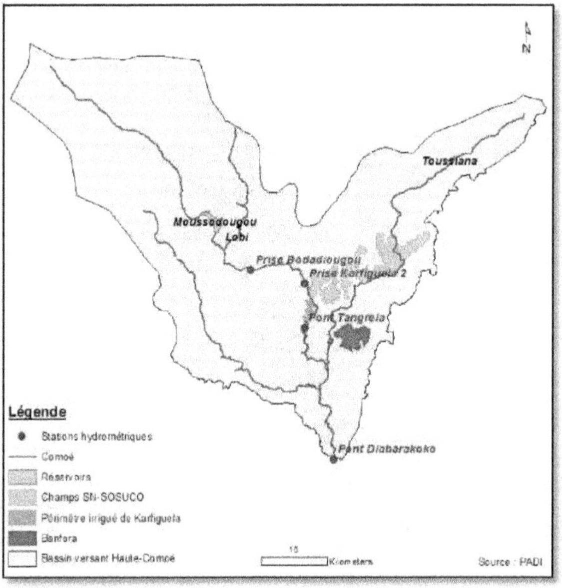

Figura 8: Localização das estações hidrométricas estudadas (Fonte: PADI)

Capítulo 3: Resultados do estudo

3.1. Práticas locais de gestão de recursos hídricos

3.1.1. Intervenientes e usos da água na sub-bacia

A escassez de água é sentida especialmente durante a estação seca de produção (Dezembro a Junho). Estamos particularmente interessados neste período do ano, durante o qual a gestão da água se baseia essencialmente nas três barragens da sub-bacia, nomeadamente as barragens de Moussodougou, Lobi e Toussiana.

> **Categorias de utilizações principais da água**

Quatro grandes grupos de utilizadores foram identificados como os principais actores na procura de água na sub-bacia. Os recursos hídricos são principalmente mobilizados para assegurar o abastecimento de água potável a grandes cidades como Banfora e Bérégadougou, e para a irrigação de parcelas agrícolas (agro-negócio, perímetro e horticultura de mercado). A utilização da água pelos industriais não está muito bem desenvolvida na zona, mas são feitas utilizações ecológicas da água para a manutenção e preservação da flora e fauna. O quadro abaixo resume os intervenientes presentes e as diferentes utilizações da água na sub-bacia.

Quadro 2: Principais actores e usos da água

Utilizadores	Tipo de utilização	Comentários
ONEA	Abastecimento de água potável e água bruta	Como empresa estatal, a ONEA é um utilizador prioritário para o abastecimento de água para fins sociais para uso doméstico
SN-SOSUCO	Agro-negócio	A SN-SOSUCO detém os 'direitos' de gerir os portões das 3 barragens. É também, nesta fase, a única instituição que fornece informações sobre a gestão das barragens.
Perímetro irrigado de Karfiguéla,	Produção de arroz com controlo total ou parcial da água	Área agrícola irrigada localizada na aldeia de Karfiguéla
Jardineiros de mercado ao longo dos rios	Jardinagem de mercado com controlo total ou parcial da água	Jardineiros informais no canal de alimentação, ao longo do Comoé (a jusante de Karfiguéla) e do rio Yannon.
Fluxos ambientais	Ecologia	Fluxos para a manutenção da fauna e da flora

> **Estado da procura de água pelos principais grupos de utilizadores**

A equipa técnica do CLE-HC para a época 2016-2017 estimou as necessidades de água (Dezembro a Junho): AEP (ONEA) a 1.044.010 m^3 , planície de arroz de Karfiguéla (248 ha) a 3.300.000 m^3 , horticultores (1.330 ha) a 5.987.000 m^3 , SN-SOSUCO a 34.032.872 m^3 ; necessidades ecológicas a 3.680.640 m^3 ou seja, uma procura de 48.044.522 m^3 para um volume de água disponível nas três barragens de 44.092.100 m^3 .

A figura 9 mostra que a maior procura de água é da SN SOSUCO (cerca de 70%), seguida por jardineiros de mercado informais ao longo dos rios. A procura de água para uso doméstico permanece baixa em comparação com os outros utilizadores (cerca de 2%). O caudal sanitário do curso de água (Qrest) para assegurar a sua sustentabilidade representa cerca de 6,87% da procura total de água.

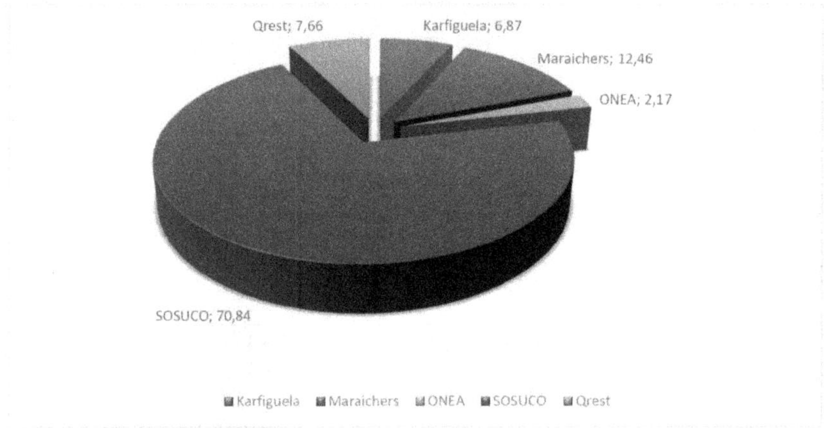

Figura 9: Percentagem da procura de água pelos principais grupos de utilizadores para a época 2016-2017

Os períodos de maior consumo são os meses de Fevereiro e Março, particularmente para os horticultores. No entanto, as exigências de outros consumidores são quase estáveis ao longo do período considerado (Figura 10).

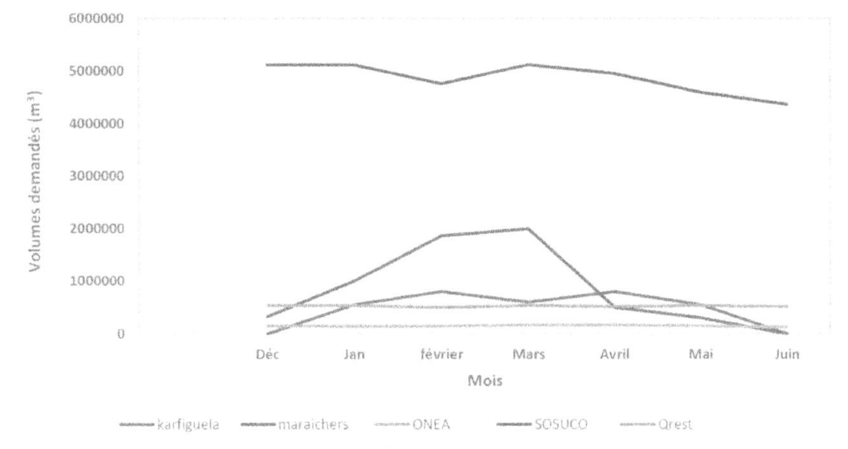

Figura 10: Evolução mensal das necessidades teóricas dos grandes utilizadores para a campanha 2016-2017

3.1.2. Órgãos de gestão da água: ferramentas e métodos de financiamento

A sub-bacia superior de Comoé faz parte da bacia nacional de Comoé, uma das quatro bacias do país. No âmbito da implementação da PAGIRE, cada bacia foi dotada de autoridades e organismos de gestão da água.

As estruturas de gestão da água presentes na bacia de Comoé são: a Agence de l'Eau des Cascades (AEC), o Comité Local de l'Eau de la Haute Comoé (CLE-HC), a Union des Coopératives des Exploitants du Périmètre Aménagé de Karfiguéla (UCEPAK) e os Comités d'Irrigants (CI) das aldeias ribeirinhas dos rios.

A operacionalização da GIRH na bacia hidrográfica foi feita essencialmente a três (03) níveis de gestão (Figura 11):

- ***a nível da bacia hidrográfica***: a organização das actividades é realizada em torno de um Comité de Gestão da Bacia (BMC) que reúne todos os representantes dos grupos de utilizadores de água, administrações públicas e autoridades locais (municípios, regiões, comunidades);
- ***a nível regional/provincial:*** os serviços estatais desconcentrados (Departamentos de Agricultura, Água e Recursos Pesqueiros) são responsáveis por fazer um balanço da situação na região. [25]

É também responsável pelo controlo da aplicação dos textos e pela prestação de apoio e aconselhamento sobre questões hídricas. A este nível, existe um quadro de consulta duas vezes por ano entre os diferentes serviços envolvidos nas questões relacionadas com a água. Contudo, há já algum tempo que as sessões do quadro de consulta inter-serviços sobre a água na bacia não se realizam.

- ***a nível local***: as ligações básicas na GIRH são os utilizadores de água e as autoridades comunitárias que participam na gestão dos recursos hídricos e das obras hidráulicas a nível local. Estes actores estão reunidos em torno da CLE-HC.

O resumo do funcionamento destas estruturas e de todos os actores envolvidos na gestão dos recursos hídricos é apresentado na Figura 11. Para cada estrutura apresentamos os instrumentos do sistema e os seus métodos de financiamento e depois analisamos a eficiência e sustentabilidade destes métodos de financiamento.

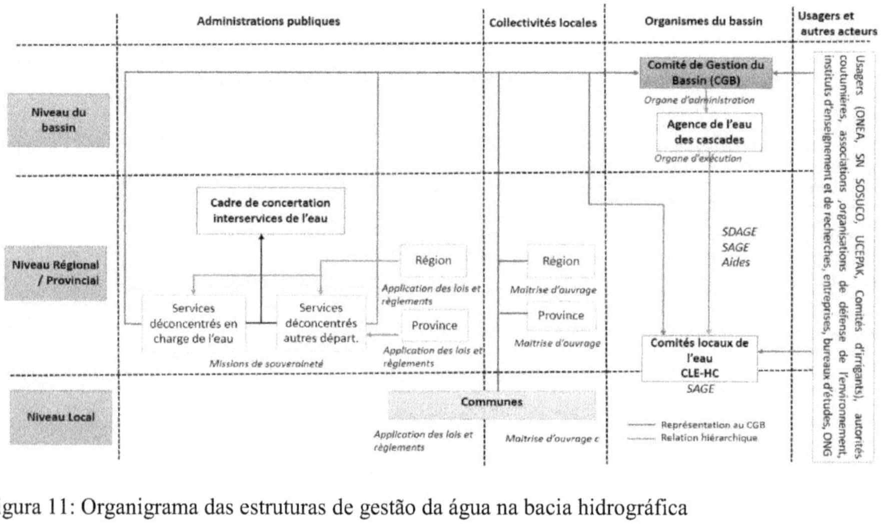

Figura 11: Organigrama das estruturas de gestão da água na bacia hidrográfica

Fonte: MAHRH, 2009 (fase PAGIRE Π)

> A Agência da Água das Cascatas (CWA)

Criada em 2010, a AEC é a estrutura mãe responsável pela gestão dos recursos hídricos a nível da bacia hidrográfica do rio Comoé no Burkina Faso. Actua como um quadro de conhecimento, planeamento e gestão da água na bacia nacional de Comoé.

O AEC está estruturado nos órgãos de gestão e administração da seguinte forma

- O Comité de Bacia (BC): Assembleia Geral da ACS, responsável pela implementação da política nacional da água na bacia de Comoé (SDAGE e SAGE);

- o Conselho de Administração (BD): este é o órgão executivo do comité da bacia. Assegura a implementação das directrizes, em particular através dos programas de intervenção plurianuais;
- A Direcção Geral (GM): assegura a organização e desenvolvimento das missões da ACS sob a direcção do Conselho de Administração e nos termos estabelecidos pelo Comité de Bacia;
- Comités Locais da Água (LWCs): estes são as ligações básicas e cruciais na implementação da GIRH a nível local.

A área de competência da AEC vai além dos limites administrativos da região de Cascades, uma vez que também abrange (embora em menor medida) duas outras regiões vizinhas, nomeadamente os Haut-Bassins e parte do Sudoeste.

As actividades da AEC são principalmente financiadas por parceiros como a SIDA (Agência Sueca de Cooperação Internacional para o Desenvolvimento) e a DANIDA (Agência Dinamarquesa de Desenvolvimento Internacional), o subsídio do Estado Burkinabe, as receitas do CFE e quaisquer outras receitas autorizadas pelo Comité de Bacia (doações e legados).

Para ter mais sucesso na sua missão no terreno, a agência deve ser financeiramente autónoma. Neste contexto, foi iniciado pelo Estado Burkinabe um sistema de auto-financiamento através de uma taxa parafiscal, denominada Contribuição Financeira para a Água (CFE), em benefício das agências. Esta taxa tem sido objecto de experimentação pela AEC. Voltaremos a este assunto em pormenor na secção 3.2.

> O Comité Local da Água da Haute Comoé (CLE-HC)

O CLE-HC foi criado em 2008 pelo decreto conjunto n.º 2008-002/RCAS/ RHBS de 5 de Março de 2008 dos Governadores das regiões Cascades e Hauts-Bassins. Foi criado na sequência de conflitos em 2007 entre utilizadores, nomeadamente agricultores de arroz na zona irrigada de Karfiguéla e alguns horticultores ao longo do rio Comoé, que tinham atacado a SN-SOSUCO. De facto, os agricultores acusaram a SN SOSUCO de utilizar a maioria das reservas de água na sub-bacia, o que colocou problemas de disponibilidade de água ao seu nível.

A CLE-HC é composta por cerca de cinquenta membros incluindo representantes dos serviços estatais, representantes das autoridades locais, representantes dos utilizadores, representantes da sociedade civil, projectos e programas e associações cujas actividades têm impactos quantitativos ou qualitativos sobre os recursos hídricos da sub-bacia.

O comité reúne uma vez por ano numa assembleia geral ordinária. Também se reúne seis (6) meses após a adopção do programa de actividades do LEC na assembleia para examinar e avaliar o seu estado de execução e tomar as medidas necessárias para a sua correcta implementação. Em caso de necessidade de qualquer outro aspecto inerente à gestão da água da sub-bacia, a comissão pode reunir-se numa assembleia geral extraordinária.

Desde a sua constituição, as assembleias gerais ordinárias da CLE têm sido realizadas regularmente. O CLE Haute Comoé tem sido capaz de organizar os actores da água (administração, utilizadores, autoridades locais, autoridades consuetudinárias e religiosas, organizações da sociedade civil) e trabalhar para a sua adesão à gestão concertada dos recursos hídricos da bacia hidrográfica. Esta adesão quase total dos utilizadores à CLE é um excelente instrumento e quadro de consulta, prevenção e gestão de conflitos relacionados com a água. O seu nível de organização faz dela uma referência a nível nacional para a criação de CLE. No entanto, encontra dificuldades para certas decisões concertadas, devido à falta de instrumentos operacionais (má distribuição das estações hidrométricas, ausência de equipamento de medição de caudal, etc.) e meios logísticos (escritório, viagens, etc.). A baixa capacidade do CLE para controlar os fluxos devido ao seu deficiente equipamento e o peso da SN SOSUCO na gestão e manutenção de obras hidro-agrícolas e o aumento da superfície dos jardins de mercado são outras grandes dificuldades.

Os recursos financeiros da CLE-HC provêm principalmente das contribuições dos seus membros e dos subsídios da Direcção Geral da Agência da Água de Cascades (DGAEC).

A contribuição para o sector agrícola é fixada em 500 F CFA (0,76 euros)/ha/ano para SN-SOSUCO e para o perímetro de Karfiguela, enquanto que uma taxa fixa é aplicada aos outros utilizadores, nomeadamente ONEA, horticultores, colectividades territoriais e associações. O quadro seguinte resume os montantes esperados de cada membro da CLE.

Quadro 3: Base para as contribuições dos membros da CLE

Membro da CLE	Base contributiva (FCFA)	Base contributiva (euros)
Conselho Regional/Cascades	50000	76,22
Comuna de Banfora	50000	76,22
Município de Toussiana	25000	38,11
Comuna de Bérégadougou	25000	38,11
Comuna de Samogohiri	25000	38,11
Comuna de Moussodougou	25000	38,11
Associação MUNYU	25000	38,11
SN-SOSUCO	2000000	3048,97
CRA/CAS	350000	533,57
ONEA	500000	762,24
Associação WOUOL/Béragadougou	25000	38,11

Grupo de Pescadores	25000	38,11
Planície de Karfiguéla	175000	266,78
SOFITEX	100000	152,45
AGEREF/CL	25000	38,11
21 Aldeias de jardim de mercado	210000	320,14
Plantação de bananeiras de Tengréla	25000	38,11
Plantação de bananeiras de Moussodougou	25000	38,11
TOTAL (FCFA)	**3685000**	**5617,72**

As contribuições da ELC são especificamente utilizadas para monitorizar o programa de libertação de água, para cobrir os custos de viagem dos membros não residentes, para realizar certas actividades de sensibilização e para proteger o recurso (reflorestação, barreiras de pedra).

O acompanhamento da recolha desde a criação da CLE até aos dias de hoje (Figura 12) mostra um nível muito progressivo de contribuições ao longo do tempo com uma recolha média de 32,72% ao longo de todo o período em que todos os membros da CLE Haute-Comoé são tidos em conta. A SN-SOSUCO é, portanto, o maior contribuinte e influencia fortemente a curva de recolha global. No entanto, como a maioria dos membros, não está regularmente actualizado com as suas quotas ou está em atraso (como foi o caso em 2015). Quanto à ONEA, é o segundo maior contribuinte, mas não paga as suas quotas desde 2012. A planície desenvolvida de Karfiguela nunca atinge os montantes esperados e só contribuiu durante duas épocas de cultivo (2011 e 2016). Os jardineiros do mercado, embora tenham uma baixa taxa de recolha (16,93% em média), parecem ser mais estáveis nas suas contribuições.

Estas dificuldades podem enfraquecer a estrutura e reduzir a eficácia e a sustentabilidade dos seus métodos de financiamento.

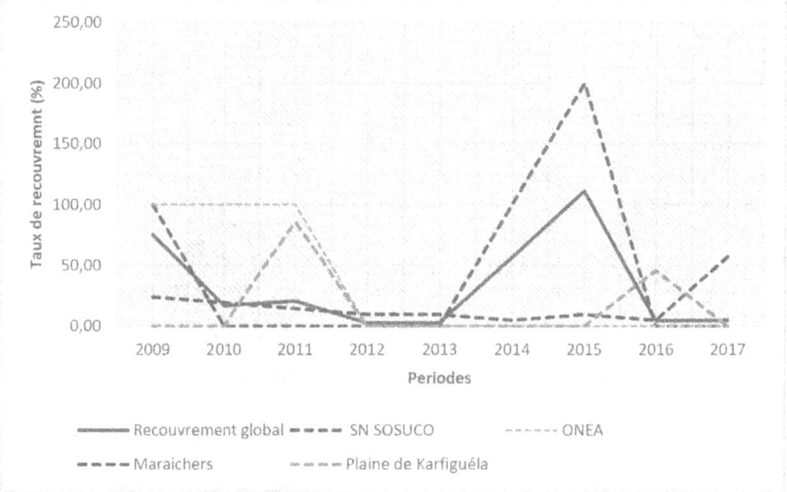

Figura 12: Evolução da recolha da contribuição dos membros para o funcionamento do CLE

> Union des Coopératives des Exploitants du Périmètre Aménagé de Karfiguéla (UCEPAK)

O perímetro irrigado de Karfiguéla cobre uma área desenvolvida de cerca de 350 hectares com um potencial de desenvolvimento de 750 hectares. Os 730 agricultores (130 mulheres e 600 homens) estão organizados em torno de cinco cooperativas de base, federadas num sindicato chamado Union des Coopératives des Exploitants du Périmètre Aménagé de Karfiguéla (UCEPAK).

A gestão da água é feita por torre de água, tendo em conta os horticultores instalados ao longo do Comoé. A UCEPAK encontra muitas dificuldades na atribuição de água na sua área de competência, porque as suas necessidades não são geralmente satisfeitas na sua totalidade. Além disso, confronta-se com o fenómeno da sifonagem a partir dos horticultores instalados a montante em torno do canal de alimentação e do [30] e o estado dilapidado da rede de irrigação. Todas estas dificuldades tornam a UCEPAK muito passiva em termos de gestão da água.

A UCEPAK é apoiada por uma equipa de consultores agrícolas designada para o perímetro pela DRAAH das Cascatas para assegurar a supervisão e organização dos produtores e para tentar encontrar soluções para as dificuldades da gestão dos recursos hídricos.

A estrutura funciona com base numa taxa individual de água fixada em 6.000 francos CFA (9,15 euros)/ha/ano. No entanto, durante as nossas entrevistas no terreno, as cooperativas mencionaram o problema da cobrança destas taxas (menos de 60% pagas), o que torna a UCEPAK impotente face a certas dificuldades e impede-a de honrar os seus compromissos para com a CLE.

> Comités de Irrigadores (ICs)

Os comités de irrigação foram criados pelo DRAAH das Cascatas entre 2007 e 2008 nas 21 aldeias limítrofes dos rios como parte das actividades para promover a produção de vegetais fora da estação.

Estas estruturas estão a tentar organizar-se melhor para partilhar a água ao nível dos locais de produção e participar na gestão do recurso ao nível da sub-bacia, mas a maior dificuldade é a extensão progressiva das superfícies devido à rentabilidade da actividade produtiva e às políticas de promoção da agricultura fora da estação. Isto tem um grande impacto no cálculo das necessidades de água e, em particular, na distribuição do recurso hídrico entre os vários utilizadores da sub-bacia.

3.1.3. Métodos de gestão da água

> Ferramentas do sistema de gestão da água

A gestão operacional da água na sub-bacia baseia-se no CLE, que depende fortemente dos seus membros e da sua organização. A gestão baseia-se em estimativas dos volumes totais que podem ser mobilizados nos reservatórios e das necessidades dos utilizadores e da distribuição dos caudais por utilização.

Em primeiro lugar, uma visita de estudo é organizada pelos membros para ver como estão cheias as barragens no final da estação das chuvas. Depois, após uma reunião com os vários utilizadores, sob a direcção da AEC, é elaborado um programa de descarga de água das barragens de acordo com um princípio que tem em conta dois níveis de prioridade de utilização:

- É dada prioridade ao fornecimento de água potável às populações para as quais a procura de água deve ser totalmente satisfeita. Assim, a ONEA recebe 100% da sua procura de água;
- Após a satisfação da ONEA, as quantidades restantes são distribuídas ao resto dos utilizadores proporcional às necessidades, respeitando a partilha justa dos défices.

É efectuado um acompanhamento pelo CLE para assegurar que o programa estabelecido é implementado e que os caudais e as voltas de água são respeitados. Em caso de incumprimento por um dos utilizadores ou de falta de satisfação, o CLE é apreendido e, por sua vez, dá prioridade ao diálogo entre os utilizadores com vista à resolução do problema.

- **As dificuldades da gestão local da água**

A nível da sub-bacia, para além das limitações acima mencionadas, as maiores dificuldades são, por um lado, a queda de chuva e, por outro lado, a baixa capacidade de armazenamento da barragem de Moussodougou devido ao mau estado do seu dique, que tem vindo a vazar e a sofrer erosão interna desde 2011. Os recursos hídricos da sub-bacia diminuem anualmente, enquanto as necessidades dos vários utilizadores aumentam consideravelmente. Isto tem consequências para a gestão do recurso.

3.2. Análise da implementação da "Contribuição Financeira para a Água

O sector da água é financiado principalmente por recursos externos (empréstimos ou subvenções ao Estado), ou seja, cerca de 89% do total dos investimentos, e é o Estado que assegura o financiamento dos investimentos (MEE, 2001). Para permitir que este sector seja mais autónomo, actualmente são os utilizadores que são solicitados a participar no financiamento; encarregando-se da manutenção e manutenção das obras e infra-estruturas hidro-agrícolas.

3.2.1. Análise dos textos sobre os impostos sobre a água

A implementação de uma política fiscal no sector da água visa mobilizar recursos financeiros no seio das Agências da Água para apoiar o financiamento de actividades de conservação de recursos hídricos no Burkina Faso. A tarifação da água foi adoptada em aplicação dos princípios "utilizador-pagador" e "poluidor-pagador" e baseia-se na existência de três textos jurídicos que estabelecem os direitos e impostos sobre a utilização da água. Em particular, estes são

> **Disposição 1:** Lei N°058-2009/AN que estabelece uma taxa parafiscal em benefício das Agências da Água. Esta é a primeira lei desde a criação das agências de água que institui uma taxa parafiscal conhecida como a "Contribuição Financeira da Água", abreviada para CFE.

> **Disposição 2:** Decreto N°2011- 445MAR que determina as modalidades de recolha/retirada de água bruta.

Este texto estabelece os preços para a captação de água bruta para diferentes utilizações:

- produção de água potável: 1 FCFA/m^3 captação de água ;
- actividades mineiras e industriais: 200 FCFA/m^3 de água captada;
- obras de engenharia civil: 10 FCFA/m^3 de aterro executado e 20 FCFA m^3 de betão implementado, todas as classes de betão incluídas;
- produção agrícola, pastoral e pesqueira: para estas utilizações, estão actualmente a ser elaborados textos.

> **Disposição 3**: Decreto N°2015-1470 /PRES-TRANS /PM/MEF/MARHASA que determina as taxas e modalidades de cobrança da taxa de captação de água bruta. A diferença em relação à disposição 2 é que este decreto diferencia entre a produção de água potável para fins sociais (1 FCFA por m^3 de água captada) e a venda de água potável por empresas comerciais (50 F CFA/m^3 para captação). Além disso, a taxa do imposto para a captação de água para fins industriais (minas

ou outros) é reduzida (125 FCFA por m^3 de água captada).

3.2.2. Aplicação do imposto sobre a captação de água bruta na sub-bacia superior de Comoé

3.2.2.1. Utilizações de água bruta sujeitas a CFE

O CFE agrupa três (03) tipos de impostos, incluindo: o imposto sobre a captação de água bruta, o imposto para modificar o regime da água e o imposto sobre a poluição da água.

Dos impostos previstos, apenas o imposto sobre a captação de água bruta com base nos volumes de água captados é aplicado na sub-bacia. As diferentes utilizações da água bruta que são objecto do pagamento do imposto sobre a captação de água são principalmente

- água bruta retirada para a produção hidroagrícola;
- água bruta para a produção de água potável ;
- água bruta captada para fins de produção de energia hidroeléctrica;
- água bruta retirada pelos sectores mineiro e industrial;
- e a água bruta retirada das obras civis.

3.2.2.2. Potencial financeiro da aplicação CFE

Na sub-bacia, para além do consumo de água potável, a água é útil para actividades agrícolas, pecuária, piscicultura, usos industriais, turismo, produção de electricidade e para a manutenção dos ecossistemas.

Para todas estas utilizações, o conhecimento de cada sector permite uma melhor compreensão dos resultados esperados da implementação do CFE.

> **Produção de água potável**

Este sector inclui o abastecimento de água potável (AEP) para fins sociais e a produção de água para fins comerciais onde a água é a matéria-prima. São principalmente empresas como a ONEA, a indústria agro-alimentar (produção de água potável SN-SOSUCO) e dez outras empresas que produzem e comercializam água mineral (água em sacos ou latas). Os locais de produção estão localizados em torno das cidades de Banfora e Niangoloko (ver lista de inquiridos, Quadro 4).

A taxa aplicada para o CFE é de 1 FCFA (0,0015 euros) por metro cúbico de água potável para fins sociais para a ONEA e 50 FCFA (0,0762 euros) para a categoria comercial.

> **O sector industrial**

São empresas que utilizam a água como factor de produção, neste caso para arrefecer motores, na lavagem e limpeza de produtos ou subprodutos ou para girar turbinas (hidroelectricidade). As indústrias presentes na área incluem a SN-SOSCUCO (industrial), algumas empresas mineiras (GRYPHON Minerals) e mineiros artesanais de ouro. Existem também duas pequenas centrais eléctricas pertencentes à Société Nationale d'Electricité du Burkina Faso (SONABEL), que produzem electricidade para abastecer as cidades de Banfora, Niangoloko, Orodara, Toussiana e Kinshasa. Orodara, Toussiana e Péni em electricidade.

Para os industriais, o imposto retido é de 125 FCFA (0,19 euros) por cada metro cúbico de água retirada.

> **Obras de engenharia civil**

São empresas de construção e obras públicas que utilizam a água principalmente em actividades de terraplanagem e betão. Estes tipos de actividades são sazonais e dependem das prioridades de desenvolvimento da área.

O regulamento prevê 10 FCFA (0,0152 euros) por metro cúbico de aterro e 20 FCFA (0,0305 euros) por m^3 de betão, todas as classes de betão incluídas.

> **Sector de produção hidro-agrícola**

A bacia hidrográfica tem um elevado potencial para a produção agrícola, particularmente na agricultura irrigada. Com um potencial que ainda pode ser desenvolvido, as áreas irrigadas estendem-se por cerca de 5480 ha, incluindo uma grande área irrigada de cana de açúcar (3800 ha), 350 ha de planície desenvolvida para produção de arroz e cerca de 1330 ha de planícies para a jardinagem de mercado, de acordo com dados do CLE-HC.

Estas áreas desenvolvidas, em particular o perímetro do açúcar explorado na agro-indústria, constituem um grande trunfo económico para a sub-bacia.

Actualmente, o imposto sobre a captação de água bruta para fins hidro-agrícolas ainda não está em vigor na bacia, mas de acordo com as nossas informações recolhidas no terreno, foi realizado um estudo pela Direcção-Geral dos Recursos Hídricos (DGRE) para elaborar um projecto de decreto sobre o assunto (ver Anexo 1). Este estudo propôs taxas fixas para a utilização de água para a produção de cereais, a jardinagem de mercado, a cana-de-açúcar, as árvores de fruto e a produção em viveiro.

> **Sector pecuário**

Para actividades pecuárias, esta é uma área pastoral onde o gado transumante ou intensivo é regado em torno de pontos de água (barragens, reservatórios de água, poços, furos, etc.). As necessidades de água dos animais são difíceis de obter e de momento não é aplicado qualquer imposto a este sector, mas nos projectos de lei (estudo da DGRE) estão previstos impostos para este sector (ver anexo).

> **Outros sectores**

Para os outros sectores, principalmente para actividades de lazer e pesca, existe um potencial, mas poucos estudos foram realizados para estimar as necessidades destas utilizações. No entanto, para a pesca, o projecto de lei prevê impostos nesta área (ver anexo).

A tabela seguinte mostra a lista de entidades declarantes em toda a jurisdição do ACS.

Quadro 4: Lista de contribuintes na bacia de Comoé

N°	Utilizadores	Localização	Domínio de actividade
1.	ONEA	Banfora	Produção de água potável
2.	SOKOMAF-SARL FARADJI	Banfora	Produção de água mineral
3.	KOUROUDJI	Toumousséni	Produção de água mineral
4.	AWADJI	Banfora	Produção de água mineral
5.	CACHOEIRAS CAMPEÃS DE ÁGUA	Niangoloko	Produção de água mineral
6.	NADHIS BURKINA SANTANI	Niangoloko	Produção de água mineral
7.	VIDA ÁGUA	Niangoloko	Produção de água mineral
8.	SN SOSUCO	Banfora	Agro Industrial
9.	Minerais GRYPHON	Niankorodougou	Mina
10.	ERRI-BF	Péni	BTP
11.	ADAE	Bobo-Dioulasso	Produção de água potável
12.	SONABHY	Péni	Hidrocarboneto
13.	SANTANI		Produção de água mineral
14.	FARADJI		Produção de água mineral
15.	DAFANI	Orodara	Produção de sumo de manga
16.	SELMON		Produção de água mineral
17.	BABALI		Produção de água mineral

Uma avaliação do potencial anual do CFE de algumas das principais categorias de actividade é apresentada no quadro abaixo. Os resultados mostram um potencial financeiro de aproximadamente 72.983.803 FCFA, ou 111.263 euros, para a taxa de captação de água bruta na bacia hidrográfica.

Quadro 5: Estimativa do potencial financeiro da taxa de captação de água bruta

Sector	Unidade	Quantidade anual	Montante unitário (FCFA)	Montante total (FCFA)
AEP	m^3	1789403	1	1 789 403
Água potável SN-SOSUCO	m3	119388	50	5 969 400
Água potável comercial	m^3	212500	50	10 625 000
Industrial	m3	436800	125	54 600 000
Total		**2558091**		**72 983 803**

Fonte: Relatórios da AEC, contribuição para as taxas de captação de água bruta

3.2.2.3. Estado de implementação do imposto de abstracção na sub-bacia

De entrevistas com as partes interessadas e a AEC, parece que a implementação do imposto ainda se encontra na fase experimental e diz respeito apenas a actividades de produção de água potável, minas, actividades industriais e obras de engenharia civil. A situação da recolha das principais categorias de utilizações é apresentada na Figura 14, desde a sua implementação em 2013 até 2016.

A ONEA e algumas empresas comerciais de venda de água com taxas médias de recolha de 50% e 46% respectivamente estão assim posicionadas como os membros mais envolvidos no processo. No entanto, a ONEA tem um baixo potencial financeiro com pequenos montantes e tem registado atrasos de pagamento desde 2015 (Figura 13). Quanto à SN-SOCUSCO (componente industrial e produção de água potável), ainda não está envolvida no processo de pagamento devido ao facto de ter um acordo com o Estado Burkinabe para assegurar a manutenção das barragens na bacia. Acontece que os custos afectados a estes trabalhos excedem largamente a sua contribuição financeira para o mesmo fim. Está à espera de clarificar a sua posição sobre o seu envolvimento no processo. Uma empresa mineira que iniciou as suas actividades começou a contribuir, mas as suas contribuições são claramente baixas em comparação com a componente industrial da SN-SOSUCO.

Para os outros membros, nomeadamente os horticultores, e a planície de Karfiguela, os textos que regem as taxas de captação de água para usos agrícolas, pastoris e florestais ainda não são eficazes, e por isso não estão preocupados com as taxas.

A taxa média de recuperação durante o período de 2013 a 2016 tende a estabilizar em torno do valor médio anual de 24%, embora tenha havido um aumento do nível de recuperação por parte das empresas de água mineral.

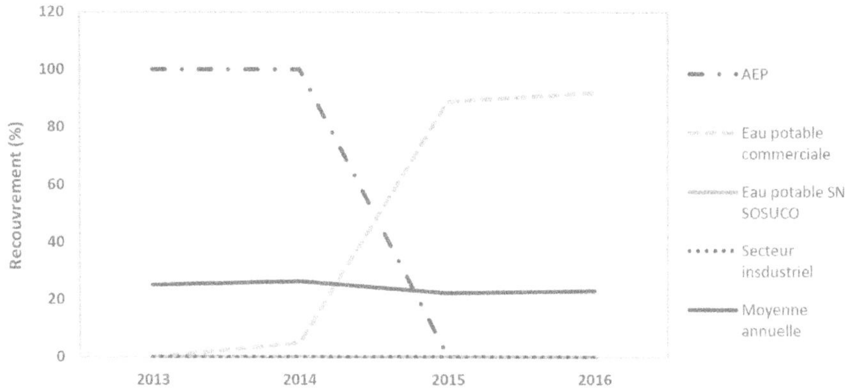

Figura 13: Nível de cobrança do imposto sobre a captação de água bruta

3.2.3. Utilização de impostos pagos pelos utilizadores

De momento, ainda não foi feita qualquer utilização dos impostos pagos pelos utilizadores na sub-bacia. No entanto, os recursos recolhidos devem ser utilizados para

- para financiar obras de interesse comum para a protecção dos recursos hídricos;
- contribuir para o funcionamento das estruturas encarregadas da água ;
- e apoiar a operação e aquisição de equipamento para a ACS.

Os benefícios destas contribuições deverão portanto ser úteis para os utilizadores dos recursos hídricos da bacia e sobretudo servir para preservar o recurso e melhorar as actividades relacionadas com a água.

O CFE é unanimemente apoiado por todos os actores e todas as pessoas entrevistadas concordaram com a necessidade de contribuir para a protecção do recurso hídrico comum. Além disso, na política de implementação do CFE, foi atribuída ao Ministério da Economia e Finanças do Burkina Faso a responsabilidade pela cobrança e gestão dos impostos com a AEC. Isto prejudicou grandemente o nível de cobrança. Para uma melhor operacionalização do mecanismo de cobrança em 2016, o Estado forneceu ao AEC um agente contabilístico que está mais próximo dos utilizadores da água e autorizado a cobrar o CFE. Isto melhorou significativamente a sensibilização dos actores e o seu compromisso com o sistema.

3.3. Monitorização dos caudais ao longo do rio Comoé na sub-bacia superior de Comoé

O objectivo deste estudo era acompanhar a evolução dos caudais em diferentes secções ao longo do rio Comoé. O estudo foi realizado em quatro estações hidrométricas representativas da sub-bacia, mas apenas apresentaremos os resultados das estações localizadas na tomada de água de Bodadiougou, na ponte de Tengrela e na ponte de Diarabakoko na saída da sub-bacia, para o período de Novembro de 2014 a Novembro de 2015. Os dados da estação de Karfiguela foram invalidados pelo PADI devido a problemas de medição.

O mês de Novembro foi escolhido para corresponder ao final da estação do Inverno, um período que determina o nível de enchimento dos reservatórios de água da bacia e, portanto, as quantidades de água disponíveis para as várias utilizações na estação baixa. Assim, iremos acompanhar a evolução dos caudais do rio desde o final da estação das chuvas até ao final da estação seca (Dezembro a Junho, o que corresponde ao período de elevada procura de água).

3.3.1. Análise dos fluxos diários em todos os locais

A figura 14 mostra as curvas das taxas de fluxo diárias para cada sítio.

A análise da situação global mostra o aparecimento de picos de fluxo nos resultados das sondas durante o período de Maio a Novembro. As taxas de fluxo da sonda Bodadiougou são mais baixas, especialmente na época das chuvas (Agosto a Setembro); isto está em contradição com os acontecimentos pluviométricos na área. Quanto ao caudal na estação de Diarabakoko, este excede largamente o caudal diário dos outros locais; esta situação é mais acentuada entre Agosto e Setembro.

O sítio Bodadiougou está localizado a montante das quedas de água e é utilizado como local de retirada de tubos para abastecer a ONEA (AEP) e o perímetro de açúcar SN-SOSUCO. Para este sítio, o caudal médio observado durante o período de Junho a Outubro de 2015 é de 0,79 m^3 /s com um mínimo de 0,19 m^3 /s e um caudal máximo de 3,01 m /s.^{3}

O caudal médio durante o período de medição (Novembro 2014 a Novembro 2015) no site de Tengrela é de 1,14 m^3 /s com um caudal máximo diário de 3,58 m^3 /s medido em Outubro 2015 e um mínimo de 0,04 m /s.^{3}

Para a estação de Diarabakoko, localizada na saída da bacia, o caudal diário mínimo é medido em Dezembro (0,02 m^3 /s) mas o caudal médio para todo o período de medição é de 5,79 m^3 /s e um caudal máximo de 22,68 m /s.^{3}

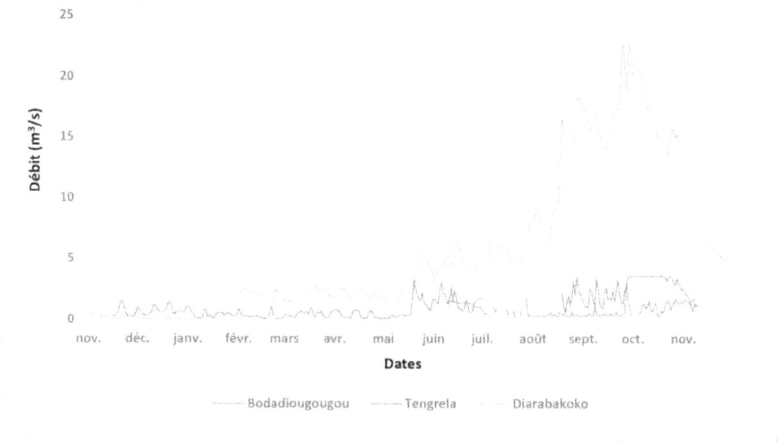

Figure 14 *Evolução dos fluxos diários em todos os locais monitorizados (fonte: dados PADI)*

Os caudais médios mensais de Novembro de 2014 a Outubro de 2015, bem como o nível médio da água da estação de Bérégadougou durante o mesmo período, são apresentados na figura abaixo. Observamos uma boa correlação entre as profundidades da água e a evolução dos caudais ao longo do tempo. De facto, os meses mais chuvosos (Agosto a Setembro) correspondem bem aos períodos em que os caudais observados são importantes, excepto para a estação de Bodadiougou onde se observa a situação oposta.

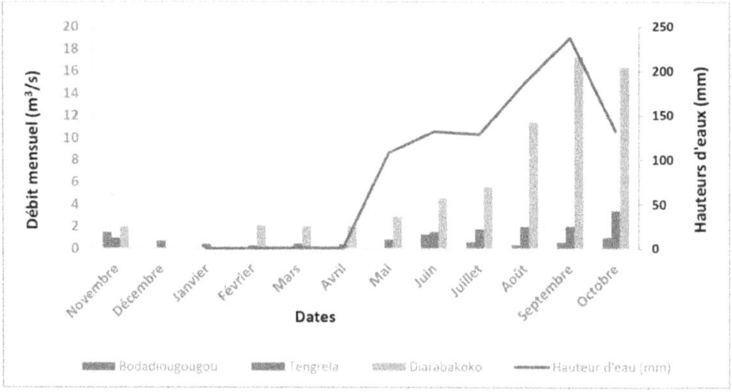

Figure 15 *Tendências em caudais mensais e profundidades da água na sub-bacia*

3.3.2. Estrutura das necessidades, usos e recursos hídricos na sub-bacia

Para melhor compreender a estrutura da competição pelos recursos hídricos na sub-bacia, organizámos os resultados obtidos acima, e as estimativas das exigências dos utilizadores da CLE-HC para o mesmo período para representar a correspondência entre as exigências e os recursos disponíveis a nível da sub-bacia, apresentadas na Figura 16.

Os resultados do fluxo mostram que, para além da ONEA, as necessidades dos outros utilizadores estão abaixo dos níveis de satisfação das exigências ao nível da CLE. O SN-SOSUCO parece ser o utilizador menos satisfeito dos recursos hídricos partilhados. A falta de dados para o período em consideração para a estação de Bodadiougou (onde o SN-SOSUCO e a ONEA tomam a sua água) e a captação de Karfiguéla (onde os jardineiros de planície e de mercado desenvolvidos tomam a sua água) significou que não foi possível avaliar os níveis reais de captação de água a partir dos medidores No entanto, os fluxos registados pela sonda Tengrela revelam que as retiradas a montante, nomeadamente pelos utilizadores da planície e certos horticultores, e mais adiante pela SN-SOSUCO e ONEA, estão abaixo dos fluxos que fluem para o Comoé. Isto permite a manutenção de um fluxo sanitário no rio.

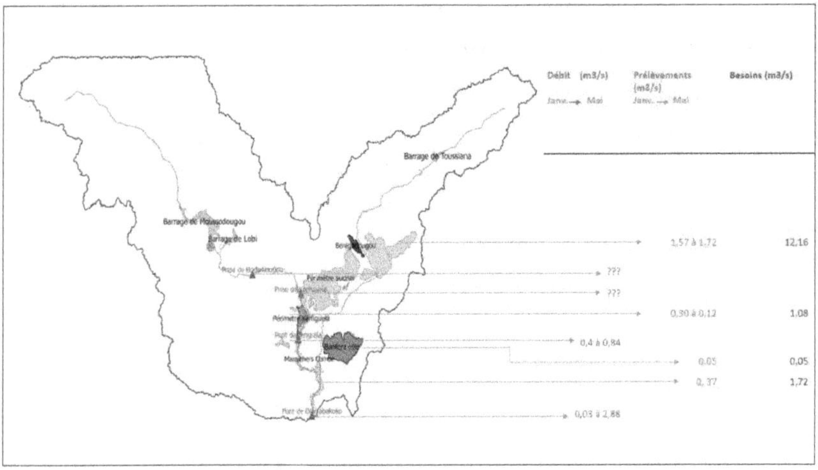

Figura 16: Estrutura do concurso para os recursos hídricos na área de estudo

CAPÍTULO 4: DISCUSSÃO E RECOMENDAÇÕES

4.1. Discussão geral do estudo

A avaliação geral do quadro institucional na sub-bacia de Comoé Superior parece mostrar que a gestão dos recursos hídricos está em conformidade com os princípios fundamentais da GIRH. Isto reflecte-se no desempenho e no nível de organização do sistema de gestão, em particular através da abordagem integrada, do forte envolvimento das comunidades de base na gestão dos recursos hídricos, mas também através do valor económico dado ao sector da água. Contudo, é de notar que o papel das mulheres na gestão dos recursos hídricos não é muito visível, mas o seu número está a aumentar progressivamente nas estruturas de gestão e, em particular, estão fortemente representadas na CLE-HC através da associação "Munyu" de Mulheres Comoé. Participam também activamente no desenvolvimento sócio-económico e cultural da zona, dada a sua elevada representação na população.

A nível da sub-bacia, entre os resultados alcançados pela PAGIRE, podemos citar a criação da Agência de Água Cascades e do Comité Local da Água do Alto Comoé, que estão a funcionar relativamente bem, e a implementação da contribuição financeira para a água (CFE).

Para a implementação do CFE, a aplicação dos textos diz respeito, em primeiro lugar, à taxa sobre a captação de água bruta para usos industriais, de engenharia civil e de produção de água potável. Isto enquadra-se bem na vontade política do Estado Burkinabe na implementação do programa IWRM de gerar recursos internos para financiar a monitorização, saneamento e distribuição de recursos hídricos para usos através de medidas legais e regulamentares sobre a água (MEE, 2001).

Os pontos fortes do sistema de financiamento residem na existência de regulamentos fiscais e no nível de organização dos actores. Contudo, existem ainda enormes desafios a enfrentar. Estes incluem o baixo nível de cobrança do CFE após 4 anos de implementação, a inexistência de textos fiscais para a componente agrícola e a baixa capacidade do AEC para controlar os níveis reais de captação de água pelos vários utilizadores na sub-bacia.

Em geral, existe um baixo potencial fiscal na sub-bacia, que poderia ser explicado pelas insuficiências dos textos do regulamento da água, mas também pela influência que esta nova taxa poderia ter nos rendimentos dos actores, particularmente dos pequenos agricultores. Kaboré e Ouattara (2000), no seu estudo sobre a avaliação das contribuições financeiras dos utilizadores, identificaram várias categorias de actores. Estes incluem utilizadores com uma certa taxa de recuperação (recuperação fácil: hidroelectricidade, hidráulica urbana e industrial, SN-SOSUCO, actividades mineiras e de construção e obras públicas), utilizadores com uma taxa de recuperação incerta (perímetros privados, jardins de mercado, hidráulica pastoril e actividades de piscicultura), utilizadores em risco (grandes perímetros estatais e utilizadores insolventes (terras baixas, pequenos perímetros cerealíferos, AEPS e PEM). Para além da SN-SOSUCO, a maioria dos utilizadores na sub-bacia têm um baixo potencial económico.

A existência de vários impostos sobre a água na sub-bacia, nomeadamente a contribuição para o funcionamento da ELC, o imposto sobre a captação de água bruta combinado com o imposto sobre a modificação do regime da água e a poluição da água que em breve estará em aplicação, pode constituir uma desvantagem para a colecta do CFE. O processo de fixação de impostos, as suas estruturas e a sua operacionalização devem, portanto, ter em conta novas considerações ambientais, financeiras e sociais que desempenham um papel muito determinante de acordo com o ADB (2000).

A partir dos resultados da estrutura do concurso para os recursos hídricos, podemos concluir que os recursos são bem geridos no seu conjunto pelos interessados na sub-bacia. De facto, embora as exigências dos vários utilizadores não pareçam ser satisfeitas, a gestão da água permite manter um fluxo sanitário no Comoé, necessário para a preservação do recurso e dos ecossistemas associados. Quanto ao SN-SOSUCO, parte das suas necessidades são satisfeitas pela barragem de Toussiana. Além disso, estão a ser feitos enormes esforços pelos principais utilizadores (SN-SOSUCO e o perímetro desenvolvido) para reduzir a procura de água. Estes esforços incluem a introdução de técnicas de produção que consomem menos água (irrigação localizada) e a redução das áreas exploradas (campanhas alternadas para as cooperativas do perímetro de Karfiguela).

Actualmente na sub-bacia, assistimos ao desenvolvimento das actividades de extracção de ouro (artesanal e industrial) e à utilização de mais pesticidas para o controlo de doenças e pragas das culturas. No entanto, os impactos de tais actividades nos recursos hídricos ainda não são actualmente bem compreendidos (MEA, 2016). Isto constitui, portanto, um forte apelo a estudos aprofundados para melhor compreender e proteger os recursos hídricos na sub-bacia.

4.2. Recomendações

Do acima exposto e para uma melhoria da gestão dos recursos hídricos na Haute-Comoé, fazemos as seguintes recomendações:

> Estabelecer um bom mecanismo de monitorização regular dos caudais ao longo do rio Comoé, através da revitalização das estações hidrométricas;

> Reforçar as capacidades da AEC, fornecendo-lhes meios técnicos e financeiros para controlar as retiradas de água reais dos utilizadores;

> fornecer ao CLE as ferramentas necessárias para monitorizar os caudais, a fim de melhor estimar as necessidades e melhorar a arbitragem da gestão da água;

> continuar a sensibilizar para o CFE e prever sanções para aqueles que não pagarem os impostos;

> organizar reuniões com utilizadores, em particular com a SN-SOCUCO, a fim de melhor compreender a sua posição e decidir sobre as modalidades da sua participação no CFE;

> harmonizar bem as taxas pagas pelos utilizadores para o funcionamento da CLE e da CFE;

> Rediscutir com os produtores as tarifas retidas para a taxa de captação de água para usos agrícolas com os utilizadores e explicar a razão de ser desta taxa a fim de obter o seu apoio antes da implementação do regulamento;

> operacionalizar as acções no terreno para que as realizações financiadas pelo CFE permitam uma melhoria das actividades dos contribuintes no terreno;

> Continuar a formar os produtores em técnicas de gestão de irrigação e reabilitar o perímetro de Karfiguela;

> Adoptar boas práticas agrícolas para reduzir o desperdício de água na rede de irrigação e na parcela e evitar o assoreamento dos reservatórios de água, protegendo as margens;

> e estudar a possibilidade de desenvolver outras fontes de extracção para colmatar a falta de água (barragens Touri e Niofila, poços de cultivo, mobilização de reservas de águas subterrâneas, etc.).

Conclusão e perspectivas

O objectivo do nosso estudo era avaliar a implementação da política de gestão integrada dos recursos hídricos na sub-bacia superior de Comoé. Podemos constatar que foram feitos enormes esforços para operacionalizar a GIRH a nível regional e através da criação de organismos e instituições relacionadas com a água que funcionam relativamente bem. Actualmente, a tendência é para uma boa gestão da água na sub-bacia, tendo em conta todos os sectores associados à partilha dos recursos hídricos e à sua utilização equilibrada, ecológica e sustentável. Também nos parece que se está a atribuir um certo valor económico à água, particularmente através da introdução de preços na captação de água. Isto prova que o sector pode assumir a responsabilidade de melhorar a gestão dos recursos hídricos.

A presença de um tal ambiente favorável à gestão do recurso limitado responde bem aos quatro princípios básicos da GIRH reconhecidos a nível internacional e entra na visão do Estado Burkinabe no que diz respeito à implementação da GIRH-NP a nível regional.

A partir das nossas hipóteses, podemos portanto confirmar que a operacionalização da GIRH é uma realidade na sub-bacia, mesmo que ainda seja necessário fazer esforços para melhorar a gestão dos recursos hídricos. Os métodos locais de gestão dos recursos hídricos pareceram-nos bastante eficazes porque o nível de organização dos utilizadores permitiu ultrapassar a crise da água em 2007 e os conflitos entre os utilizadores de água na sub-bacia. Relativamente à implementação das "Contribuições Financeiras da Água" a nível local, o processo está ainda na sua infância. Não existe um mecanismo muito bom para monitorizar as captações de água dos utilizadores. O sistema de financiamento ainda não é muito eficaz e precisa de ser apoiado por todos os utilizadores para uma maior eficiência e sustentabilidade.

O nosso trabalho permitiu-nos apresentar o ponto da situação da GIRH na bacia hidrográfica de Comoé Superior, contudo a falta de dados de caudal na área não permitiu uma melhor compreensão do estado da captação de água por parte das principais categorias de utilizadores.

Dada a complexidade deste assunto, é necessário um trabalho adicional para melhorar a modelação da gestão das águas superficiais na sub-bacia. O nosso trabalho poderia assim servir de base para outros estudos, particularmente no contexto da utilização de novas tecnologias de informação geográfica (ferramentas de detecção remota, ferramentas de planeamento integrado de recursos hídricos, etc.) nos processos de optimização da gestão dos recursos hídricos. Além disso, os impactos das actividades agrícolas e industriais sobre os recursos hídricos ainda não são bem conhecidos. Um estudo aprofundado permitirá um melhor conhecimento dos recursos hídricos e responder à questão de como conciliar a mobilização dos recursos hídricos e as opções de resposta com a abordagem da GIRH.

Bibliografia

AEDE, 2012. Ponto da situação - Cascatas, Novembro de 2012

Câmara de Comércio, Indústria e Artesanato do Burkina Faso, 2006. Données économiques et sociales du Burkina Faso, Ouagadougou, CCIA.

RCAP, 1994. Operação Arroz Comoé. Rapport de synthèse. DRA de la Comoé. 70 p.

Declarações da Conferência Internacional sobre a Água e o Ambiente em Dublin em Janeiro de 1992 (http://www.wmo.mt/pages/prog/hwrp/documents/francais/icwedecf.html) acedidas em 07 de Julho de 2017

Gautier Y., "CONFERÊNCIA DO RIO DE (1992)". Em Universalis education [online]. Encyclopædia Universalis, acedida a 7 de Julho de 2017. Disponível em http://www.universalis-edu.com/encyclopedie/rio-conference-de/

INSD, 2005. Inquérito anual às condições de vida dos agregados familiares (EA/QUIBB)

INSD, 2007. Resultados preliminares do recenseamento geral da população e habitação (RGPH) de 2006 do Burkina Faso, 51p.

Institut National de la Statistique et de le Démographie (INSD), 1996.

Lankoande O. e Sebego M., 2005. Monographie de la province de la Comoé, Ministério da Economia e do Desenvolvimento, Burina Faso, 136p.

Lankoande O. e Sebego M., 2008. Monographie de la région des Cascades; Ministério da Economia e do Desenvolvimento, Burina Faso, 148p.

MAHRH, 2003. Plano de Acção para a Gestão Integrada dos Recursos Hídricos no Burkina Faso Faso (PAGIRE), 62p. (http://documentation.2ieedu.org/cdi2ie/opac css/doc num.php?explnum id=41 acedido em 20 de Outubro de 2016)

MAHRH, 2009. Plano de Acção para a Gestão dos Recursos Hídricos (PAGIRE), segunda fase, Burkina Faso, 65p. (http://www.eauburkina.org/images/GIRE/PAGIRE phase%202 vf.pdf acedido em 20 de Outubro de 2016)

MCA, 2012. Rapport d'Etat des lieux du bassin de la Comoé. MCA-BF- AD9.1, 215pp

MEF/BF - SNAT, 2007. Perspectivas e cenários; 15p.

Ministério da Água e do Saneamento, 2016. Programme National pour la Gestion Intégrée des Ressources en Eau 2016-2030 - PNGIRE - (Versão Definitiva), Burkina Faso, 43p e anexos (http://www.pseau.org/outils/ouvrages/mea PNGIRE 2016 2030.pdf consultado em 30/07/2016)

Banco Africano de Desenvolvimento, 2000. La politique de gestion intégrée de ressources en eau,

Abidjan, éd Ocod, 46p
(https://www.afdb.org/fileadmin/uploads/afdb/Documents/Policy-Documents/10000016-FR-POLITIQUE-DE-GESTION-INTEGREE-DES-RESSOURCES-EAU.PDF acedido em 22 de Novembro de 2016)

Programas de Desenvolvimento Económico Local (PPDEL), 2003. Relatório da Economia Local de Cascades

Traoré, F., Cornet, Y., Denis, A., Wellens, J., Tychon, B., 2013. Monitorização da evolução das áreas irrigadas com imagens Landsat utilizando análise de detecção de alterações para a frente e para trás na bacia hidrográfica de Kou, Burkina Faso. Geocarto International, 28 (8), 733-752.

UM HABITAT, 2003. Resultados de um estudo sobre a economia local de Banfora

UN HABITAT, 2005. Elaboração do relatório nacional habitat iii de burkina faso http://habitat3.org/wp-content/uploads/National-Report-Africa-Burkina-Faso-Final-in- French.pdf consultado em 25 de Junho de 2017

VREO, 2010. Schéma Directeur d'Aménagement et de Gestion des Ressources en Eau du Bassin de la Comoé (SDAGE - Comoé). (Preliminary draft). Volume n°1. Análise e diagnóstico do estado dos recursos hídricos da bacia. Versão final/ Fevereiro de 2010. Projecto n°8 acpbk 038- 9acp bkk 010. 8º Fundo Europeu de Desenvolvimento.

WAIPRO-CILS-IWMI. (2009). Diagnóstico participativo e planeamento da acção do perímetro irrigado de Kartiguela, Província de Comoé-Burkina Faso, CILS-IWMI- projecto USAID, 42p.

WAIPRO-CILS-IWMI, 2010. Gestion des eaux du bassin de la Haute Comoé, CILS- IWMI-USAID project, 21p.

Wellens, J., 2014. Um quadro para a utilização de ferramentas de apoio à decisão em várias escalas espaciais para a gestão da agricultura irrigada no semi-árido.África Ocidental. Dissertação de doutoramento. Université de Liège, Liège, Bélgica. 106 p.

Wellens, J., Raes, D., Traoré, F., Denis, A., Djaby, B., 2013. Avaliação do desempenho do modelo AquaCrop da FAO para couve irrigada em parcelas de agricultores num ambiente semi-árido. Gestão da Água Agrícola, 127, 40-47.

Yeo W. E., 2008. Contribuição para a gestão integrada dos recursos hídricos na bacia hidrográfica de Comoé. Tese de Mestrado, Instituto de Engenharia e Água, Burkina Faso, 34p.

Anexos

Annexi: Tabela de taxas de taxa de captação de água bruta para actividades agrícolas, pastoris, piscicultura, aquacultura e silvicultura

1. Actividades hidro-agrícolas	*Taxa base*	*Taxa aplicável*		**Propostas das partes interessadas**	**Montante retido pela comissão**	
		Rega por gravidade ou similar	*Aspersão e micro-irrigação ou similar*		*Rega por gravidade ou similar*	*Aspersão e microirrigação ou similar*
o Cereais *(arroz. milho. ...) e tubérculos (batatas. inhame...)	500 FCFA /ha / ciclo de produção	-tipo de família: 500 FCFA/ha/ ciclo de produção -tipo empresarial: 1000FCFA/ha/ ciclo de produção	-tipo familiar: 300 FCFA/ha/ciclo de produção -tipo empresarial: 600 FCFA/ha/ciclo de produção	**500f/ha/ano**	**-Tipo de família: 500F/ha/ciclo de produção Tipo de empresa: 1000FCFA/ha/ciclo de produção**	**-tipo de família: 300 FCFA/ha/ciclo de produção -tipo empresarial: 600 FCFA/ha/ ciclo de produção**
o Jardinagem de mercado	2 000 FCFA /ha / ciclo de produção	-tipo de família: 2,000 FCFA/ha/ ciclo de produção -tipo empresarial: 4000FCF A/ha/ ciclo de produção	-família : 1,500FCF A/ha/ ciclo de produção -tipo empresarial: 3000FCF A/ha/ ciclo de produção	**4600 FCFA/ha/ ciclo de produção**	**-tipo de família: 2,000 FCFA/ha/ ciclo de produção -tipo empresarial: 4000FCFA/ha/ ciclo de produção**	**-tipo de família: 1,500FCFA/ha/ ciclo de produção -tipo empresarial: 3000FCFA/ha/ ciclo de produção**
o Cana-de-açúcar	2 000 FCFA /ha / ciclo de produção	-tipo de família: 2,000 FCFA/ha/ ciclo de produção	-tipo de família:! 500 FCFA/ha/ ciclo de produção -tipo empresarial : 3000FCF A/ha/ ciclo de produção	**2000FCFA/ha/chave de produção**	**-tipo de família: 1500 FCFA/ha/ciclo de produção**	**-Tipo de família: 1250 FCFA/ha/ ciclo de produção**
		-tipo empresarial: 4000FCF A/ha/ ciclo de produção			**-tipo empresarial: 3000FCFA/ha/ ciclo de produção**	**-tipo empresarial: 2500FCFA/ha/ ciclo de produção**
o Árvores de fruto	2 000 FCFA /ha/ ciclo de produção	2 000 FCFA/ha/ ciclo de produção	1 500 FCFA/ha/ ciclo de produção	**5 000 FCFA/ha/ ciclo de produção**	**2 000 FCFA/ha/ ciclo de produção**	**1 500 FCFA/ha/ ciclo de produção**
o Enfermeiros	1500 FCFA/ fazenda/ano	1500 FCFA/ fazenda/ano	1500 FCFA/ fazenda/ano	**2000 FCFA/ fazenda/ano**	**1500 FCFA/ fazenda/ano**	**1500 FCFA/ fazenda/ano**
2. Actividades pastorais	*Taxa base*	*Taxa aplicável*		**Proposta das partes interessadas (CFAF)**	**Montante retido pela comissão (FCFA)**	
		Bem, furo de sondagem, AEPS, lago artificial	*Lago natural, rio (desvio ou retirada directa)*		*Poço moderno, furo de sondagem, AEPS, lago artificial*	*Lago natural, rio (desvio ou retirada directa)*
o Criação tradicional	200 FCFA/UBT/ano	150FCFA/UBT/AN	200 FCFA/UBT/ANO	**357 FCFA /cattle**	**150FCFA/UBT/AN**	**200 FCFA/UBT/ANO**
o Pecuária intensiva (engorda,	200 FCFA/UBT/se	100 FCFA/UBT/semestre	200 FCFA/UBT/ANO		**100FCFA/UBT/AN**	**200 FCFA/UBT/ANO**

lacticínios, etc.) (estratégica)*	mestre					
o Criação de aves de capoeira (pelo menos 1000 cabeças)	500FCFA /100 0 cabeças/ano	500FCFA /1000 cabeças/An	500FCFA/1000 cabeças/An	**750 FCFA /1000 aves/ano**	**500 FCFA /1000 aves/ano**	**500 FCFA /1000 aves/ano**
o Agricultura da vida selvagem	200FCFA /head/ Um	200FCFA/head/An	200FCFA/head/An	**nenhuma**	**200FCFA/head/An**	**200FCFA/head/An**
o 3- Aquicultura	*Taxa base*	*Taxa aplicável*		**Propostas das partes interessadas (CFAF)**	**Montante retido pela comissão (FCFA)**	
		Sistemas sem recuperação de esgotos	*Sistemas com recuperação de esgotos*		*Sistemas sem recuperação de esgotos*	*Sistemas com recuperação de esgotos*
o Aquacultura em terra * (Lagoas, lagoas com infiltração de água)	Não definido	**-Tipo de família: 1000 FCFA/ha/ ciclo de produção -tipo empresarial: 2000 FCFA//ha/ ciclo de produção**	**-Tipo de família: 600 FCFA/ha/ ciclo de produção -tipo empresarial: 1200 FCFA/ha/ ciclo de produção**	**Nenhum**	**-Tipo de família: 1000 FCFA/ha/ ciclo de produção -tipo empresarial: 2000 FCFA//ha/ ciclo de produção**	**-Tipo de família: 600 FCFA/ha/ ciclo de produção -tipo empresarial: 1200 FCFA/ha/ ciclo de produção**
o Aquacultura * aquacultura em terra (tanques, e tanques que não são de sítio) a partir de IOOOm3	Não definido	**2 FCFA/m3/ciclo de produção**	**1 FCFA/m3/ciclo de produção**	**Nenhum**	**2FCFA/m3/ciclo de produção**	**1 FCFA/m3/ciclo de produção**

Fonte: Relatório de síntese dos trabalhos da comissão encarregada de finalizar o estudo sobre a elaboração dos textos relativos à aplicação do imposto sobre as captações de água bruta (DGRE)

Anexo 2: Ficha de inquérito sobre o estado da gestão integrada dos recursos hídricos na sub-bacia superior de Comoé, Burkina Faso

BURKINA FASO

Unidade-Progress-Justiça

MINISTÉRIO DA AGRICULTURA E GESTÃO DA ÁGUA

SECRETARIADO GERAL

DIRECÇÃO-GERAL DE GESTÃO DA ÁGUA E DESENVOLVIMENTO DA IRRIGAÇÃO (DGAHDI)

Ficha de inquérito sobre o estado da gestão integrada dos recursos hídricos na sub-bacia superior de Comoé,
Burkina Faso

Folha N°....

Data do inquérito :	Estrutura:
Nome do investigador :	

Parte A: Captações de água, volumes exigidos e contribuições *financeiras para a água*

NB: parte reservada apenas para os vários utilizadores de água

1. Necessidades hídricas

Qual tem sido a evolução das suas necessidades anuais de água ao longo dos últimos 5 anos?

Recurso mobilizável na estação seca												
PERÍODO	**2012**		**2013**		**2014**		**2015**		**2016**		**2017**	
	l/s	m3	l/s	m3	l/s	m3	l/s	m3	l/s	m3	l/s	m3
Janeiro												
Fevereiro												
Março												
Abril												
Maio												
Junho												
Julho												
TOTAL												

2. Qual é o volume ou taxa de fluxo real libertado para satisfazer as suas várias necessidades

Recurso real mobilizado na estação seca												
PERÍODO	**2012**		**2013**		**2014**		**2015**		**2016**		**2017**	
	l/s	m3	l/s	m3	l/s	m3	l/s	m3	l/s	m3	l/s	m3
Janeiro												
Fevereiro												
Março												
Abril												
Maio												
Junho												
Julho												
TOTAL												

3. Nível de conhecimentos sobre o estado dos recursos hídricos

Como avalia as suas necessidades e as quantidades de água disponíveis para as suas actividades (caudal, quantidade, qualidade da água)?

4. De que tanque se extrai a água

Reservatório de Moussodougou
Reservatório de Toussiana
Reservatório do Lobi
Entrada de água a montante das quedas
Entrada de água ao pé das quedas
Bombagem ao longo de riachos a jusante de reservatórios
Outros (por favor especifique)

5. Como é gerido o recurso hídrico com outros utilizadores

6. **Consulta das partes interessadas para a gestão da água**

Existe um quadro de consulta para a gestão da água a nível global?

Forte participação em grupos directivos, reuniões de trabalho com um grande número de interessados
Participação em reuniões com um número mais limitado de interessados
Participação mais ocasional em grupos directivos, algumas reuniões com outros interessados na água
Poucas ligações com outros actores
Sem consulta com outros interessados na água

7. Que meios e competências técnicas utiliza para a gestão da água? Isto permite-lhe implementar uma gestão integrada?

8. Quais são as principais dificuldades relacionadas com a gestão da água na sua área de intervenção?
1)
2)
3)
4)
5)

9. Quais são os desafios da gestão da água a nível global?

1) ....

2) ....

3) ....

4) ....

5) ....

10. Como se gerem os conflitos entre diferentes utilizadores para a procura de água?

....

....

....

....

11. Existe uma taxa para a captação de água bruta?

Sim Não

12. Em caso afirmativo, quanto é que paga por m3 de água utilizada?

....

....

13. Qual é o nível de recuperação para esta campanha?

....

....

14. Quais foram os níveis de recuperação nos anos anteriores de 2012 a 2016?

Campanhas	2011-2012	2012-2013	2013-2014	2014-2015	2015-2017
Taxa de recuperação (%)					

15. Como se explica o nível de recuperação?

....

....

....

16. Que estrutura é responsável pela recolha e gestão dos seus fundos?

Agência Cascades CLE/HC

Membro do comité da cooperativa Irrigator's Outro (por favor especificar)

17. Sabe para que é que é utilizado o financiamento?

Sim Não

18. Se sim, na sua opinião, quais são as utilizações dos seus recursos recolhidos (em percentagem) obras de manutenção Investimento para novas obras remuneração do pessoal de apoio remuneração do pessoal da LEC Outros (especificar)..

Outros (por favor especifique)..

Outros (por favor especifique)..

19. O que pensa sobre o preço da água bruta

20. Que propostas tem para melhorar a gestão da água a nível da bacia?

21. Que papel desempenham os serviços técnicos (AEC, DRAAH, DPAAH) na gestão dos recursos hídricos e as suas propostas de melhoramento?

Papéis

Sugestões

22. Qual é o papel das autoridades locais na gestão de recursos e as suas propostas de melhoramento?

Papéis

Sugestões

Parte B: Funcionamento da CLE/HC e do ACS

NB: Esta parte é reservada apenas para a CLE /HC e para a ACS

23. Estrutura do gestor

Em duas palavras, qual é a sua estrutura? Qual é o papel institucional da sua estrutura na gestão dos recursos (responsabilidade, legitimidade, competências alargadas)?

24. Quem são os seus membros

25. Quais foram os fluxos previstos para esta época?

Recurso mobilizável a partir de												
PERÍODO	**KARFIGUÉLA**		**MARAICHERS**		**ONEA**		**SOSUCO**		**Qrest**		**TOTAL**	
	l/s	m3	l/s	m3	l/s	m3	l/s	m3	l/s	m3	l/s	m3
Setembro												
Outubro												
Novembro												
Dezembro												
Janeiro												
Fevereiro												
Março												
Abril												
Maio												
Junho												
Julho												
Agosto												

TOTAL												

1. Que rendimento real foi libertado para estes diferentes utilizadores

Recursos mobilizados a partir de												
PERÍODO	**KARFIGUÉLA**		**MARAICHERS**		**ONEA**		**SOSUCO**		**Qrest**		**TOTAL**	
	l/s	m3	l/s	m3	l/s	m3	l/s	m3	l/s	m3	l/s	m3
Setembro												
Outubro												
Novembro												
Dezembro												
Janeiro												
Fevereiro												
Março												
Abril												
Maio												
Junho												
Julho												
Agosto												
TOTAL												

2.

3. Qual é a situação de recuperação (percentagem) de cada membro de 2012 a 2017?

Campanhas	2011-2012	2012-2013	2013-2014	2014-2015	2015-2017	Comentários
KARFIGUELA						
SN SOSUCO						
ONEA						
MARAICHERS						
OUTROS						

4. Como satisfazer as necessidades dos diferentes utilizadores

5. Que ferramentas de planeamento utiliza?

6. Quais são os desafios globais da gestão da água?

7. Como lidar com os conflitos entre diferentes agricultores sobre a procura de água

8. Quais são as suas dificuldades operacionais?

9. Quais são as vossas propostas para melhorar a gestão da água à escala global?

Anexo 3: Lista de pessoas consultadas

NOME, nome próprio	Função
Sr. SANON Fousseni	DR, ponto focal
Sr. COULIBALY Ousman	DP, Chefe do Departamento de Gestão da Água e Produção Agrícola
Sr. OULE Jean-Marcel	RD, Director DRAAH Cascades
Sr. BADO Mathias	DP, Director DPAAH Comoé
Sr TOU Moussa	Presidente dos Comités de Irrigadores
Sr. TOU Siriki	SG planície de Karfiguéla
Sr. OUATTARA Daouda	Chefe da Irrigação e Fertilização SN SOCUCO
Sr. BAKYONO Pierre Damien	AEC, Director-Geral
Sr. BAGAYA Ousséni	ACS, Director de Prospectiva e Planeamento
Sr. BADO Francis	ACS, Socioeconomista
Sr. KABORE Rémi	AEC, Técnico de Hidráulica
Sra. Sawadogo Céline	DREA, Director
Sr. ACKA Alexandre	DREA, Chefe do Departamento de Água Potável
Sr. NEBIE Babou	Chefe da equipa técnica da planície de Karfiguéla
Sr. SAWADOGO Jean-Ferdinand	SG CLE-HC
Senhora Tagnan Hema Djeneba	DP, Responsável Técnico Comum, Cooperativa 3
Sra. Compoaré Koné Awa	DP, Responsável Técnico Comum, Cooperativa 4
Sra. Seré Djénéba	DP, Responsável Técnico Comum, Cooperativas 1 e 5
Sr. Hien Michel	ONEA, chefe do centro de agrupamento Banfora
Sr. Coulibaly	ONEA, chefe da estação Banfora

CPSIA information can be obtained
at www.ICGtesting.com
Printed in the USA
LVHW042117310323
743153LV00020B/468

Touring Spa

Portugal

Also available:

Published by The Caravan and Motorhome Club Limited
East Grinstead House, East Grinstead
West Sussex RH19 1UA

General Enquiries: 01342 326944
Travel Service Reservations: 01342 316101
Red Pennant Overseas Holiday Insurance: 01342 336633
Website: camc.com

Editor: Kate Walters
Publishing service provided by Fyooz Ltd
Printed by Stephens & George Ltd
Merthyr Tydfil
ISBN 978-1-9993236-1-5

Maps and distance charts generated from Collins Bartholomew Digital Database

Welcome...

...to another year of touring across Europe!

Wherever and however you travel, people who camp (whether in a caravan, motorhome, campervan or tent) share a sense of adventure and freedom that you don't get from other types of holidays.

However as we enter 2019 we must be aware that there is a great period of change ahead for the UK which may affect the way we travel to the continent. While it is unlikely that Brexit will stop us from touring the EU, you might need to plan further ahead and do more research before you travel.

At the time of this book going to press much of what will change once the UK leaves the EU is unknown, and therefore it is very difficult to advise on issues such as border controls, customs, visa requirements and pet passports. Therefore you're strongly advised to check this information before you travel, especially if the UK has already left or will leave the EU while you're travelling. Check camc.com/overseas for our most up-to-date advice.

So as you start another year of touring adventures, I would like to thank you for continuing to buy and contribute to these guides. If you can, please spare five minutes to fill in one of the site report forms at the back of this book or visit camc.com/europereport to let us know what you think about the sites you've stayed on this year. Happy touring!

Kate Walters

Kate Walters, Editor

Contents

Continental Campsites

Site Listings

How to use this guide

The information contained within Touring Spain & Portugal is presented in the following categories:

The Handbook

This includes general information about touring in France and Andorra, such as legal requirements, advice and regulations. The Handbook chapters are at the front of the guide and are separated as follows:

Planning Your Trip	Information you'll need before you travel including information on documents and insurance, advice on money, customs regulations and planning your channel crossings.
Motoring Advice	Advice on motoring overseas, essential equipment and roads in Europe including mountain passes and tunnels.
During Your Stay	Information for while you're away including telephone, internet and TV advice, medical information and advice on staying safe.

Country Introduction

Following on from the Handbook chapters you will find the Country Introductions containing information, regulations and advice specific to each country. You should read the Country Introduction in conjunction with the Handbook chapters before you set off on your holiday.

Campsite Entries

After the country introduction you will find the campsite entries listed alphabetically under their nearest town or village. Where there are several campsites shown in and around the same town they will be listed in clockwise order from the north.

To find a campsite all you need to do is look for the town or village of where you would like to stay, or use the maps at the back of the book to find a town where sites are listed. In order to provide you with the details of as many site as possible in Touring Spain & Portugal we use abbreviations in the site entries.

For a full list of these abbreviations please see the following pages of this section.

We have also included some of the most regularly used abbreviations, as well as an explanation of a campsite entry, on the fold-out on the back cover.

Campsite Fees

Campsite entries show high season fees per night for an outfit plus two adults, as at the year of the last report. Prices given may not include electricity or showers, unless indicated. Outside of the main holiday season many sites offer discounts on the prices shown and some sites may also offer a reduction for longer stays.

Campsite fees may vary to the prices stated in the site entries, especially if the site has not been reported on for a few years. You are advised to always check fees when booking, or at least before pitching, as those shown in site entries should be used as a guide only.

Site Maps

Each town and village listed alphabetically in the site entry pages has a map grid reference number, e.g. 3B4. The map grid reference number is shown on each site entry.

The maps can be found at the end of the book. The reference number will show you where each town or village is located, and the site entry will tell you how far the site is from that town. Place names are shown on the maps in two colours:

Red where we list a site which is open all year (or for at least eleven months of the year)

Black where we only list seasonal sites which close in winter.

These maps are intended for general campsite location purposes only; a detailed road map or atlas is essential for route planning and touring.

Town names in capital letters (RED, BLACK or in ITALICS) correspond with towns listed on the Distance Chart.

The scale of the map means that it isn't possible to show every town or village where a campsite is listed, so some sites in small villages may be listed under a nearby larger town instead.

Satellite Navigation

Most campsite entries now show a GPS (sat nav) reference. There are several different formats of writing co-ordinates, and in this guide we use decimal degrees, for example 48.85661 (latitude north) and 2.35222 (longitude east).

Minus readings, shown as -1.23456, indicate that the longitude is west of the Greenwich meridian. This will only apply to sites in the west of France, most of Spain and all of Portugal as the majority of Europe are east of the Greenwich meridian.

Manufacturers of sat navs all use different formats of co-ordinates so you may need to convert the co-ordinates before using them with your device. There are plenty of online conversion tools which enable you to do this quickly and easily - just type 'co-ordinate converter' into your search engine.

Please be aware if you are using a sat nav device some routes may take you on roads that are narrow and/or are not suitable for caravans or large outfits.

The GPS co-ordinates given in this guide are provided by members and checked wherever possible, however we cannot guarantee their accuracy due to the rural nature of most of the sites. The Caravan and Motorhome Club cannot accept responsibility for any inaccuracies, errors or omissions or for their effects.

Site Report Forms

With the exception of campsites in The Club's Overseas Site Booking Service (SBS) network, The Caravan and Motorhome Club does not inspect sites listed in this guide. Virtually all of the sites listed in Touring Spain & Portugal are from site reports submitted by users of these guides. You can use the forms at the back of the book or visit camc.com/europereport tell us about great sites you have found or update the details of sites already within the books.

Sites which are not reported on for five years are deleted from the guide, so even if you visit a site and find nothing different from the site listing we'd appreciate an update to tell us as much.

You will find site report forms towards the back of this guide which we hope you will complete and return to us. Use the abbreviated site report form if you are reporting no changes, or only minor changes, to a site entry. The full report form should be used for new sites or sites which have changed a lot since the last report.

You can complete both the full and abbreviated versions of the site report forms by visiting camc.com/europereport.

Please submit reports as soon as possible. Information received by mid August 2019 will be used wherever possible in the next edition of Touring Spain & Portugal. Reports received after that date are still very welcome and will appear in the following edition. The editor is unable to respond individually to site reports submitted due to the large quantity that we receive.

Tips for Completing Site Reports

- If possible fill in a site report form while at the campsite. Once back at home it can be difficult to remember details of individual sites, especially if you visited several during your trip.
- When giving directions to a site, remember to include the direction of travel, e.g. 'from north on D137, turn left onto D794 signposted Combourg' or 'on N83 from Poligny turn right at petrol station in village'. Wherever possible give road numbers, junction numbers and/or kilometre post numbers, where you exit from motorways or main roads. It is also helpful to mention useful landmarks such as bridges, roundabouts, traffic lights or prominent buildings.

We very much appreciate the time and trouble you take submitting reports on campsites that you have visited; without your valuable contributions it would be impossible to update this guide.

Acknowledgements

Thanks go to the AIT/FIA Information Centre (OTA), the Alliance Internationale de Tourisme (AIT), the Fédération International de Camping et de Caravaning (FICC) and to the national clubs and tourist offices of those countries who have assisted with this publication.

Every effort is made to ensure that information provided in this publication is accurate. The Caravan and Motorhome Club Ltd has not checked these details by inspection or other investigation and cannot accept responsibility for the accuracy of these reports as provided by members and non-members, or for errors, omissions or their effects. In addition The Caravan and Motorhome Club Ltd cannot be held accountable for the quality, safety or operation of the sites concerned, or for the fact that conditions, facilities, management or prices may have changed since the last recorded visit. Any recommendations, additional comments or opinions have been contributed by people staying on the site and are not those of The Caravan and Motorhome Club.

The inclusion of advertisements or other inserted material does not imply any form of approval or recognition, nor can The Caravan and Motorhome Club Ltd undertake any responsibility for checking the accuracy of advertising material.

Explanation of a Campsite Entry

The town under which the campsite is listed, as shown on the relevant Sites Location Map at the end of each country's site entry pages

Distance and direction of the site from the centre of the town the site is listed under in kilometres (or metres), together with site's aspect

Site Location Map grid reference

Indicates that the site is open all year

Telephone and fax numbers including national code where applicable

Campsite name

Contact email and website address

Description of the campsite and its facilities

Directions to the campsite

Charge per night in high season for car, caravan + 2 adults as at year of last report

Unspecified facilities for disabled guests.

The year in which the site was last reported on by a visitor

> 🅣 **BLANES** *3C3* (1km S Coastal) *41.65933, 2.77000* **Camping Blanes, Avda Vila de Madrid 33, 17300 Blanes (Gerona) [972-33 15 91; fax 972-33 70 63; info@campingblanes.com;www.campingblanes.com]** Fr N on AP7/E15 exit junc 9 onto NII dir Barcelona & foll sp Blanes. Fr S to end of C32, then NII dir Blanes. On app Blanes, foll camping sps & Playa S'Abanell - all campsites are sp at rndabts; all sites along same rd. Site adj Hotel Blau-Mar. Lge, mkd pitch, shd; wc; chem disp; mv service pnt; shwrs inc; shop; el pnts (5A) inc; gas; lndtte; supmkt; snacks high ssn; bar; playgrnd; pool; solarium; dir access to sand beach; watersports; cycle hire; games rm; wifi; entmnt; dogs; phone; bus; poss cr; Eng spkn; quiet; ccard acc; red low ssn. "Excel site, espec low ssn; helpful owner; easy walk to town cent; trains to Barcelona & Gerona." ♦
> € 36.95 2011*

Campsite address

GPS co-ordinates – latitude and longitude in decimal degrees. Minus figures indicate that the site is west of the Greenwich meridian

> **NOJA** *1A4* (700m N Coastal) *43.49011, -3.53636* **Camping Playa Joyel, Playa del Ris, 39180 Noja (Cantabria) [942-63 00 81; fax 942-63 12 94; playajoyel@telefonica.net; www.playajoyel.com]** Fr Santander or Bilbao foll sp A8/E70 (toll-free). Approx 15km E of Solares exit m'way junc 185 at Beranga onto CA147 N twd Noja & coast. On o'skirts of Noja turn L sp Playa del Ris, (sm brown sp) foll rd approx 1.5km to rndabt, site sp to L, 500m fr rndabt. Fr Santander take S10 for approx 8km, then join A8/E70. V lge, mkd pitch, pt sl, pt shd; wc; chem disp; mv service pnt; baby facs; shwrs inc; el pnts (6A) inc; gas; lndtte (inc dryer); supmkt; tradsmn; rest; snacks; bar; BBQ (gas/charcoal); playgrnd; pool; paddling pool; jacuzzi; direct access to sand beach adj; windsurfing; sailing; tennis; hairdresser; car wash; cash dispenser; wifi; entmnt; games/TV rm; 15% statics; no dogs; no c'vans/m'vans over 8m high ssn; phone; recep 0800-2200; poss v cr w/end & high ssn; Eng spkn; adv bkg; ccard acc; quiet at night; red low ssn/snr citizens; CCI. "Well-organised site on sheltered bay; v busy high ssn; pleasant staff; gd, clean facs; superb pool & beach; some narr site rds with kerbs; midnight silence enforced; Wed mkt outside site; highly rec." ♦ 15 Apr-1 Oct. (CChq acc) € 47.40 SBS - E05 2011*

Comments and opinions of caravanners who have visited the site (within inverted commas)

The site accepts Camping Cheques, see the Continental Campsites chapter for details

Booking reference for a site the Club's Overseas Travel Service work with, i.e. bookable via The Club.

Opening dates

Site Description Abbreviations

Each site entry assumes the following unless stated otherwise:

Level ground, open grass pitches, drinking water on site, clean wc unless otherwise stated (own sanitation required if wc not listed), site is suitable for any length of stay within the dates shown.

aspect

- urban – within a city or town, or on its outskirts
- rural – within or on edge of a village or in open countryside
- coastal – within one kilometre of the coast

size of site

- sm – max 50 pitches
- med – 51 to 150 pitches
- lge – 151 to 500 pitches
- v lge – 501+ pitches

pitches

- hdg pitch – hedged pitches
- mkd pitch – marked or numbered pitches
- hdstg – some hard standing or gravel

levels

- sl – sloping site
- pt sl – sloping in parts
- terr – terraced site

shade

- shd – plenty of shade
- pt shd – part shaded
- unshd – no shade

Site Facilities

adv bkg -

- acc - advance booking accepted
- rec – advance booking recommended
- req - advance booking required

beach

Beach for swimming nearby;

- 1km – distance to beach
- sand beach – sandy beach
- shgl beach – shingle beach

bus/metro/tram

Public transport within 5km

chem disp

Dedicated chemical toilet disposal facilities;

chem disp (wc) – no dedicated point; disposal via wc only

CKE/CCI

Camping Key Europe and/or Camping Card International accepted

CL-type

Very small, privately-owned, informal and usually basic, farm or country site similar to those in the Caravan and Motorhome Club's network of Certificated Locations

dogs

Dogs allowed on site with appropriate certification (a daily fee may be quoted and conditions may apply)

el pnts

Mains electric hook-ups available for a fee;

- inc – cost included in site fee quoted
- 10A – amperage provided
- conn fee – one-off charge for connection to metered electricity supply
- rev pol – reversed polarity may be present

(see Electricity and Gas in the section DURING YOUR STAY)

Eng spkn

English spoken by campsite reception staff

entmnt

Entertainment facilities or organised entertainment for adults and/or children

fam bthrm

Bathroom for use by families with small children

gas

Supplies of bottled gas available on site or nearby

Wi-fi

Wireless local area network available

lndry

Washing machine(s) with or without tumble dryers, sometimes other equipment available, eg ironing boards

Mairie

Town hall (France); will usually make municipal campsite reservations

mv service pnt

Special low level waste discharge point for motor caravans; fresh water tap and rinse facilities should also be available

NH
Suitable as a night halt

open 1 Apr-15 Oct
Where no specific dates are given, opening dates are assumed to be inclusive, ie Apr-Oct – beginning April to end October
(NB: opening dates may vary from those shown; check before travelling, particularly when travelling out of the main holiday season)

phone
Public payphone on or adjacent to site

playgrnd
Children's playground

pool
Swimming pool (may be open high season only);
htd – heated pool
covrd – indoor pool or one with retractable cover

quiet site
peaceful, tranquil site

red CCI/CCS
Reduction in fees on production of a Camping Card International or Camping Card Scandinavia

rest
Restaurant on site or nearby;
bar – bar on site or nearby
BBQ – barbecues allowed (may be restricted to a separate, designated area)
cooking facs – communal kitchen area
snacks – snack bar or takeaway on site

SBS
Site Booking Service (pitch reservation can be made through the Club's Travel Service)

serviced pitch
Electric hook-ups and mains water inlet and grey water waste outlet to pitch;
all – to all pitches
50% – percentage of pitches

shop
on site – shop(s) on site
nearby – within 2km
supmkt – supermarket
hypmkt – hypermarket

shwrs
Hot showers available for a fee;
inc – cost included in site fee quoted

ssn
Season;
high ssn – peak holiday season
low ssn – out of peak season

50% statics
Percentage of static caravans/mobile homes/chalets/fixed tents/cabins or long term seasonal pitches on site, including those run by tour operators

sw
Swimming nearby;
1km – nearest swimming
lake – in lake
rv – in river

TV
TV available for viewing by visitors;
room – separate TV room (often also a games room)
pitch – cable or satellite connections to pitches

wc
Clean flushing toilets on site;
(cont) – continental type with floor-level hole
htd – sanitary block centrally heated in winter
own san – use of own sanitation facilities recommended

Other Abbreviations

AIT	Alliance Internationale de Tourisme
a'bahn	Autobahn
a'pista	Autopista
a'route	Autoroute
a'strada	Autostrada
adj	Adjacent, nearby
alt	Alternative
app	Approach, on approaching
arr	Arrival, arriving
avail	Available
Ave	Avenue
bdge	Bridge
bef	Before
bet	Between
Blvd	Boulevard
C	Century, eg 16thC
c'van	Caravan
CC	Caravan and Motorhome Club
ccard acc	Credit and/or debit cards accepted (check with site for specific details)
CChq acc	Camping Cheques accepted
cent	Centre or central
clsd	Closed
conn	Connection
cont	Continue or continental (wc)
conv	Convenient
covrd	Covered
dep	Departure
diff	Difficult, with difficulty

dir	Direction
dist	Distance
dual c'way	Dual carriageway
E	East
ent	Entrance/entry to
espec	Especially
ess	Essential
excel	Excellent
facs	Facilities
FIA	Fédération Internationale de l'Automobile
FICC	Fédération Internationale de Camping & de Caravaning
FFCC	Fédération Française de Camping et de Caravaning
FKK/FNF	Naturist federation, ie naturist site
foll	Follow
fr	From
g'ge	Garage
gd	Good
grnd(s)	Ground(s)
hr(s)	Hour(s)
immac	Immaculate
immed	Immediate(ly)
inc	Included/inclusive
indus est	Industrial estate
INF	Naturist federation, ie naturist site
int'l	International
irreg	Irregular
junc	Junction
km	Kilometre
L	Left
LH	Left-hand
LS	Low season
ltd	Limited
mkd	Marked
mkt	Market
mob	Mobile (phone)
m'van	Motor caravan
m'way	Motorway
N	North
narr	Narrow
nr, nrby	Near, nearby
opp	Opposite
o'fits	Outfits
o'look(ing)	Overlook(ing)
o'night	Overnight
o'skts	Outskirts
PO	Post office
poss	Possible, possibly
pt	Part
R	Right
rd	Road or street

rec	Recommend/ed
recep	Reception
red	Reduced, reduction (for)
reg	Regular
req	Required
RH	Right-hand
rlwy	Railway line
rm	Room
rndabt	Roundabout
rte	Route
RV	Recreational vehicle, ie large motor caravan
rv/rvside	River/riverside
S	South
san facs	Sanitary facilities ie wc, showers, etc
snr citizens	Senior citizens
sep	Separate
sh	Short
sp	Sign post, signposted
sq	Square
ssn	Season
stn	Station
strt	Straight, straight ahead
sw	Swimming
thro	Through
TO	Tourist Office
tour ops	Tour operators
traff lts	Traffic lights
twd	Toward(s)
unrel	Unreliable
vg	Very good
vill	Village
W	West
w/end	Weekend
x-ing	Crossing
x-rds	Cross roads

Symbols Used

◆ Unspecified facilities for disabled guests check before arrival

 Open all year

* Last year site report received (see Campsite Entries in Introduction)

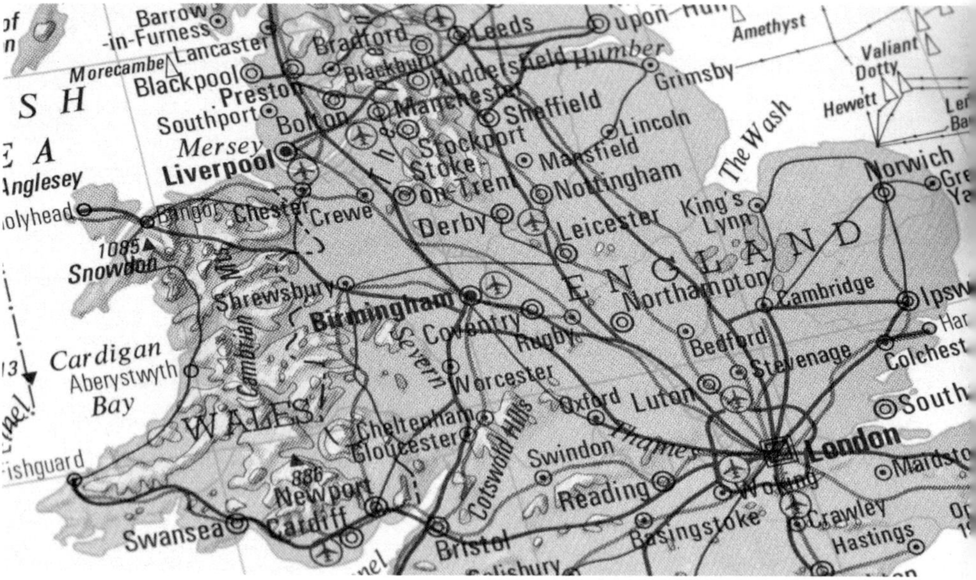

Documents

Camping Card Schemes

Camping Key Europe (CKE) is a useful touring companion. Not only does it serve as ID at campsites, meaning that you don't have to leave your passport at reception, it also entitles you to discounts at over 2200 sites.

CKE also offers third-party liability insurance for families including up to three children, which provides cover for loss or damage that occurs while on site. For more information on the scheme and all its benefits visit www.campingkey.com.

You can purchase the CKE from the Club by calling 01342 336633, or it is provided free for Red Pennant Overseas Holiday Insurance customers taking out 'Motoring' cover.

An alternative scheme is Camping Card International (CCI) - to find out more visit www.campingcardinternational.com.

If you are using a CKE or CCI card as ID at a site, make sure that you collect your card when checking out. Also check that you have been given your own card instead of someone else's.

Driving Licence

A full (not provisional), valid driving licence should be carried at all times when driving abroad. You must produce it when asked to do so by the police and other authorities, or you may be liable for an immediate fine and confiscation of your vehicle(s).

If your driving licence is due to expire while you are away it can be renewed up to three months before expiry - contact the DVLA if you need to renew more than three months ahead.

All EU countries recognise the photocard driving licence introduced in the UK in 1990, subject to the minimum age requirements (normally 18 years for a vehicle with a maximum weight of 3,500 kg carrying no more than 8 people).

Old-style green paper licences or Northern Irish licences issued before 1991 should be updated before travelling as they may not be recognised by local authorities.

Selected post offices and DVLA local offices offer a premium checking service for photocard applications but the service is not available for online applications.

MOT Certificate

Carry your vehicle's MOT certificate (if applicable) when driving on the Continent. You may need to show it to the authorities if your vehicle is involved in an accident, or in the event of random vehicle checks. If your MOT certificate is due to expire while you are away you should have the vehicle tested before you leave home.

Passport

In many EU countries everyone is required to carry photographic ID at all times. Enter next-of-kin details in the back of your passport and keep a separate photocopy. It's also a good idea to leave a photocopy of it with a relative or friend at home.

The following information applies to British passport holders only. For information on passports issued by other countries you should contact the local embassy.

Applying for a Passport

Each person travelling out of the UK (including babies) must hold a valid passport - it is no longer possible to include children on a parent's passport. A standard British passport is valid for ten years, or 5 years for under 16s.

All newly issued UK passports are now biometric, also known as e-passports, which contain a microchip with information which can be used to authenticate the holder's identity.

Full information and application forms are available from main post offices or from the Identity & Passport Service's website, www.gov.uk where you can complete an online application. Allow at least six weeks for first-time passport applications, for which you may need to attend an interview at your nearest Identity and Passport Service (IPS) regional office. Allow three weeks for a renewal application or replacement of a lost, stolen or damaged passport.

Post offices offer a 'Check & Send' service for passport applications which can prevent delays due to errors on your application form. To find your nearest 'Check & Send' post office call 0345 611 2970 or see www.postoffice.co.uk.

Passport Validity

Most countries in the EU only require your passport to be valid for the duration of your stay. However, in case your return home is delayed it is a good idea make sure you have six month's validity remaining. Any time left on a passport (up to a maximum of nine months) will be added to the validity of your new passport on renewal.

Schengen Agreement

The Schengen Agreement allows people and vehicles to pass freely without border checks from country to country within the Schengen area (a total of 26 countries). Where there are no longer any border checks you should still not attempt to cross land borders without a full, valid passport. It is likely that random identity checks will continue to be made for the foreseeable future in areas surrounding land borders. The United Kingdom and Republic of Ireland do not fully participate in the Schengen Agreement.

Regulations for Pets

Some campsites do not accept dogs at all and some have restrictions on the number and breed of dogs allowed. Visit camc.com/overseasadvice for more information and country specific advice.

In popular tourist areas local regulations may ban dogs from beaches during the summer months.

Pet Travel Scheme (PETS)

The Pet Travel Scheme (PETS) allows owners of dogs, cats and ferrets from qualifying European countries, to bring their pets into the UK (up to a limit of five per person) without quarantine. The animal must have an EU pet passport, be microchipped and be vaccinated against rabies. Dogs must also have been treated for tapeworm. It also allows pets to travel from the UK to other EU qualifying countries.

There are country specific regulations regarding certain breeds of dogs. You can't import breeds classed as dangerous dogs to many countries, and other breeds will require additional documentation. For more information or to find out which breeds are banned or restricted visit camc.com/pets or call us on 01342 336766.

Pets resident anywhere in the British Isles (excluding the Republic of Ireland) are able to travel freely within the British Isles and are not subject to PETS rules.

For details of how to obtain a Pet Passport visit www.defra.gov.uk of call 0370 241 1710.

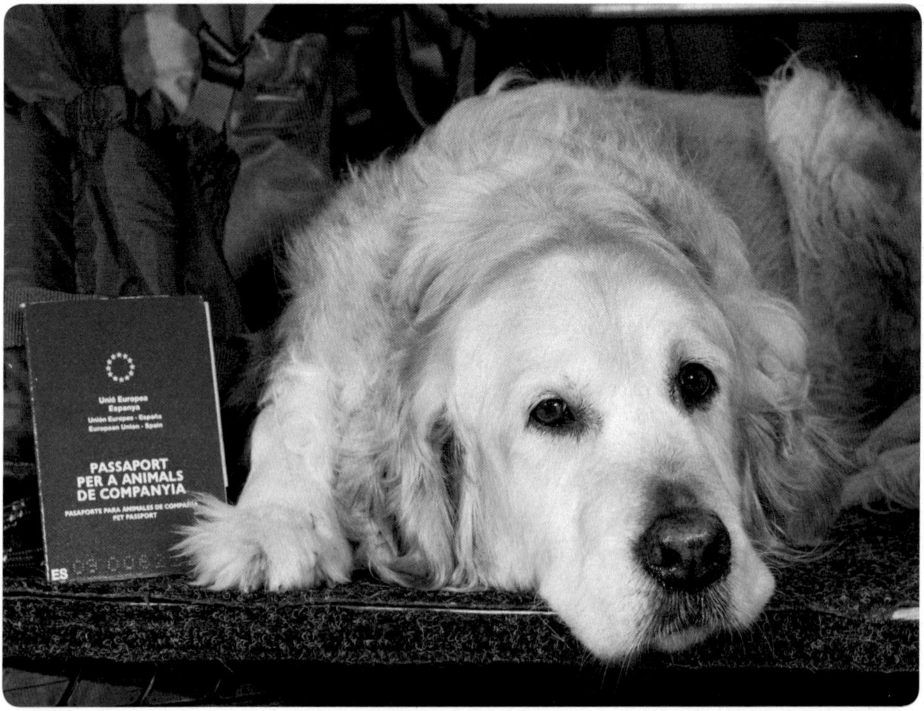

Returning to the UK

On your return to the UK with your pet you will need to visit a vet between 24 and 120 hours prior to your return journey in order for your pet to be treated for tapeworm. The vet will need to sign your pet passport - ensure that they put the correct date against their signature or you may not fall within the correct time range for travel. Ask your campsite to recommend a local vet, or research vets near to the port you will be returning from before you travel.

Travelling with Children

Some countries require evidence of parental responsibility for people travelling alone with children, especially those who have a different surname to them (including lone parents and grandparent). The authorities may want to see a birth certificate, a letter of consent from the child's parent (or other parent if you are travelling alone with your own child) and some evidence as to your responsibility for the child.

For further information on exactly what will be required at immigration contact the Embassy or Consulate of the countries you intend to visit.

Vehicle Tax

While driving abroad you still need to have current UK vehicle tax. If your vehicle's tax is due to expire while you are abroad you may apply to re-license the vehicle at a post office, by post, or in person at a DVLA local office, up to two months in advance.

Since October 2014 the DVLA have no longer issued paper tax discs - EU Authorities are aware of this change.

Vehicle Registration Certificate (V5C)

You must always carry your Vehicle Registration Certificate (V5C) and MOT Certificate (if applicable) when taking your vehicle abroad. If yours has been lost, stolen or destroyed you should apply to a DVLA local office on form V62. Call DVLA Customer Enquiries on 0300 790 6802 for more information.

Caravan – Proof of Ownership (CRIS)

In Britain and Ireland, unlike most other European countries, caravans are not formally registered in the same way as cars. This may not be fully understood by police and other authorities on the Continent. You are strongly advised, therefore, to carry a copy of your Caravan Registration Identification Scheme (CRIS) document.

Hired or Borrowed Vehicles

If using a borrowed vehicle you must obtain a letter of authority to use the vehicle from the registered owner. You should also carry the Vehicle Registration Certificate (V5C).

In the case of hired or leased vehicles, including company cars, when the user does not normally possess the V5C, ask the company which owns the vehicle to supply a Vehicle On Hire Certificate, form VE103, which is the only legal substitute for a V5C. The BVRLA, the trade body for the vehicle rental and leasing sector, provide advice on hired or leased vehicles - see www.bvrla.co.uk or call them on 01494 434747 for more information.

If you are caught driving a hired vehicle abroad without this certificate you may be fined and/or the vehicle impounded.

Visas

British citizens holding a full UK passport do not require a visa for entry into any EU countries, although you may require a permit for stays of more than three months. Contact the relevant country's UK embassy before you travel for information.

British subjects, British overseas citizens, British dependent territories citizens and citizens of other countries may need visas that are not required by British citizens. Again check with the authorities of the country you are due to visit at their UK embassy or consulate. Citizens of other countries should apply to their own embassy, consulate or High Commission.

Insurance

Car, Motorhome and Caravan Insurance

It is important to make sure your outfit is covered whilst you are travelling abroad. Your car or motorhome insurance should cover you for driving in the EU, but check what you are covered for before you travel. If you are travelling outside the EU or associated countries you'll need to inform your insurer and may have to pay an additional premium.

Make sure your caravan insurance includes travel outside of the UK, speak to your provider to check this. You may need to notify them of your dates of travel and may be charged an extra premium dependent on your level of cover.

The Caravan and Motorhome Club's Car, Caravan Cover and Motorhome Insurance schemes extend to provide policy cover for travel within the EU free of charge, provided the total period of foreign travel in any one year does not exceed 270 days for Car and Motorhome Insurance and 182 for Caravan Insurance. It may be possible to extend this period, although a charge may apply.

Should you be delayed beyond these limits notify your broker or insurer immediately in order to maintain your cover until you can return to the UK.

If your outfit is damaged during ferry travel (including while loading or unloading) it must be reported to the carrier at the time of the incident. Most insurance policies will cover short sea crossings (up to 65 hours) but check with your insurer before travelling.

Visit camc.com/insurance or call 01342 336610 for full details of our Caravan Cover or for Car or Motorhome Insurance call 0345 504 0334.

European Accident Statement

Your car or motorhome insurer may provide you with a European Accident Statement form (EAS), or you may be given one if you are involved in an accident abroad. The EAS is a standard form, available in different languages, which gives all parties involved in an accident the opportunity to agree on the facts. Signing the form doesn't mean that you are accepting liability, just that you agree

with what has been stated on the form. Only sign an EAS if you are completely sure that you understand what has been written and always make sure that you take a copy of the completed EAS.

Vehicles Left Behind Abroad

If you are involved in an accident or breakdown abroad which prevents you taking your vehicle home, you must ensure that your normal insurance will cover your vehicle if left overseas while you return home. Also check if you're covered for the cost of recovering it to your home address.

In this event you should remove all items of baggage and personal belongings from your vehicles before leaving them unattended. If this isn't possible you should check with your insurer if extended cover can be provided. In all circumstances, you must remove any valuables and items liable for customs duty, including wine, beer, spirits and cigarettes.

Legal Costs Abroad

If an accident abroad leads to you being taken to court you may find yourself liable for legal costs – even if you are not found to be at fault. Most UK vehicle insurance policies include cover for legal costs or have the option to add cover for a small additional cost – check if you are covered before you travel.

Holiday Travel Insurance

A standard motor insurance policy won't cover you for all eventualities, for example vehicle breakdown, medical expenses or accommodation so it's important to also take out adequate travel insurance. Make sure that the travel insurance you take out is suitable for a caravan or motorhome holiday.

Remember to check exemptions and exclusions, especially those relating to pre-existing medical conditions or the use of alcohol. Be sure to declare any pre-existing medical conditions to your insurer.

The Club's Red Pennant Overseas Holiday Insurance is designed specifically for touring holidays and can cover both motoring and personal use. Depending on the level of cover chosen the policy will cover you for vehicle recovery and repair, holiday continuation, medical expenses and accommodation.

Visit camc.com/redpennant for full details or call us on 01342 336633.

Holiday Insurance for Pets

Taking your pet with you? Make sure they're covered too. Some holiday insurance policies, including The Club's Red Pennant, can be extended to cover pet expenses relating to an incident normally covered under the policy – such as pet repatriation in the event that your vehicle is written off.

However in order to provide cover for pet injury or illness you will need a separate pet insurance policy which covers your pet while out of the UK. For details of The Club's Pet Insurance scheme visit camc.com/petins or call 0345 504 0336.

Home Insurance

Your home insurer may require advance notification if you are leaving your home unoccupied for 30 days or more. There may be specific requirements, such as turning off mains services (except electricity), draining water down and having somebody check your home periodically. Read your policy documents or speak to your provider.

The Club's Home Insurance policy provides full cover for up to 90 days when you are away from home (for instance when touring) and requires only common sense precautions for longer periods of unoccupancy. See camc.com/homeins or call 0345 504 0335 for details.

Personal Belongings

The majority of travellers are able to cover their valuables such as jewellery, watches, cameras, laptops, and bikes under a home insurance policy. This includes the Club's Home Insurance scheme.

Specialist gadget insurance is now commonly available and can provide valuable benefits if you are taking smart phones, tablets, laptops or other gadets on holiday with you. The Club offers a Gadget Insurance policy - visit camc.com/gadget or call 01342 779413 to find out more.

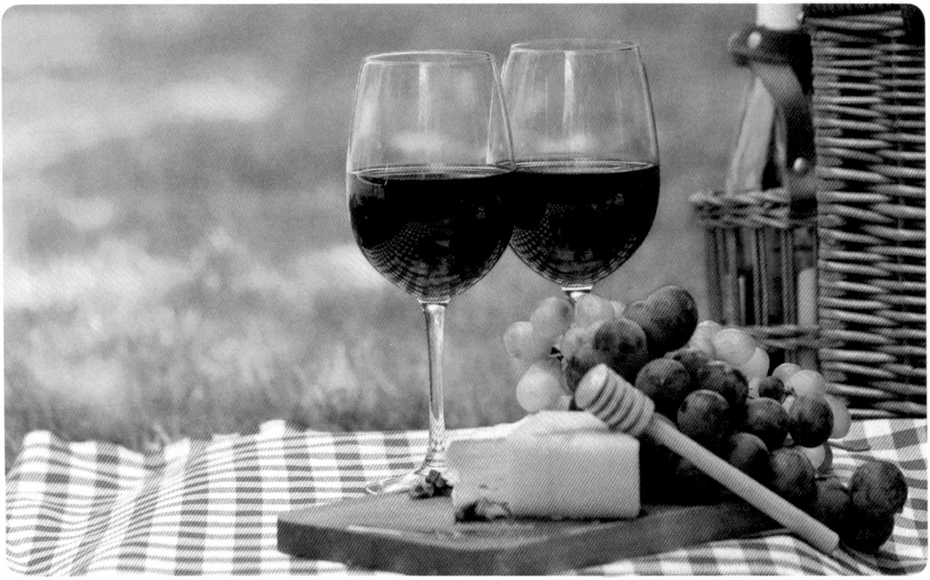

Customs Regulations

Caravans and Vehicles

You can temporarily import a caravan, trailer tent or vehicle from one EU country to another without any Customs formalities. Vehicles and caravans may be temporarily imported into non-EU countries generally for a maximum of six months in any twelve month period, provided they are not hired, sold or otherwise disposed of in that country.

If you intend to stay longer than six months, dispose of a vehicle while in another country or leave your vehicle there in storage you should seek advice well before your departure from the UK.

Borrowed Vehicles

If you are borrowing a vehicle from a friend or relative, or loaning yours to someone, you should be aware of the following:

- The total time the vehicle spends abroad must not exceed the limit for temporary importation (generally six months).
- The owner of the caravan must provide the other person with a letter of authority.
- The owner cannot accept a hire fee or reward.
- The number plate on the caravan must match the number plate on the tow car.
- Both drivers' insurers must be informed if a caravan is being towed and any additional premium must be paid.

Currency

You must declare cash of €10,000 (or equivalent in other currencies) or more when travelling between the UK and a non-EU country. The term 'cash' includes cheques, travellers' cheques, bankers' drafts, notes and coins. You don't need to declare cash when travelling within the EU.

For further information contact HMRC Excise & Customs Helpline on 0300 200 3700.

Customs Allowances

Travelling within the European Union

If you are travelling to the UK from within the EU you can bring an unlimited amount of most goods without being liable for any duty or tax, but certain rules apply. The goods must be

for your own personal use, which can include use as a gift (if the person you are gifting the goods to reimburses you this is not classed as a gift), and you must have paid duty and tax in the country where you purchased the goods. If a customs official suspects that any goods are not for your own personal use they can question you, make further checks and ultimately seize both the goods and the vehicle used to transport them. Although no limits are in place, customs officials are less likely to question you regarding your goods if they are under the following limits:

- 800 cigarettes
- 400 cigarillos
- 200 cigars
- 1kg tobacco
- 10 litres of spirits
- 20 litres of fortified wine (e.g. port or sherry)
- 90 litres of wine
- 110 litres of beer

The same rules and recommended limits apply for travel between other EU countries.

Travelling Outside the EU

There are set limits to the amount of goods you bring back into the UK from countries outside of the EU. All goods must be for your own personal use. Each person aged 17 and over is entitled to the following allowance:

- 200 cigarettes, or 100 cigarillos, or 50 cigars, or 250g tobacco
- 1 litre of spirits or strong liqueurs over 22% volume, or 2 litres of fortified wine, sparkling wine or any other alcoholic drink that's less than 22% volume
- 4 litres of still wine
- 16 litres of beer
- £390 worth of all other goods including perfume, gifts and souvenirs without having to pay tax and/or duty
- For further information contact HMRC National Advice Service on 0300 200 3700.

Medicines

There is no limit to the amount of medicines you can take abroad if they are obtained without prescription (i.e. over the counter medicines). Medicines prescribed by your doctor may contain controlled drugs (e.g. morphine), for which you will need a licence if you're leaving the UK for 3 months or more. Visit www.gov.uk/travelling-controlled-drugs or call 020 7035 0771 for a list of controlled drugs and to apply for a licence.

You don't need a licence if you carry less than 3 months' supply or your medication doesn't contain controlled drugs, but you should carry a letter from your doctor stating your name, a list of your prescribed drugs and dosages for each drug. You may have to show this letter when going through customs.

Personal Possessions

Visitors to countries within the EU are free to carry reasonable quantities of any personal possessions such as jewellery, cameras, and electrical equipment required for the duration of their stay. It is sensible to carry sales receipts for new items in case you need to prove that tax has already been paid.

Prohibited and Restricted Goods

Regardless of where you are travelling from the importation of some goods into the UK is restricted or banned, mainly to protect health and the environment. These include:

- Endangered animals or plants including live animals, birds and plants, ivory, skins, coral, hides, shells and goods made from them such as jewellery, shoes, bags and belts.
- Controlled, unlicensed or dangerous drugs.
- Counterfeit or pirated goods such as watches, CDs and clothes; goods bearing a false indication of their place of manufacture or in breach of UK copyright.
- Offensive weapons such as firearms, flick knives, knuckledusters, push daggers, self-defence sprays and stun guns.
- Pornographic material depicting extreme violence or featuring children

This list is not exhaustive; if in doubt contact HMRC on 0300 200 3700 (+44 2920 501 261 from outside the UK) or go through the red Customs channel and ask a Customs officer when returning to the UK.

Plants and Food

Travellers from within the EU may bring into the UK any fruit, vegetable or plant products without restriction as long as they are grown in the EU, are free from pests or disease and are for your own consumption. For food products Andorra, the Channel Islands, the Isle of Man, San Marino and Switzerland are treated as part of the EU.

From most countries outside the EU you are not allowed to bring into the UK any meat or dairy products. Other animal products may be severely restricted or banned and it is important that you declare any such products on entering the UK.

For up to date information contact the Department for Environment, Food and Rural Affairs (Defra) on 0345 33 55 77 or +44 20 7238 6951 from outside the UK. You can also visit www.defra.gov.uk to find out more.

Money

Being able to safely access your money while you're away is a necessity for you to enjoy your break. It isn't a good idea to rely on one method of payment, so always have a backup plan. A mixture of a small amount of cash plus one or two electronic means of payment are a good idea.

Traveller's cheques have become less popular in recent years as fewer banks and hotels are willing or able to cash them. There are alternative options which offer the same level of security but are easier to use, such as prepaid credit cards.

Local Currency

It is a good idea to take enough foreign currency for your journey and immediate needs on arrival, don't forget you may need change for tolls or parking on your journey. Currency exchange facilities will be available at ports and on ferries but rates offered may not be as good as you would find elsewhere.

The Post Office, banks, exchange offices and travel agents offer foreign exchange. All should stock Euros but during peak holiday times or if you need a large amount it may be sensible to pre-order your currency. You should also pre-order any less common currencies. Shop around and compare commission and exchange rates, together with minimum charges.

Banks and money exchanges in central and eastern Europe won't usually accept Scottish and Northern Irish bank notes and may be reluctant to change any sterling which has been written on or is creased or worn.

Foreign Currency Bank Accounts

Frequent travellers or those who spend long periods abroad may find a Euro bank account useful. Most such accounts impose no currency conversion charges for debit or credit card use and allow fee-free cash withdrawals at ATMs. Some banks may also allow you to spread your account across different currencies, depending on your circumstances. Speak to your bank about the services they offer.

Prepaid Travel Cards

Prepaid travel money cards are issued by various providers including the Post Office, Travelex, Lloyds Bank and American Express.

They are increasingly popular as the PIN protected travel money card offers the security of Traveller's Cheques, with the convenience of paying by card. You load the card with the amount you need before leaving home, and then use cash machines to make withdrawals or use the card to pay for goods and services as you would a credit or debit card. You can top the card up over the telephone or online while you are abroad. However there can be issues with using them with some automated payment systems, such as pay-at-pump petrol stations and toll booths, so you should always have an alternative payment method available.

These cards can be cheaper to use than credit or debit cards for both cash withdrawals and purchases as there are usually no loading or transaction fees to pay. In addition, because they are separate from your bank account, if the card is lost or stolen you bank account will still be secure.

Credit and Debit Cards

Credit and debit cards offer a convenient way of spending abroad. For the use of cards abroad most banks impose a foreign currency conversion charge of up to 3% per transaction. If you use your card to withdraw cash there will be a further commission charge of up to 3% and you will be charged interest (possibly at a higher rate than normal) as soon as you withdraw the money.

There are credit cards available which are specifically designed for spending overseas and will give you the best available rates. However they often have high interest rates so are only economical if you're able to pay them off in full each month.

If you have several cards, take at least two in case you encounter problems. Credit and debit 'Chip and PIN' cards issued by UK banks may not be universally accepted abroad so check that your card will be accepted if using it in restaurants or other situations where you pay after you have received goods or services

Contact your credit or debit card issuer before you leave home to let them know that you will be travelling abroad. In the battle against card fraud, card issuers frequently query transactions which they regard as unusual or suspicious, causing your card to be declined or temporarily stopped. You should always carry your card issuer's helpline number with you so that you can contact them if this happens. You will also need this number should you need to report the loss or theft of your card.

Dynamic Currency Conversion

When you pay with a credit or debit card, retailers may offer you the choice of currency for payment, e.g. a euro amount will be converted into sterling and then charged to your card account. This is known as a 'Dynamic Currency Conversion' but the exchange rate used is likely to be worse than the rate offered by your card issuer, so will work out more expensive than paying in the local currency.

Emergency Cash

If an emergency or theft means that you need cash in a hurry, then friends or relatives at home can send you emergency cash via money transfer services. The Post Office, MoneyGram and Western Union all offer services which, allows the transfer of money to over 233,000 money transfer agents around the world. Transfers take approximately ten minutes and charges are levied on a sliding scale.

Ferries & the Channel Tunnel

Booking Your Ferry

If travelling at peak times, such as Easter or school holidays, make reservations as early as possible. Each ferry will have limited room for caravans and large vehicles so spaces can fill up quickly, especially on cheaper crossings. If you need any special assistance or arrangements request this at the time of booking.

When booking any ferry crossing, make sure you give the correct measurements for your outfit including bikes, roof boxes or anything which may add to the length or height of your vehicle - if you underestimate your vehicle's size you may be turned away or charged an additional fee.

The Caravan and Motorhome Club is an agent for most major ferry companies operating services. Call The Club's Travel Service on 01342 316 101 or see camc.com/ferries to book.

The table at the end of this section shows ferry routes from the UK to the Continent and Ireland. Some ferry routes may not be operational all year, and during peak periods there may be a limit to the number of caravans or motorhomes accepted. For the most up-to-date information visit camc.com/ferries or call the Club's Travel Services team.

On the Ferry

Arrive at the port with plenty of time before your boarding time. Motorhomes and car/caravan outfits will usually either be the first or last vehicles boarded onto the ferry. Almost all ferries are now 'drive on – drive off' so you won't be required to do any complicated manoeuvres. You may be required to show ferry staff that your gas is switched off before boarding the ferry.

Be careful using the ferry access ramps, as they are often very steep which can mean there is a risk of grounding the tow bar or caravan hitch. Drive slowly and, if your ground clearance is low, consider whether removing your jockey wheel and any stabilising devices would help.

Vehicles are often parked close together on ferries, meaning that if you have towing extension mirrors they could get knocked or damaged by people trying to get past your

vehicle. If you leave them attached during the ferry crossing then make sure you check their position on returning to your vehicle.

Channel Tunnel

The Channel Tunnel operator, Eurotunnel, accepts cars, caravans and motorhomes (except those running on LPG) on their service between Folkestone and Calais. You can just turn up and see if there is availability on the day, however prices increase as it gets closer to the departure time so if you know your plans in advance it is best to book as early as possible.

On the Journey

You will be asked to open your roof vents prior to travel and you will also need to apply the caravan brake once you have parked your vehicle on the train. You will not be able to use your caravan until arrival.

Pets

It is possible to transport your pet on a number of ferry routes to the Continent and Ireland, as well as on Eurotunnel services from Folkestone to Calais. Advance booking is essential as restrictions apply to the number of animals allowed on any one crossing. Make sure you understand the carrier's terms and conditions for transporting pets. Brittany Ferries ask for all dogs to be muzzled when out of the vehicle but this varies for other operators so please check at the time of booking.

Once on board pets are normally required to remain in their owner's vehicle or in kennels on the car deck and you won't be able to access your vehicle to check on your pet while the ferry is at sea. On longer crossings you should make arrangements at the on-board information desk for permission to visit your pet in order to check its well-being. You should always make sure that ferry staff know your vehicle has a pet on board.

Information and advice on the welfare of animals before and during a journey is available on the website of the Department for Environment, Food and Rural Affairs (Defra), www.defra.gov.uk.

Gas

UK based ferry companies usually allow up to three gas cylinders per caravan, including the cylinder currently in use, however some may restrict this to a maximum of two cylinders. Some operators may ask you to hand over your gas cylinders to a member of the crew so that they can be safely stored during the crossing. Check that you know the rules of your ferry operator before you travel.

Cylinder valves should be fully closed and covered with a cap, if provided, and should remain closed during the crossing. Cylinders should be fixed securely in or on the caravan in the position specified by the manufacturer.

Gas cylinders must be declared at check-in and the crew may ask to inspect each cylinder for leakage before travel.

The carriage of spare petrol cans, whether full or empty, is not permitted on ferries or through the Channel Tunnel.

LPG Vehicles

Vehicles fully or partially powered by LPG can't be carried through the Channel Tunnel. Gas for domestic use (e.g. heating, lighting or cooking) can be carried, but the maximum limit is 47kg for a single bottle or 50kg in multiple bottles. Tanks must be switched off before boarding and must be less than 80% full; you will be asked to demonstrate this before you travel.

Most ferry companies will accept LPG-powered vehicles but you must let them know at the time of booking. During the crossing the tank must be no more than 75% full and it must be turned off. In the case of vehicles converted to use LPG, some ferry companies also require a certificate showing that the conversion has been carried out by a professional - before you book speak to the ferry company to see what their requirements are.

Club Sites Near Ports

If you've got a long drive to the ferry port, or want to catch an early ferry then an overnight stop near to the port gives you a relaxing start to your holiday. The following table lists Club sites which are close to ports.

Club Members can book online at camc.com or call 01342 327490. Non-members can book by calling the sites directly on the telephone numbers below when the sites are open.

Please note that Commons Wood, Fairlight Wood, Hunter's Moon, Mildenhall and Old Hartley are open to Club members only. Non-members are welcome at all other sites listed below.

Port	Nearest Club Site	Tel No.
Cairnryan	New England Bay	01776 860275
Dover, Folkestone, Channel Tunnel	Bearsted	01622 730018
	Black Horse Farm*	01303 892665
	Daleacres	01303 267679
	Fairlight Wood	01424 812333
Fishguard, Pembroke	Freshwater East	01646 672341
Harwich	Cambridge Cherry Hinton*	01223 244088
	Commons Wood*	01707 260786
	Mildenhall	01638 713089
Holyhead	Penrhos	01248 852617
Hull	York Beechwood Grange*	01904 424637
	York Rowntree Park*	01904 658997
Newcastle upon Tyne	Old Hartley	0191 237 0256
Newhaven	Brighton*	01273 626546
Plymouth	Plymouth Sound	01752 862325
Poole	Hunter's Moon*	01929 556605
Portsmouth	Rookesbury Park	01329 834085
Rosslare	River Valley	00353 (0)404 41647
Weymouth	Crossways	01305 852032

** Site open all year*

Ferry Routes and Operators

Route	Operator	Approximate Crossing Time	Maximum Frequency
Belgium			
Hull – Zeebrugge	P & O Ferries	12-14 hrs	1 daily
France			
Dover – Calais	P & O Ferries	1½ hrs	22 daily
Dover – Calais	DFDS Seaways	1½ hrs	10 daily
Dover – Dunkerque	DFDS Seaways	2 hrs	12 daily
Folkestone – Calais	Eurotunnel	35 mins	3 per hour
Newhaven – Dieppe	DFDS Seaways	4 hrs	2 daily
Plymouth – Roscoff	Brittany Ferries	6 hrs	2 daily
Poole – St Malo (via Channel Islands)*	Condor Ferries	5 hrs	1 daily (May to Sep)
Portsmouth – Caen	Brittany Ferries	6 / 7 hrs	3 daily (maximum)
Portsmouth – Cherbourg	Brittany Ferries	3 hrs	2 daily (maximum)
Portsmouth – Le Havre	Brittany Ferries	3¼ / 8 hrs	1 daily (minimum)
Portsmouth – St Malo	Brittany Ferries	9 hrs	1 daily
Ireland – Northern			
Cairnryan – Larne	P & O Irish Sea	1 / 2 hrs	7 daily
Liverpool (Birkenhead) – Belfast	Stena Line	8 hrs	2 daily
Cairnryan – Belfast	Stena Line	2 / 3 hrs	7 daily
Ireland – Republic			
Cork – Roscoff*	Brittany Ferries	14 hrs	1 per week
Dublin - Cherbourg	Irish Ferries	19 hrs	1 per week
Fishguard – Rosslare	Stena Line	3½ hrs	2 daily
Holyhead – Dublin	Irish Ferries	2-4 hrs	Max 4 daily
Holyhead – Dublin	Stena Line	2-4 hrs	Max 4 daily
Liverpool – Dublin	P & O Irish Sea	8 hrs	2 daily
Pembroke – Rosslare	Irish Ferries	4 hrs	2 daily
Rosslare – Cherbourg*	Irish Ferries	19½ hrs	3 per week
Rosslare – Cherbourg	Stena Line	19 hrs	3 per week
Rosslare – Roscoff*	Irish Ferries	19½ hrs	4 per week
Netherlands			
Harwich – Hook of Holland	Stena Line	7 hrs	2 daily
Hull – Rotterdam	P & O Ferries	11-12 hrs	1 daily
Newcastle – Ijmuiden (Amsterdam)	DFDS Seaways	15½ hrs	1 daily
Spain			
Portsmouth – Bilbao	Brittany Ferries	24 / 32 hrs	1 - 3 per week
Portsmouth or Plymouth – Santander	Brittany Ferries	20 / 32 hrs	4 per week

**Not bookable through The Club's Travel Service.*
Note: Services and routes correct at time of publication but subject to change.

Motoring Advice

Preparing for Your Journey

The first priority in preparing your outfit for your journey should be to make sure it has a full service. Make sure that you have a fully equipped spares kit, and a spare wheel and tyre for your caravan – it is easier to get hold of them from your local dealer than to have to spend time searching for spares where you don't know the local area.

Club members should carry their UK Sites Directory & Handbook with them, as it contains a section of technical advice which may be useful when travelling.

The Club also has a free advice service covering a wide range of technical topics – download free information leaflets at camc.com/advice or contact the team by calling 01342 336611 or emailing technical@caravanclub.co.uk.

For advice on issues specific to countries other than the UK, Club members can contact the Travel Service Information Officer, email: travelserviceinfo@caravanclub.co.uk or call 01342 336766.

Weight Limits

From both a legal and a safety point of view, it is essential not to exceed vehicle weight limits. It is advisable to carry documentation confirming your vehicle's maximum permitted laden weight - if your Vehicle Registration Certificate (V5C) does not state this, you will need to produce alternative certification, e.g. from a weighbridge.

If you are pulled over by the police and don't have certification you will be taken to a weighbridge. If your vehicle(s) are then found to be overweight you will be liable to a fine and may have to discard items to lower the weight before you can continue on your journey.

Some Final Checks

Before you start any journey make sure you complete the following checks:

- All car and caravan or motorhome lights are working and sets of spare bulbs are packed
- The coupling is correctly seated on the towball and the breakaway cable is attached
- Windows, vents, hatches and doors are shut
- On-board water systems are drained

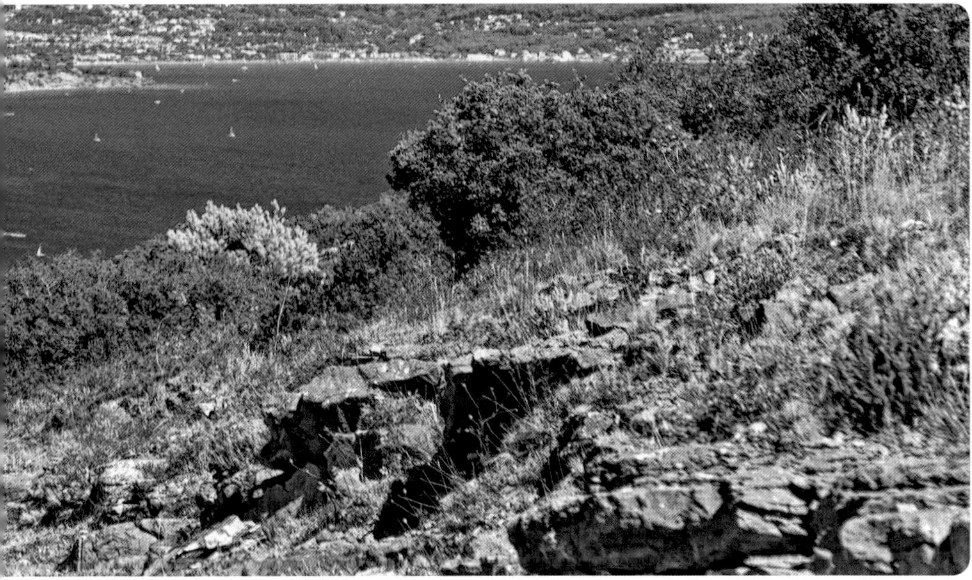

- Mirrors are adjusted for maximum visibility
- Corner steadies are fully wound up and the brace is handy for your arrival on site
- Any fires or flames are extinguished and the gas cylinder tap is turned off. Fire extinguishers are fully charged and close at hand
- The over-run brake is working correctly
- The jockey wheel is raised and secured, the handbrake is released.

Driving in Europe

Driving abroad for the first time can be a daunting prospect, especially when towing a caravan. Here are a few tips to make the transition easier:

- Remember that Sat Navs may take you on unsuitable roads, so have a map or atlas to hand to help you find an alternative route.
- It can be tempting to try and get to your destination as quickly as possible but we recommend travelling a maximum of 250 miles a day when towing.
- Share the driving if possible, and on long journeys plan an overnight stop.
- Remember that if you need to overtake or pull out around an obstruction you will not be able to see clearly from the driver's seat. If possible, always have a responsible adult in the passenger seat who can advise you when it is clear to pull out. If that is not possible then stay well back to get a better view and pull out slowly.
- If traffic builds up behind you, pull over safely and let it pass.
- Driving on the right should become second nature after a while, but pay particular attention when turning left, after leaving a rest area, petrol station or site or after a one-way system.
- Stop at least every two hours to stretch your legs and take a break.

Fuel

Grades of petrol sold on the Continent are comparable to those sold in the UK; 95 octane is frequently known as 'Essence' and 98 octane as 'Super'. Diesel may be called 'Gasoil' and is widely available across Europe.

E10 petrol (containing 10% Ethanol) can be found in certain countries in Europe. Most modern cars are E10 compatible, but those which aren't could be damaged by filling up with E10. Check your vehicle handbook or visit www.acea.be and search for 'E10' to find 'Vehicle compatibility with new fuel standards'.

Members of The Caravan Club can check current average fuel prices by country at camc.com/overseasadvice.

Away from major roads and towns it is a good idea not to let your fuel tank run too low as you may have difficulty finding a petrol station, especially at night or on Sundays. Petrol stations offering a 24-hour service may involve an automated process, in some cases only accepting credit cards issued in the country you are in.

Automotive Liquefied Petroleum Gas (LPG)

The increasing popularity of dual-fuelled vehicles means that the availability of LPG – also known as 'autogas' or GPL – has become an important issue for more drivers.

There are different tank-filling openings in use in different countries. Currently there is no common European filling system, and you might find a variety of systems. Most Continental motorway services will have adaptors but these should be used with care – see www.autogas.ltd.uk for more information.

Low Emission Zones

Many cities in countries around Europe have introduced 'Low Emission Zones' (LEZ's) in order to regulate vehicle pollution levels. Some schemes require you to buy a windscreen sticker, pay a fee or register your vehicle before entering the zone. You may also need to provide proof that your vehicle's emissions meet the required standard. Before you travel visit www.lowemissionzones.eu for maps and details of LEZ's across Europe. Also see the Country Introductions later in this guide for country specific information.

Motorhomes Towing Cars

If you are towing a car behind a motorhome, our advice would be to use a trailer with all four wheels of the car off the ground. Although most countries don't have specific laws banning A-frames, there may be laws in place which prohibit motor vehicle towing another motor vehicle.

Priority and Roundabouts

When driving on the Continent it can be difficult to work out which vehicles have priority in different situations. Watch out for road signs which indicate priority and read the Country Introductions later in this guide for country specific information.

Take care at intersections – you should never rely on being given right of way, even if you have priority; especially in small towns and villages where local traffic may take right of way. Always give way to public service and military vehicles and to buses and trams.

In some countries in Europe priority at roundabouts is given to vehicles entering the roundabout (i.e. on the right) unless the road signs say otherwise.

Public Transport

In general in built-up areas be prepared to stop to allow a bus to pull out from a bus stop when the driver is signalling his intention to do so.

Take particular care when school buses have stopped and passengers are getting on and off.

Overtaking trams in motion is normally only allowed on the right, unless on a one way street where you can overtake on the left if there is not enough space on the right. Do not overtake a tram near a tram stop. These may be in the centre of the road. When a tram or bus stops to allow passengers on and off, you should stop to allow them to cross to the pavement. Give way to trams which are turning across your carriageway. Don't park or stop across tram lines; trams cannot steer round obstructions!

Pedestrian Crossings

Stopping to allow pedestrians to cross at zebra crossings is not always common practice on the Continent as it is in the UK. Pedestrians expect to wait until the road is clear before crossing, while motorists behind may be taken by surprise by your stopping. The result may be a rear-end shunt or vehicles overtaking you at the crossing and putting pedestrians at risk.

Traffic Lights

Traffic lights may not be as easily visible as they are in the UK, for instance they may be smaller or suspended across the road with a smaller set on a post at the roadside. You may find that lights change directly from red to green, bypassing amber completely. Flashing amber lights generally indicate that you may proceed with caution if it is safe to do so but you must give way to pedestrians and other vehicles.

A green filter light should be treated with caution as you may still have to give way to pedestrians who have a green light to cross the road. If a light turns red as approached, continental drivers will often speed up to get through the light instead of stopping. Be aware that if you brake sharply because a traffic light has turned red as you approached, the driver behind might not be expecting it.

Motoring Equipment

Essential Equipment

The equipment that you legally have to carry differs by country. For a full list see the Essential Equipment table at the end of this chapter. Please note equipment requirements and regulations can change frequently. To keep up to date with the latest equipment information visit camc.com/overseasadvice.

Child Restraint Systems

Children under 10 years of age are not permitted to travel in front seats of vehicles, unless there are no rear seats in the vehicle, the rear seats are already occupied with other children, or there are no seat belts in the rear. In these situations a child must not be placed in the front seats in a rear-facing child seat, unless any airbag is deactivated. Children up to 10 must travel in an approved child seat or restraint system, adapted to their size. A baby up to 13kg in weight must be carried in a rear facing baby seat. A child between 9kg and 18kg in weight must be seated in a child seat. A child from 15kg in weight up to the age of 10 can use a booster seat with a seat belt.

Children must not travel in the front of a vehicle if there are rear seats available. If they travel in the front the airbag must be deactivated and again they must use an EU approved restraint system adapted to their size.

Fire Extinguisher

As a safety precaution, an approved fire extinguisher should be carried in all vehicles. This is a legal requirement in several countries in Europe.

Lights

When driving in on the right headlights should be adjusted if they are likely to dazzle other road users. You can do this by applying beam deflectors, or some newer vehicles have a built-in adjustment system. Some high-density discharge (HID), xenon or halogen lights, may need to be taken to a dealer to make the necessary adjustment.

Remember also to adjust headlights according to the load being carried and to compensate for the weight of the caravan on the back of

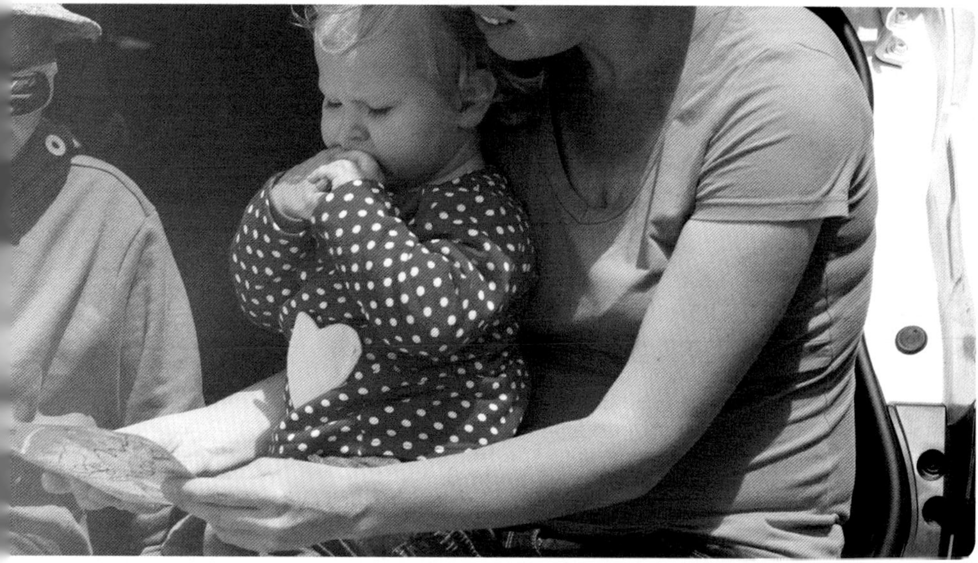

your car. Even if you do not intend to drive at night, it is important to ensure that your headlights are correctly adjusted as you may need to use them in heavy rain, fog or in tunnels. If using tape or a pre-cut adhesive mask remember to remove it on your return home.

All vehicle lights must be in working condition. If your lights are not in working order you may be liable for a fine of up to €450 and confiscation of your vehicle is a possibility in some European countries.

Headlight-Flashing

On the Continent headlight-flashing is used as a warning of approach or as an overtaking signal at night, and not, as is commonly the case in the UK, an indication that you are giving way. Be more cautious with both flashing your headlights and when another driver flashes you. If a driver flashes his headlights they are generally indicating that he has priority and you should give way, contrary to standard practice in the UK.

Hazard Warning Lights

Hazard warning lights should not be used in place of a warning triangle, but should be used in addition to it.

Nationality Plate (GB/IRL)

A nationality plate must be fixed to the rear of both your car or motorhome and caravan. Checks are made and a fine may be imposed for failure to display a nationality plate correctly. If your number plates have the Euro-Symbol on them there is no requirement to display an additional GB sticker within the EU and Switzerland. If your number plate doesn't have the EU symbol or you are planning to travel outside of the EU you will need a GB sticker.

GB is the only national identification code allowed for cars registered in the UK.

Reflective Jackets/ Waistcoats

If you break down outside of a built-up area it is normally a legal requirement that anyone leaving the vehicle must be wearing a reflective jacket or waistcoat. Make sure that your jacket is accessible from inside the car as you will need to put it on before exiting the vehicle. Carry one for each passenger as well as the driver.

Route Planning

It is always a good idea to carry a road atlas or map of the countries you plan to visit, even if you have Satellite Navigation. You can find information on UK roads from Keep Moving – www.keepmoving.co.uk or call 09003 401100. Websites offering a European route mapping service include www.google.co.uk/maps, www.mappy.com or www.viamichelin.com.

Satellite Navigation/GPS

European postcodes don't cover just one street or part of a street in the same way as UK postcodes, they can cover a very large area. GPS co-ordinates and full addresses are given for site entries in this guide wherever possible, so that you can programme your device as accurately as possible.

It is important to remember that sat nav devices don't usually allow for towing or driving a large motorhome and may try to send you down unsuitable roads. Always use your common sense, and if a road looks unsuitable find an alternative route.

Use your sat nav in conjunction with the directions given in the site entries, which have been provided by members who have actually visited. Please note that directions given in site entries have not been checked by the Caravan and Motorhome Club.

In nearly all European countries it is illegal to use car navigation systems which actively search for mobile speed cameras or interfere with police equipment (laser or radar detection).

Car navigation systems which give a warning of fixed speed camera locations are legal in most countries with the exception of France, Germany, and Switzerland where this function must be de-activated.

Seat Belts

The wearing of seat belts is compulsory throughout Europe. On-the-spot fines will be incurred for failure to wear them and, in the event of an accident failure to wear a seat belt may reduce any claim for injury. See the country introductions for specific regulations on both seat belts and car seats.

Caravan Spares

It will generally be much harder to get hold of spare parts for caravans on the continent, especially for UK manufactured caravans. It is therefore advisable to carry any commonly required spares (such as light bulbs) with you.

Take contact details of your UK dealer or manufacturer with you, as they may be able to assist in getting spares delivered to you in an emergency.

Car Spares

Some car manufacturers produce spares kits; contact your dealer for details. The choice of spares will depend on the vehicle and how long you are away, but the following is a list of basic items which should cover the most common causes of breakdown:

- Radiator top hose
- Fan belt
- Fuses and bulbs
- Windscreen wiper blade
- Length of 12V electrical cable
- Tools, torch and WD40 or equivalent water repellent/ dispersant spray

Spare Wheel

Your local caravan dealer should be able to supply an appropriate spare wheel. If you have any difficulty in obtaining one, the Club's Technical Department can provide Club members with a list of suppliers on request.

Tyre legislation across Europe is more or less consistent and, while the Club has no specific knowledge of laws on the Continent regarding the use of space-saver spare wheels, there should be no problems in using such a wheel provided its use is in accordance with the manufacturer's instructions. Space-saver spare wheels are designed for short journeys to get to a place where it can be repaired and there will usually be restrictions on the distance and speed at which the vehicle should be driven.

Towbar

The vast majority of cars registered after 1 August 1998 are legally required to have a European Type approved towbar (complying with European Directive 94/20) carrying a plate giving its approval number and various technical details, including the maximum noseweight. Your car dealer or specialist towbar fitter will be able to give further advice.

All new motorhomes will need some form of type approval before they can be registered in the UK and as such can only be fitted with a type approved towbar. Older vehicles can continue to be fitted with non-approved towing brackets.

Tyres

Tyre condition has a major effect on the safe handling of your outfit. Caravan tyres must be suitable for the highest speed at which you can legally tow, even if you choose to drive slower.

Most countries require a minimum tread depth of 1.6mm but motoring organisations recommend at least 3mm. If you are planning a long journey, consider if they will still be above the legal minimum by the end of your journey.

Tyre Pressure

Tyre pressure should be checked and adjusted when the tyres are cold; checking warm tyres will result in a higher pressure reading. The correct pressures will be found in your car handbook, but unless it states otherwise to add an extra 4 - 6 pounds per square inch to the rear tyres of a car when towing to improve handling. Make sure you know what pressure your caravan tyres should be. Some require a pressure much higher than that normally used for cars. Check your caravan handbook for details.

Tyre Sizes

It is worth noting that some sizes of radial tyre to fit the 13" wheels commonly used on older UK caravans are virtually impossible to find in stock at retailers abroad, e.g. 175R13C.

After a Puncture

A lot of cars now have a liquid sealant puncture repair kit instead of a spare wheel. These should not be considered a permanent repair, and in some cases have been known to make repair of the tyre impossible. If you need to use a liquid sealant you should get the tyre repaired or replaced as soon as possible.

Following a caravan tyre puncture, especially on a single-axle caravan, it is advisable to have the non-punctured tyre removed from its wheel and checked inside and out for signs of damage resulting from overloading during the deflation of the punctured tyre.

Winter Driving

Snow chains must be fitted to vehicles using snow-covered roads in compliance with the relevant road signs. Fines may be imposed for non-compliance. Vehicles fitted with chains must not exceed 50 km/h (31mph).

They are not difficult to fit but it's a good idea to carry sturdy gloves to protect your hands when handling the chains in freezing conditions. Polar Automotive Ltd sells and hires out snow chains, contact them on 01892 519933, www.snowchains.com, or email: polar@snowchains.com.

In Andorra winter tyres are recommended. Snow chains must be used when road conditions necessitate their use and/or when road signs indicate.

Warning Triangles

In almost all European countries it is compulsory to carry a warning triangle which, in the event of vehicle breakdown or accident, must be placed (providing it is safe to do so) on the carriageway at least 30 metres from the vehicle. In some instances it is not compulsory to use the triangle but only when this action would endanger the driver.

A warning triangle should be placed on the road approximately 30 metres (100 metres on motorways) behind the broken down vehicle on the same side of the road. Always assemble the triangle before leaving your vehicle and walk with it so that the red, reflective surface is facing oncoming traffic. If a breakdown occurs round a blind corner, place the triangle in advance of the corner. Hazard warning lights may be used in conjunction with the triangle but they do not replace it.

Essential Equipment Table

The table below shows the essential equipment required for each country. Please note that this information was correct at the time of going to print but is subject to change.

For up to date information on equipment requirements for countries in Europe visit camc.com/overseasadvice.

Country	Warning Triangle	Spare Bulbs	First Aid Kit	Reflective Jacket	Additional Equipment to be Carried/Used
Andorra	Yes (2)	Yes	Rec	Yes	Dipped headlights in poor daytime visibility. Winter tyres recommended; snow chains when road conditions or signs dictate.
Austria	Yes	Rec	Yes	Yes	Winter tyres from 1 Nov to 15 April.*
Belgium	Yes	Rec	Rec	Yes	Dipped headlights in poor daytime visibility.
Croatia	Yes (2 for vehicle with trailer)	Yes	Yes	Yes	Dipped headlights at all times from last Sunday in Oct - last Sunday in Mar. Spare bulbs compulsory if lights are xenon, neon or LED. Snow chains compulsory in winter in certain regions.*
Czech Rep	Yes	Yes	Yes	Yes	Dipped headlights at all times. Replacement fuses. Winter tyres or snow chains from 1 Nov - 31st March.*
Denmark	Yes	Rec	Rec	Rec	Dipped headlights at all times. On motorways use hazard warning lights when queues or danger ahead.
Finland	Yes	Rec	Rec	Yes	Dipped headlights at all times. Winter tyres Dec - Feb.*
France	Yes	Rec	Rec	Yes	Dipped headlights recommended at all times. Legal requirement to carry a breathalyser, but no penalty for non-compliance.

Country	Warning Triangle	Spare Bulbs	First Aid Kit	Reflective Jacket	Additional Equipment to be Carried/Used
Germany	Rec	Rec	Rec	Rec	Dipped headlights recommended at all times. Winter tyres to be used in winter weather conditions.*
Greece	Yes	Rec	Yes	Rec	Fire extinguisher compulsory. Dipped headlights in towns at night and in poor daytime visibility.
Hungary	Yes	Rec	Yes	Yes	Dipped headlights at all times outside built-up areas and in built-up areas at night. Snow chains compulsory on some roads in winter conditions.*
Italy	Yes	Rec	Rec	Yes	Dipped headlights at all times outside built-up areas and in poor visibility. Snow chains from15 Oct - 15 April.*
Luxembourg	Yes	Rec	Rec	Yes	Dipped headlights at night and daytime in bad weather.
Netherlands	Yes	Rec	Rec	Rec	Dipped headlights at night and in bad weather and recommended during the day.
Norway	Yes	Rec	Rec	Rec	Dipped headlights at all times. Winter tyres compulsory when snow or ice on the roads.*
Poland	Yes	Rec	Rec	Rec	Dipped headlights at all times. Fire extinguisher compulsory.
Portugal	Yes	Rec	Rec	Rec	Dipped headlights in poor daytime visibility, in tunnels and in lanes where traffic flow is reversible.
Slovakia	Yes	Rec	Yes	Yes	Dipped headlights at all times. Winter tyres compulsory when compact snow or ice on the road.*
Slovenia	Yes (2 for vehicle with trailer)	Yes	Rec	Yes	Dipped headlights at all times. Hazard warning lights when reversing. Use winter tyres or carry snow chains 15 Nov - 15 Mar.
Spain	Yes (2 Rec)	Rec	Rec	Yes	Dipped headlights at night, in tunnels and on 'special' roads (roadworks).
Sweden	Yes	Rec	Rec	Rec	Dipped headlights at all times. Winter tyres 1 Dec to 31 March.
Switzerland (inc Liechtenstein)	Yes	Rec	Rec	Rec	Dipped headlights recommended at all times, compulsory in tunnels. Snow chains where indicated by signs.

NOTES:
1) All countries: seat belts (if fitted) must be worn by all passengers.
2) Rec: not compulsory for foreign-registered vehicles, but strongly recommended
3) Headlamp converters, spare bulbs, fire extinguisher, first aid kit and reflective waistcoat are strongly recommended for all countries.
4) In some countries drivers who wear prescription glasses must carry a spare pair.
5) Please check information for any country before you travel. This information is to be used as a guide only and it is your responsibility to make sure you have the correct equipment.

* For more information and regulations on winter driving please see the Country Introduction.

Route Planning

Den Haag
Rotterdam
NETHERLANDS
Utrecht
Apeldoorn
Arnhem
Bocholt
Osnabrück
Magdeburg
POLAND
Bruges
Dunkerque
Antwerpen
Essen
Dortmund
Düsseldorf
Leipzig
Wrocław
BRUXELLES
Lille
Troisdorf
Erfurt
Dresden
Opole
Mons
Liège
BELGIUM
GERMANY
Liberec
Hradec Králové
Amiens
Karlovy Vary
Wiesbaden
Mainz
Frankfurt am Main
PRAHA
Plzeň
CZECHIA
(CZECH REPUBLIC)
Olomouc
LUXEMBOURG
Zlín
Nürnberg
Saarbrücken
České Budějovice
Brno
Trenčín
Metz
Châlons-en-Champagne
Karlsruhe
Sindelfingen
Stuttgart
SLOVAKIA
Trnava
Strasbourg
Linz
WIEN
BRATISLAVA
München
Salzburg
Sopron
Győr
Szombathely
Dijon
Basel
Zürich
Innsbruck
AUSTRIA
Graz
HUNGARY
BERN
Luzern
LIECHTENSTEIN
FRANCE
SWITZERLAND
Klagonfurt
Maribor
Nagykanizsa
Lausanne
Bolzano
Genève
SLOVENIA
LJUBLJANA
ZAGREB
Lyon
Trento
Trieste
Karlovac
Milano
Verona
Venezia
Grenoble
Torino
Rijeka
CROATIA
Banja Luka
Pula
BOSNIA AND HERZEGOVINA
Bologna
Genova
Zadar
SAN MARINO
Nice
MONACO
Firenze
Ancona
Montpellier
Split
Livorno
Marseille
Perugia
Perpignan
Bastia
ITALY
Adriatic Sea
L'Aquila
Pescara
Corse
(Corsica)
(France)
Ajaccio
ROMA
Campobasso
Bari
Olbia
Napoli
Potenza
Sassari
Mahón
Menorca
(Minorca)
Sardegne
(Sardinia)
(Italy)
Tyrrhenian Sea
Mallorca
(Majorca)
Baleares Islands
Cagliari
Nicastro
Catanzaro
Messina
Mediterranean Sea
Palermo
Reggio di Calabria
Sicilia
(Sicily)
Catania
ALGERIA
Bizerte
Menzel Bourguiba
TUNISIA
Tizi Ouzou
Bejaïa
Jijel
Skikda
Annaba
Siracusa
Tunis
© Collins Bartholomew Ltd 2019

Mountain Passes & Tunnels

Mountain Passes

Mountain passes can create difficult driving conditions, especially when towing or driving a large vehicle. You should only use them if you have a good power to weight ratio and in good driving conditions. If in any doubt as to your outfit's suitability or the weather then stick to motorway routes across mountain ranges if possible.

The tables on the following pages show which passes are not suitable for caravans, and those where caravans are not permitted. Motorhomes aren't usually included in these restrictions, but relatively low powered or very large vehicles should find an alternative route. Road signs at the foot of a pass may restrict access or offer advice, especially for heavy vehicles. Warning notices are usually posted at the foot of a pass if it is closed, or if chains or winter tyres must be used.

Caravanners are particularly sensitive to gradients and traffic/road conditions on passes. The maximum gradient is usually on the inside of bends but exercise caution if it is necessary to pull out. Always engage a lower gear before taking a hairpin bend and give priority to vehicles ascending. On mountain roads it is not the gradient which puts strain on your car but the duration of the climb and the loss of power at high altitudes: approximately 10% at 915 metres (3,000 feet) and even more as you get higher. To minimise the risk of the engine overheating, take high passes in the cool part of the day, don't climb any faster than necessary and keep the engine pulling steadily. To prevent a radiator boiling, pull off the road safely, turn the heater and blower full on and switch off air conditioning. Keep an eye on water and oil levels. Never put cold water into a boiling radiator or it may crack. Check that the radiator is not obstructed by debris sucked up during the journey.

A long descent may result in overheating brakes; select the correct gear for the gradient and avoid excessive use of brakes. Even if you are using engine braking to control speed, caravan brakes may activate due to the overrun mechanism, which may cause them to overheat.

Travelling at altitude can cause a pressure build up in tanks and water pipes. You can prevent this by slightly opening the blade valve of your portable toilet and opening a tap a fraction.

Tunnels

Long tunnels are a much more commonly seen feature in Europe than in the UK, especially in mountainous regions. Tolls are usually charged for the use of major tunnels.

Dipped headlights are usually required by law even in well-lit tunnels, so switch them on before you enter. Snow chains, if used, must be removed before entering a tunnel in lay-bys provided for this purpose.

'No overtaking' signs must be strictly observed. Never cross central single or double lines. If overtaking is permitted in twin-tube tunnels, bear in mind that it is very easy to underestimate distances and speed once inside. In order to minimise the effects of exhaust fumes close all car windows and set the ventilator to circulate air, or operate the air conditioning system coupled with the recycled air option.

If you break down, try to reach the next lay-by and call for help from an emergency phone. If you cannot reach a lay-by, place your warning triangle at least 100 metres behind your vehicle. Modern tunnels have video surveillance systems to ensure prompt assistance in an emergency. Some tunnels can extend for miles and a high number of breakdowns are due to running out of fuel so make sure you have enough before entering the tunnel.

Mountain Pass Information

The dates of opening and closing given in the following tables are approximate. Before attempting late afternoon or early morning journeys across borders, check their opening times as some borders close at night.

Gradients listed are the maximum which may be encountered on the pass and may be steeper at the inside of curves, particularly on older roads.

Gravel surfaces (such as dirt and stone chips) vary considerably; they can be dusty when dry and slippery when wet. Where known to exist, this type of surface has been noted.

In fine weather winter tyres or snow chains will only be required on very high passes, or for short periods in early or late summer. In winter conditions you will probably need to use them at altitudes exceeding 600 metres (approximately 2,000 feet).

Converting Gradients

20% = 1 in 5	11% = 1 in 9
16% = 1 in 6	10% = 1 in 8
14% = 1 in 7	8% = 1 in 12
12% = 1 in 8	6% = 1 in 16

Tables and Maps

Much of the information contained in the following tables was originally supplied by The Automobile Association and other motoring and tourist organisations. The Caravan and Motorhome Club haven't checked this information and cannot accept responsibility for the accuracy or for errors or omissions to these tables.

The mountain passes, rail and road tunnels listed in the tables are shown on the following maps. Numbers and letters against each pass or tunnel in the tables correspond with the numbers and letters on the maps.

Abbreviations

MHV	Maximum height of vehicle
MLV	Maximum length of vehicle
MWV	Maximum width of vehicle
MWR	Minimum width of road
OC	Occasionally closed between dates
UC	Usually closed between dates
UO	Usually open between dates, although a fall of snow may obstruct the road for 24-48 hours.

Major Mountain Passes – Pyrenees and Northern Spain

Before using any of these passes, please read the advice at the beginning of this chapter.

	Pass Height In Metres (Feet)	From To	Max Gradient	Conditions and Comments
1	**Aubisque** (France) 1710 (5610)	Eaux Bonnes *Argelés-Gazost*	10%	UC mid Oct-Jun. MWR 3.5m (11'6") Very winding; continuous on D918 but easy ascent; descent including Col-d'Aubisque 1709m (5607 feet) and Col-du-Soulor 1450m (4757 feet); 8km (5 miles) of very narrow, rough, unguarded road with steep drop. **Not recommended for caravans**.
2	**Bonaigua** (Spain) 2072 (6797)	Viella (Vielha) *Esterri-d'Aneu*	8.5%	UC Nov-Apr. MWR 4.3m (14'1") Twisting, narrow road (C28) with many hairpins and some precipitous drops. **Not recommended for caravans**. Alternative route to Lerida (Lleida) through Viella (Vielha) Tunnel is open all year.
3	**Cabrejas** (Spain) 1167 (3829)	Tarancon *Cuenca*	14%	UO. On N400/A40. Sometimes blocked by snow for 24 hours. MWR 5m (16')
4	**Col-d'Haltza and Col-de-Burdincurutcheta** (France) 782 (2565) and 1135 (3724)	St Jean-Pied-de-Port *Larrau*	11%	UO. A narrow road (D18/D19) leading to Iraty skiing area. Narrow with some tight hairpin bends; rarely has central white line and stretches are unguarded. Not for the faint-hearted. **Not recommended for caravans**.
5	**Envalira** (France – Andorra) 2407 (7897)	Pas-de-la-Casa *Andorra*	12.5%	OC Nov-Apr. MWR 6m (19'8") Good road (N22/CG2) with wide bends on ascent and descent; fine views. MHV 3.5m (11'6") on N approach near l'Hospitalet. Early start rec in summer to avoid border delays. Envalira Tunnel (toll) reduces congestion and avoids highest part of pass.
6	**Escudo** (Spain) 1011 (3317)	Santander *Burgos*	17%	UO. MWR probably 5m (16'5") Asphalt surface but many bends and steep gradients. **Not recommended in winter**. On N632; A67/N611 easier route.
7	**Guadarrama** (Spain) 1511 (4957)	Guadarrama *San Rafael*	14%	UO. MWR 6m (19'8") On NVI to the NW of Madrid but may be avoided by using AP6 motorway from Villalba to San Rafael or Villacastin (toll).
8	**Ibañeta (Roncevalles)** (France – Spain) 1057 (3468)	St Jean-Pied-de-Port *Pamplona*	10%	UO. MWR 4m (13'1") Slow and winding, scenic route on N135.
9	**Manzanal** (Spain) 1221 (4005)	Madrid *La Coruña*	7%	UO. Sometimes blocked by snow for 24 hours. On A6.
10	**Navacerrada** (Spain) 1860 (6102)	Madrid *Segovia*	17%	OC Nov-Mar. On M601/CL601. Sharp hairpins. Possible but **not recommended for caravans**.

Before using any of these passes, please read the advice at the beginning of this chapter.

	Pass Height In Metres (Feet)	From To	Max Gradient	Conditions and Comments
11	**Orduna** (Spain) 900 (2953)	Bilbao *Burgos*	15%	UO. On A625/BU556; sometimes blocked by snow for 24 hours. Avoid by using AP68 motorway.
12	**Pajares** (Spain) 1270 (4167)	Oviedo *Léon*	16%	UO. On N630; sometimes blocked by snow for 24 hours. **Not recommended for caravans**. Avoid by using AP66 motorway.
13	**Paramo-de-Masa** 1050 (3445)	Santander *Burgos*	8%	UO. On N623; sometimes blocked by snow for 24 hours.
14	**Peyresourde** (France) 1563 (5128)	Arreau *Bagnères-de-Luchon*	10%	UO. MWR 4m (13'1") D618 somewhat narrow with several hairpin bends, though not difficult. **Not recommended for caravans**.
15	**Picos-de-Europa: Puerto-de-San Glorio, Puerto-de-Pontón, Puerto-de-Pandetrave** (Spain), 1609 (5279)	Unquera *Riaño*	12%	UO. MWR probably 4m (13'1") Desfiladero de la Hermida on N621 good condition. Puerto-de-San-Glorio steep with many hairpin bends. For confident drivers only.
		Riaño *Cangas-de-Onis*		Puerto-de-Ponton on N625, height 1280 metres (4200 feet). Best approach fr S as from N is very long uphill pull with many tight turns.
		Portilla-de-la-Reina *Santa Marina-de-Valdeón*		Puerto-de-Pandetrave, height 1562 metres (5124 feet) on LE245 not rec when towing as main street of Santa Marina steep & narrow.
16	**Piqueras** (Spain) 1710 (5610)	Logroño *Soria*	7%	UO. On N111; sometimes blocked by snow for 24 hours.
17	**Port** (France) 1249 (4098)	Tarascon-sur-Ariège *Massat*	10%	OC Nov-Mar. MWR 4m (13'1") A fairly easy, scenic road (D618), but narrow on some bends.
18	**Portet-d'Aspet** (France) 1069 (3507)	Audressein *Fronsac*	14%	UO. MWR 3.5m (11'6") Approached from W by the easy Col-des-Ares and Col-de-Buret; well-engineered but narrow road (D618); care needed on hairpin bends. **Not recommended for caravans**.
19	**Pourtalet** (France – Spain) 1792 (5879)	Laruns *Biescas*	10%	UC late Oct-early Jun. MWR 3.5m (11'6") A fairly easy, unguarded road, but narrow in places. Easier from Spain (A136), steeper in France (D934). **Not recommended for caravans**.
20	**Puymorens** (France) 1915 (6283)	Ax-les-Thermes *Bourg-Madame*	10%	OC Nov-Apr. MWR 5.5m (18') MHV 3.5m (11'6") A generally easy, modern tarmac road (N20). Parallel toll road tunnel available.

Before using any of these passes, please read the advice at the beginning of this chapter.

	Pass Height In Metres (Feet)	From To	Max Gradient	Conditions and Comments
21	**Quillane** (France) 1714 (5623)	Axat *Mont-Louis*	8.5%	OC Nov-Mar. MWR 5m (16'5") An easy, straightforward ascent and descent on D118.
22	**Somosierra** (Spain) 1444 (4738)	Madrid *Burgos*	10%	OC Mar-Dec. MWR 7m (23') On A1/E5; may be blocked following snowfalls. Snow-plough swept during winter months but wheel chains compulsory after snowfalls. Well-surfaced dual carriageway, tunnel at summit.
23	**Somport** (France – Spain) 1632 (5354)	Accous *Jaca*	10%	UO. MWR 3.5m (11'6") A favoured, old-established route; not particularly easy and narrow in places with many unguarded bends on French side (N134); excellent road on Spanish side (N330). Use of road tunnel advised – see ***Pyrenean Road Tunnels*** in this section. NB Visitors advise re-fuelling no later than Sabiñánigo when travelling south to north.
24	**Toses (Tosas)** (Spain) 1800 (5906)	Puigcerda *Ribes-de-Freser*	10%	UO MWR 5m (16'5") A fairly straightforward, but continuously winding, two-lane road (N152) with with a good surface but many sharp bends; some unguarded edges. Difficult in winter.
25	**Tourmalet** (France) 2114 (6936)	Ste Marie-de-Campan *Luz-St Sauveur*	12.5%	UC Oct-mid Jun. MWR 4m (13'1") The highest French Pyrenean route (D918); approaches good, though winding, narrow in places and exacting over summit; sufficiently guarded. Rough surface & uneven edges on west side. **Not recommended for caravans.**
26	**Urquiola** (Spain) 713 (2340)	Durango (Bilbao) *Vitoria/Gasteiz*	16%	UO. Sometimes closed by snow for 24 hours. On BI623/A623. **Not recommended for caravans.**

Major Pyrenean Road Tunnels

Before using any of these tunnels, please read the advice at the beginning of this chapter.

	Tunnel	Route and Height Above Sea Level	General Information and Comments
(AA)	**Bielsa** (France – Spain) 3.2 km (2 miles)	**Aragnouet to Bielsa** 1830m (6000')	Open 24 hours but possibly closed October-Easter. On French side (D173) generally good road surface but narrow with steep hairpin bends and steep gradients near summit. Often no middle white line. Spanish side (A138) has good width and is less steep and winding. Used by heavy vehicles. No tolls.
(BB)	**Cadi** (Spain) 5 km (3 miles)	**Bellver de Cerdanya to Berga** 1220m (4000')	W of Toses (Tosas) pass on E9/C16; link from La Seo de Urgel to Andorra; excellent approach roads; heavy weekend traffic. **Tolls charged.**
(CC)	**Envalira** (France – Spain via Andorra) 2.8 km (1.75 miles)	**Pas de la Casa to El Grau Roig** 2000m (6562')	Tunnel width 8.25m. On N22/CG2 France to Andorra. **Tolls charged.**
(DD)	**Puymorens** (France –Spain) 4.8 km (2.9 miles)	**Ax-les-Thermes to Puigcerda** 1515m (4970')	MHV 3.5m (11'6"). Part of Puymorens pass on N20/E9. **Tolls charged.**
(EE)	**Somport** (France – Spain) 8.6 km (5.3 miles)	**Urdos to Canfranc** 1116m (3661')	Tunnel height 4.55m (14'9"), width 10.5m (34'). Max speed 90 km/h (56 mph); leave 100m between vehicles. On N134 (France), N330 (Spain). No tolls.
(FF)	**Vielha (Viella)** (Spain) 5.2 km (3.2 miles)	**Vielha (Viella) to Pont de Suert** 1635m (5390')	Tunnel height 5.3m, width 12m. Max speed 80km/h. 3 lane, well lit, modern tunnel on N230. Gentle gradients on both sides. Good road surface. Narrow on approach from Vielha. No tolls.

Pyrenees & Northern Spain

FRANCE
ANDORRA
ANDORRA LA VELLA
Mont-de-Marsan
Dax
Bayonne
Montauban
Albi
Millau
Auch
Toulouse
Castres
Col de Sié
Pau
Tarbes
Muret
Mazamet
Col de Rodomouls
Col d'Ispéguy
Col de la Fenille
Béziers
Col d'Aspin
Carcassonne
Col du Portillon
Narbonne
Port de la Bonaigua
Durban-Corbières
Port de Venasque
Port de Saula
Port de Pailhères
Col de Jau
Perpignan
Col du Perthus
Huesca
Coll de Merolla
Figueres
Olot
Zaragoza
Vic
Salt
Girona
Lleida
Manresa
Blanes
Igualada
Mataró
Vilafranca del Penedès
Barcelona
Reus
El Prat de Llobregat
Tarragona
Vilanova i la Geltrú
Tortosa
Puerto de Cabrillas
Puerto de Escandón
Villareal
Castellón de la Plana/ Castelló de la Plana
Burriana
Sagunto
Valencia
Inca
Palma de Mallorca
Manacor
Sueca
Algemesí
Cullera
Xátiva
Gandía
Oliva
Ontinyent
Alcoy/ Alcoi
Puerto de Ares
Denia
Ibi
Eivissa
Benidorm
Motorway
Motorway (Proposed)
Motorway Road Tunnel
Major/Main Roads
Minor Mountain Passes (suitability for caravans not checked)
Major Mountain Passes Suitable for Caravans
Major Mountain Passes Unsuitable for Caravans
Major Road Tunnels
0 10 20 30 40 50 km
These maps should be used in conjunction with the information in the Mountain Passes and Tunnels tables in this chapter.
2000m – +3000m
1000m – 2000m
100m – 1000m
0 – 100m
© Collins Bartholomew Ltd 2019

Keeping in Touch

Telephones and Calling

Most people need to use a telephone at some point while they're away, whether to keep in touch with family and friends back home or call ahead to sites. Even if you don't plan to use a phone while you're away, it is best to make sure you have access to one in case of emergencies.

International Direct Dial Calls

Each country has a unique dialing code you must use if phoning from outside that country. You can find the international dialing code for any country by visiting www.thephonebook.bt.com. First dial the code then the local number. If the area code starts with a zero this should be omitted.

The international access code to dial the UK from anywhere in the world is 0044.

Ringing Tones

Ringing tones vary from country to country, so may sound very different to UK tones. Some ringing tones sound similar to error or engaged tones that you would hear on a UK line.

Using Mobile Phones

Mobile phones have an international calling option called 'roaming' which will automatically search for a local network when you switch your phone on. The EU abolished roaming charges in 2017, but you should contact your service provider to check if there are any chardges for your tariff.

Storing telephone numbers in your phone's contact list in international format (i.e. use the prefix of +44 and omit the initial '0') will mean that your contacts will automatically work abroad as well as in the UK.

Global SIM Cards

If you're planning on travelling to more than one country consider buying a global SIM card. This will mean your mobile phone can operate on foreign mobile networks, which will be more cost effective than your service provider's roaming charges. For details of SIM cards available, speak to your service provider or visit www.0044.co.uk or www.globalsimcard.co.uk.

You may find it simpler to buy a SIM card or cheap 'pay-as-you-go' phone abroad if you plan to make a lot for local calls, e.g. to book campsites or restaurants. This may mean

that you still have higher call charges for international calls (such as calling the UK). Before buying a different SIM card, check with you provider whether your phone is locked against use on other networks.

Hands-Free

Legislation in Europe forbids the use of mobile or car phones while driving except when using hands-free equipment. In some European countries it is now also illegal to drive while wearing headphones or a headset - including hands-free kits.

If you are involved in an accident whilst driving and using a hand-held mobile phone, your insurance company may refuse to honour the claim.

Accessing the internet

Accessing the internet via your mobile (data roaming) while outside of the UK can be very expensive. It is recommended that you disable your internet access by switching 'data roaming' off to avoid a large mobile phone bill.

Internet Access

Wi-Fi is available on lots of campsites in Europe, the cost may be an additional charge or included in your pitch fee. Most larger towns may have internet cafés or libraries where you can access the internet, however lots of fast food restaurants and coffee chains now offer free Wi-Fi for customers so you can get access for the price of a coffee or bite to eat.

Many people now use their smartphones for internet access. Another option is a dongle – a device which connects to your laptop to give internet access using a mobile phone network. While these methods are economical in the UK, overseas you will be charged data roaming charges which can run into hundreds or thousands of pounds depending on how much data you use. If you plan on using your smartphone or a dongle abroad speak to your service provider before you leave the UK to make sure you understand the costs or add an overseas data roaming package to your phone contract.

Making Calls from your Laptop

If you download Skype to your laptop you can make free calls to other Skype users anywhere in the world using a Wi-Fi connection. Rates for calls to non-Skype users (landline or mobile phone) are also very competitively-priced. You will need a computer with a microphone and speakers, and a webcam is handy too. It is also possible to download Skype to an internet-enabled mobile phone to take advantage of the same low-cost calls – see www.skype.com.

Club Together

If you want to chat to other members either at home or while you're away, you can do so on The Club's online community Club Together. You can ask questions and gather opinions on the forums at camc.com/together.

Radio and Television

The BBC World Service broadcasts radio programmes 24 hours a day worldwide and you can listen on a number of platforms: online, via satellite or cable, DRM digital radio, internet radio or mobile phone. You can find detailed information and programme schedules at www.bbc.co.uk/worldservice.

Whereas analogue television signals were switched off in the UK during 2012, no date has yet been fixed for the switch off of analogue radio signals.

Digital Terrestrial Television

As in the UK, television transmissions in most of Europe have been converted to digital. The UK's high definition transmission technology may be more advanced than any currently implemented or planned in Europe. This means that digital televisions intended for use in the UK might not be able to receive HD terrestrial signals in some countries.

Satellite Television

For English-language TV programmes the only realistic option is satellite, and satellite dishes are a common sight on campsites all over Europe. A satellite dish mounted on the caravan roof or clamped to a pole fixed to the drawbar, or one mounted on a foldable free-standing tripod, will provide good reception and minimal interference. Remember however that obstructions to the south east (such as tall trees or even mountains) or heavy rain, can interrupt the signals. A specialist dealer will be able to advise you on the best way of mounting your dish. You will also need a satellite receiver and ideally a satellite-finding meter.

The main entertainment channels such as BBC1, ITV1 and Channel 4 can be difficult to pick up in mainland Europe as they are now being transmitted by new narrow-beam satellites. A 60cm dish should pick up these channels in most of France, Belgium and the Netherlands but as you travel further afield, you'll need a progressively larger dish. See www.satelliteforcaravans.co.uk (created and operated by a Club member) for the latest changes and developments, and for information on how to set up your equipment.

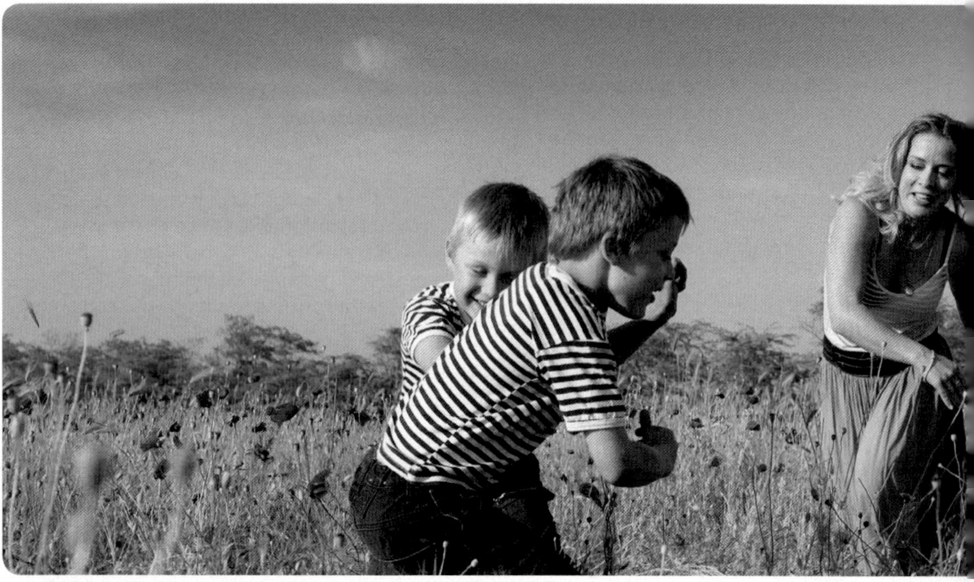

Medical Matters

Before You Travel

You can find country specific medical advice, including any vaccinations you may need, from www.nhs.uk/healthcareabroad, or speak to your GP surgery. For general enquiries about medical care abroad contact NHS England on 0300 311 22 33 or email england.contactus@nhs.uk.

If you have any pre-existing medical conditions you should check with your GP that you are fit to travel. Ask your doctor for a written summary of any medical problems and a list of medications , which is especially imporant for those who use controlled drugs or hypodermic syringes.

Always make sure that you have enough medication for the duration of your holiday and some extra in case your return is delayed. Take details of the generic name of any drugs you use, as brand names may be different abroad, your blood group and details of any allergies (translations may be useful for restaurants).

An emergency dental kit is available from High Street chemists which will allow you temporarily to restore a crown, bridge or filling or to dress a broken tooth until you can get to a dentist.

A good website to check before you travel is www.nathnac.org/travel which gives general health and safety advice, as well as highlighting potential health risks by country.

European Heath Insurance Card (EHIC)

Before leaving home apply for a European Health Insurance Card (EHIC). British residents temporarily visiting another EU country are entitled to receive state-provided emergency treatment during their stay on the same terms as residents of those countries, but you must have a valid EHIC to claim these services.

To apply for your EHIC visit www.ehic.org.uk, call 0300 330 1350 or pick up an application form from a post office. An EHIC is required by each family member, with children under 16 included in a parent or guardian's application. The EHIC is free of charge, is valid for up to five years and can be renewed up to six months before its expiry date. Before you travel remember to check that your EHIC is still valid.

An EHIC is not a substitute for travel insurance and it is strongly recommended that you arrange full travel insurance before leaving home regardless of the cover provided by your EHIC. Some insurance companies require you to have an EHIC and some will waive the policy excess if an EHIC has been used.

If your EHIC is stolen or lost while you are abroad contact 0044 191 2127500 for help. If you experience difficulties in getting your EHIC accepted, telephone the Department for Work & Pensions for assistance on the overseas healthcare team line 0044 (0)191 218 1999 between 8am to 5pm Monday to Friday. Residents of the Republic of Ireland, the Isle of Man and Channel Islands, should check with their own health authorities about reciprocal arrangements with other countries.

Holiday Travel Insurance

Despite having an EHIC you may incur high medical costs if you fall ill or have an accident. The cost of bringing a person back to the UK in the event of illness or death is never covered by the EHIC.

Separate additional travel insurance adequate for your destination is essential, such as the Club's Red Pennant Overseas Holiday Insurance – see camc.com/redpennant.

First Aid

A first aid kit containing at least the basic requirements is an essential item, and in some countries it is compulsory to carry one in your vehicle (see the Essential Equipment Table in the chapter Motoring – Equipment). Kits should contain items such as sterile pads, assorted dressings, bandages and plasters, antiseptic wipes or cream, cotton wool, scissors, eye bath and tweezers. Also make sure you carry something for upset stomachs, painkillers and an antihistamine in case of hay fever or mild allergic reactions.

If you're travelling to remote areas then you may find it useful to carry a good first aid manual. The British Red Cross publishes a comprehensive First Aid Manual in conjunction with St John Ambulance and St Andrew's Ambulance Association.

Accidents and Emergencies

If you are involved in or witness a road accident the police may want to question you about it. If possible take photographs or make sketches of the scene, and write a few notes about what happened as it may be more difficult to remember the details at a later date.

For sports activities such as skiing and mountaineering, travel insurance must include provision for covering the cost of mountain and helicopter rescue. Visitors to the Savoie and Haute-Savoie areas should be aware that an accident or illness may result in a transfer to Switzerland for hospital treatment. There is a reciprocal healthcare agreement for British citizens visiting Switzerland but you will be required to pay the full costs of treatment and afterwards apply for a refund.

Sun Protection

Never under-estimate how ill exposure to the sun can make you. If you are not used to the heat it is very easy to fall victim to heat exhaustion or heat stroke. Avoid sitting in the sun between 11am and 3pm and cover your head if sitting or walking in the sun. Use a high sun protection factor (SPF) and re-apply frequently. Make sure you drink plenty of fluids.

Tick-Borne Encephalitis (TBE) and Lyme Disease

Hikers and outdoor sports enthusiasts planning trips to forested, rural areas should be aware of tick-borne encephalitis, which is transmitted by the bite of an infected tick. If you think you may be at risk, seek medical advice on prevention and immunisation before you leave the UK.

There is no vaccine against Lyme disease, an equally serious tick-borne infection, which, if left untreated, can attack the nervous system and joints. You can minimise the risk by using an insect repellent containing DEET, wearing long sleeves and long trousers, and checking for ticks after outdoor activity.

Avoid unpasteurised dairy products in risk areas. See www.tickalert.org or telephone 01943 468010 for more information.

Water and food

Water from mains supplies throughout Europe is generally safe, but may be treated with chemicals which make it taste different to tap water in the UK. If in any doubt, always drink bottled water or boil it before drinking.

Food poisoning is potential anywhere, and a complete change of diet may upset your stomach as well. In hot conditions avoid any food that hasn't been refrigerated or hot food that has been left to cool. Be sensible about the food that you eat – don't eat unpasteurised or undercooked food and if you aren't sure about the freshness of meat or seafood then it is best avoided.

Returning Home

If you become ill on your return home tell your doctor that you have been abroad and which countries you have visited. Even if you have received medical treatment in another country, always consult your doctor if you have been bitten or scratched by an animal while on holiday. If you were given any medicines in another country, it may be illegal to bring them back into the UK. If in doubt, declare them at Customs when you return.

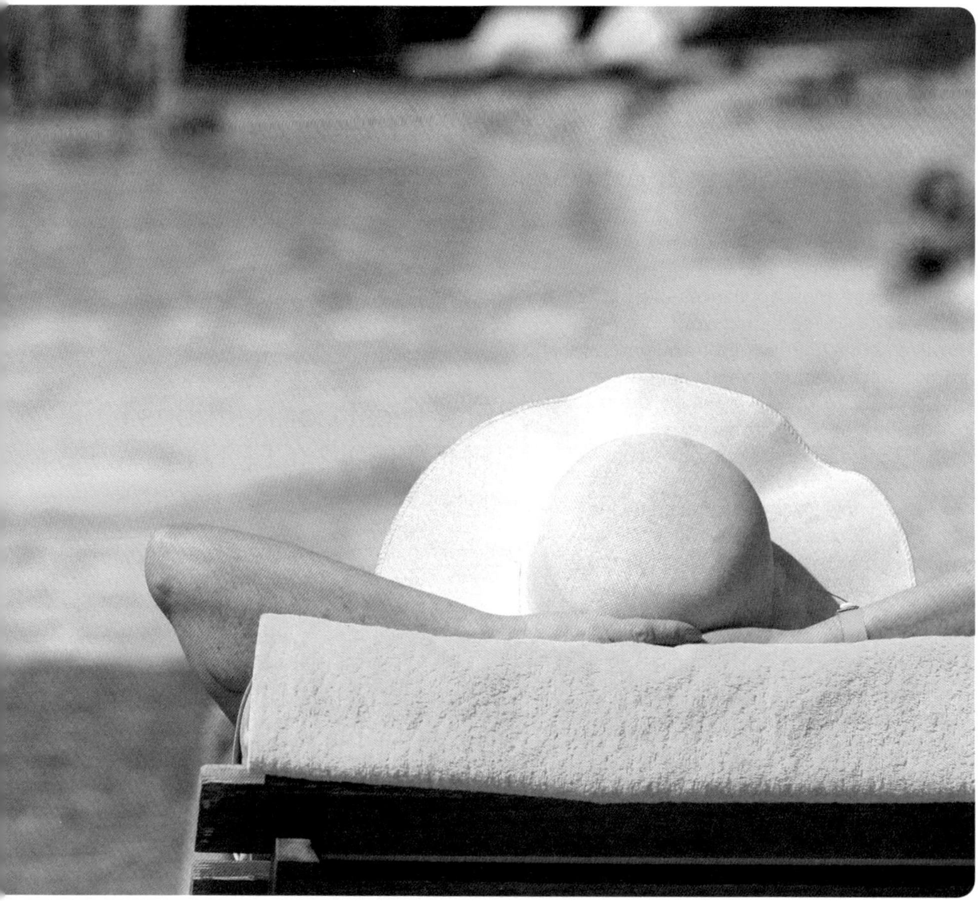

Electricity and Gas

Electricity – General Advice

The voltage for mains electricity is 230V across the EU, but varying degrees of 'acceptable tolerance' mean you may find variations in the actual voltage. Most appliances sold in the UK are 220-240V so should work correctly. However, some high-powered equipment, such as microwave ovens, may not function well – check your instruction manual for any specific instructions. Appliances marked with 'CE' have been designed to meet the requirements of relevant European directives.

The table below gives an approximate idea of which appliances can be used based on the amperage which is being supplied (although not all appliances should be used at the same time). You can work it out more accurately by making a note of the wattage of each appliance in your caravan. The wattages given are based on appliances designed for use in caravans and motorhomes. Household kettles, for example, have at least a 2000W element. Each caravan circuit will also have a maximum amp rating which should not be exceeded.

Electrical Connections – EN60309-2 (CEE17)

EN60309-2 (formerly known as CEE17) is the European Standard for all newly fitted connectors. Most sites should now have these connectors, however there is no requirement

Amps	Wattage (Approx)	Fridge	Battery Charger	Air Conditioning	LCD TV	Water Heater	Kettle (750W)	Heater (1kW)
2	400	✓	✓					
4	900	✓	✓		✓	✓		
6	1300	✓	✓	*	✓	✓	✓	
8	1800	✓	✓	✓**	✓	✓	✓	✓**
10	2300	✓	✓	✓**	✓	✓	✓	✓**
16	3600	✓	✓	✓	✓	✓	✓	✓**

* Usage possible, depending on wattage of appliance in question

** Not to be used at the same time as other high-wattage equipment

to replace connectors which were installed before this was standardised so you may still find some sites where your UK 3 pin connector doesn't fit. For this reason it is a good idea to carry a 2-pin adapter. If you are already on site and find your connector doesn't fit, ask campsite staff to borrow or hire an adaptor. You may still encounter a poor electrical supply on site even with an EN60309-2 connection.

If the campsite does not have a modern EN60309-2 (CEE17) supply, ask to see the electrical protection for the socket outlet. If there is a device marked with IDn = 30mA, then the risk is minimised.

Hooking Up to the Mains

Connection should always be made in the following order:

- Check your outfit isolating switch is at 'off'.
- Uncoil the connecting cable from the drum. A coiled cable with current flowing through it may overheat. Take your cable and insert the connector (female end) into your outfit inlet.
- Insert the plug (male end) into the site outlet socket.
- Switch outfit isolating switch to 'on'.
- Use a polarity tester in one of the 13A sockets in the outfit to check all connections are correctly wired. Never leave it in the socket.

Some caravans have these devices built in as standard.

It is recommended that the supply is not used if the polarity is incorrect (see Reversed Polarity overleaf).

Warnings:

If you are in any doubt of the safety of the system, if you don't receive electricity once connected or if the supply stops then contact the site staff.

If the fault is found to be with your outfit then call a qualified electrician rather than trying to fix the problem yourself.

To ensure your safety you should never use an electrical system which you can't confirm to be safe. Use a mains tester such as the one shown below to test the electrical supply.

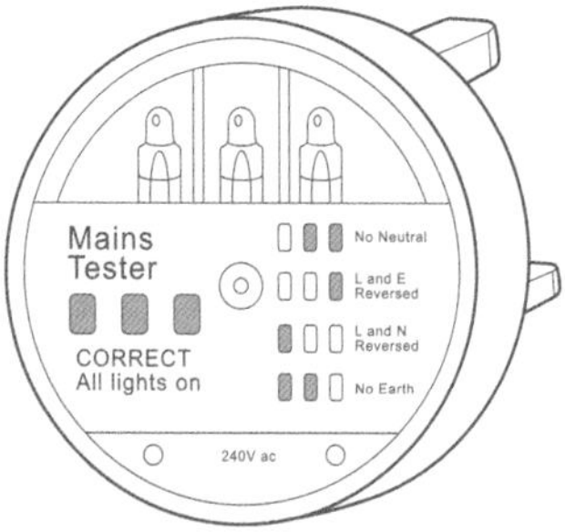

Always check that a proper earth connection exists before using the electrics. Please note that these testers may not pick up all earth faults so if there is any doubt as to the integrity of the earth system do not use the electrical supply.

Disconnection

- Switch your outfit isolating switch to 'off'.
- At the site supply socket withdraw the plug.
- Disconnect the cable from your outfit.

Motorhomes – if leaving your pitch during the day, don't leave your mains cable plugged into the site supply, as this creates a hazard if the exposed live connections in the plug are touched or if the cable is not seen during grass-cutting.

Reversed Polarity

Even if the site connector meets European Standard EN60309-2 (CEE17), British caravanners are still likely to encounter the problem known as reversed polarity. This is where the site supply 'live' line connects to the outfit's 'neutral' and vice versa. You should always check the polarity immediately on connection, using a polarity tester available from caravan accessory shops. If polarity is reversed the caravan mains electricity should not be used. Try using another nearby socket instead. Frequent travellers to the Continent can make up an adaptor themselves, or ask an electrician to make one for you, with the live and neutral wires reversed. Using a reversed polarity socket will probably not affect how an electrical appliance works, however your protection is greatly reduced. For example, a lamp socket may still be live as you touch it while replacing a blown bulb, even if the light switch is turned off.

Shaver Sockets

Most campsites provide shaver sockets with a voltage of 220V or 110V. Using an incorrect voltage may cause the shaver to become hot or break. The 2-pin adaptor available in the UK may not fit Continental sockets so it is advisable to buy 2-pin adaptors on the Continent. Many modern shavers will work on a range of voltages which make them suitable for travelling abroad. Check you instruction manual to see if this is the case.

Gas – General Advice

Gas usage can be difficult to predict as so many factors, such as temperature and how often you eat out, can affect the amount you need. As a rough guide allow 0.45kg of gas a day for normal summer usage.

With the exception of Campingaz, LPG cylinders normally available in the UK cannot be exchanged abroad. If possible, take enough gas with you and bring back the empty cylinders. Always check how many you can take with you as ferry and tunnel operators may restrict the number of cylinders you are permitted to carry for safety reasons.

The full range of Campingaz cylinders is widely available from large supermarkets and hypermarkets, although at the end of the holiday season stocks may be low. Other popular brands of gas are Primagaz, Butagaz,

Site Hooking Up Adaptor

ADAPTATEUR DE PRISE AU SITE (SECTEUR)
CAMPINGPLATZ-ANSCHLUSS (NETZ)

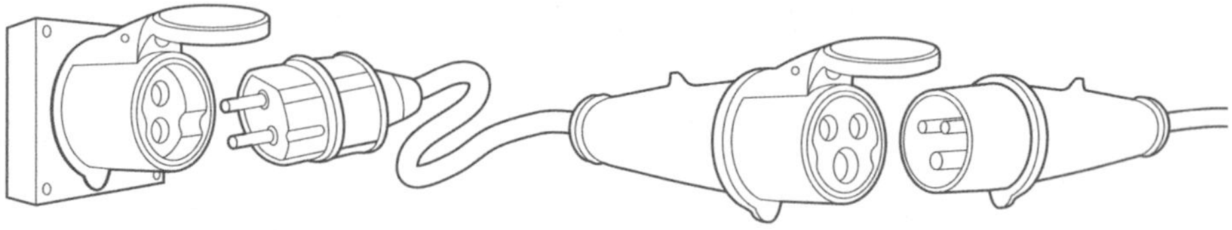

Totalgaz and Le Cube. A loan deposit is required and if you are buying a cylinder for the first time you may also need to buy the appropriate regulator or adaptor hose.

If you are touring in cold weather conditions use propane gas instead of butane. Many other brands of gas are available in different countries and, as long as you have the correct regulator, adaptor and hose and the cylinders fit in your gas locker these local brands can also be used.

Gas cylinders are now standardised with a pressure of 30mbar for both butane and propane within the EU. On UK-specification caravans and motorhomes (2004 models and later) a 30mbar regulator suited to both propane and butane use is fitted to the bulkhead of the gas locker. This is connected to the cylinder with a connecting hose (and sometimes an adaptor) to suit different brands or types of gas. Older outfits and some foreign-built ones may use a cylinder-mounted regulator, which may need to be changed to suit different brands or types of gas.

Warnings:

- Refilling gas cylinders intended to be exchanged is against the law in most countries, however you may still find that some sites and dealers will offer to refill cylinders for you. Never take them up on this service as it can be dangerous; the cylinders haven't been designed for user-refilling and it is possible to overfill them with catastrophic consequences.
- Regular servicing of gas appliances is important as a faulty appliance can emit carbon monoxide, which could prove fatal. Check your vehicle or appliance handbook for service recommendations.
- Never use a hob or oven as a space heater.

The Caravan and Motorhome Club publishes a range of technical leaflets for its members including detailed advice on the use of electricity and gas – you can request copies or see camc.com/advice-and-training

Safety and Security

EU countries have good legislation in place to protect your safety wherever possible. However accidents and crime will still occur and taking sensible precautions can help to minimise your risk of being involved.

Beaches, Lakes and Rivers

Check for any warning signs or flags before you swim and ensure that you know what they mean. Check the depth of water before diving and avoid diving or jumping into murky water as submerged objects may not be visible. Familiarise yourself with the location of safety apparatus and/or lifeguards.

Use only the designated areas for swimming, watersports and boating and always use life jackets where appropriate. Watch out for tides, undertows, currents and wind strength and direction before swimming in the sea. This applies in particular when using inflatables, windsurfing equipment, body boards, kayaks or sailing boats. Sudden changes of wave and weather conditions combined with fast tides and currents are particularly dangerous.

Campsite Safety

Once you've settled in, take a walk around the site to familiarise yourself with its layout and locate the nearest safety equipment. Ensure that children know their way around and where your pitch is.

Natural disasters are rare, but always think about what could happen. A combination of heavy rain and a riverside pitch could lead to flash flooding, for example, so make yourself aware of site evacuation procedures.

Be aware of sources of electricity and cabling on and around your pitch – electrical safety might not be up to the same standards as in the UK.

Poison for rodent control is sometimes used on sites or surrounding farmland. Warning notices are not always posted and you are strongly advised to check if staying on a rural site with dogs or children.

Incidents of theft on campsites are rare but when leaving your caravan unattended make sure you lock all doors and shut windows. Conceal valuables from sight and lock up any bicycles.

Children

Watch out for children as you drive around the site and don't exceed walking pace.

Children's play areas are generally unsupervised, so check which are suitable for your children's ages and abilities. Read and respect the displayed rules. Remember it is your responsibility to supervise your children at all times.

Be aware of any campsite rules concerning ball games or use of play equipment, such as roller blades and skateboards. When your children attend organised activities, arrange when and where to meet afterwards. You should never leave children alone inside a caravan.

Fire

Fire prevention is important on sites, as fire can spread quickly between outfits. Certain areas of southern Europe experience severe water shortages in summer months leading to an increased fire risk. This may result in some local authorities imposing restrictions at short notice on the use of barbecues and open flames.

Fires can be a regular occurrence in forested areas, especially along the Mediterranean coast during summer months. They are generally extinguished quickly and efficiently but short term evacuations are sometimes necessary. If visiting forested areas familiarise yourself with local emergency procedures in the event of fire. Never use paraffin or gas heaters inside your caravan. Gas heaters should only be fitted when air is taken from outside the caravan. Don't change your gas cylinder inside the caravan. If you smell gas turn off the cylinder immediately, extinguish all naked flames and seek professional help.

Make sure you know where the fire points and telephones are on site and know the site fire drill. Make sure everyone in your party knows how to call the emergency services.

Where site rules permit the use of barbecues, take the following precautions to prevent fire:

- Never locate a barbecue near trees or hedges.
- Have a bucket of water to hand in case of sparks.
- Only use recommended fire-lighting materials.
- Don't leave a barbecue unattended when lit and dispose of hot ash safely.
- Never take a barbecue into an enclosed area or awning – even when cooling they continue to release carbon monoxide which can lead to fatal poisoning.

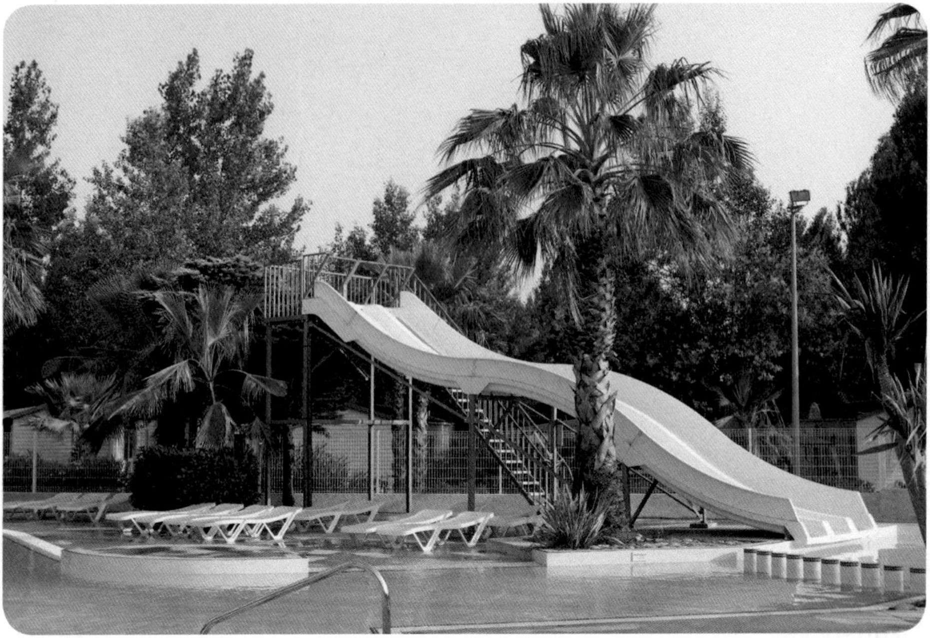

Swimming Pools

Familiarize yourself with the pool area before you venture in for a swim - check the pool layout and identify shallow and deep ends and the location of safety equipment. Check the gradient of the pool bottom as pools which shelve off sharply can catch weak or non-swimmers unawares.

Never dive or jump into a pool without knowing the depth – if there is a no diving rule it usually means the pool isn't deep enough for safe diving.

For pools with a supervisor or lifeguard, note any times or dates when the pool is not supervised. Read safety notices and rules posted around the pool.

On the Road

Don't leave valuables on view in cars or caravans, even if they are locked. Make sure items on roof racks or cycle carriers are locked securely.

Near to ports British owned cars have been targeted by thieves, both while parked and on the move, e.g. by flagging drivers down or indicating that a vehicle has a flat tyre. If you stop in such circumstances be wary of anyone offering help, ensure that car keys are not left in the ignition and that vehicle doors are locked while you investigate.

Always keep car doors locked and windows closed when driving in populated areas. Beware of a 'snatch' through open car windows at traffic lights, filling stations or in traffic jams. When driving through towns and cities keep your doors locked. Keep handbags, valuables and documents out of sight at all times.

If flagged down by another motorist for whatever reason, take care that your own car is locked and windows closed while you check outside, even if someone is left inside.

Be particularly careful on long, empty stretches of motorway and when you stop for fuel. Even if the people flagging you down appear to be officials (e.g. wearing yellow reflective jackets or dark, 'uniform-type' clothing) lock your vehicle doors. They may appear to be friendly and helpful, but could be opportunistic thieves. Have a mobile phone to hand and, if necessary, be seen to use it.

Road accidents are a increased risk in some countries where traffic laws may be inadequately enforced, roads may be poorly maintained, road signs and lighting inadequate, and driving standards poor. It's a good idea to keep a fully-charged mobile phone with you in your car with the number of your breakdown organisation saved into it.

On your return to the UK there are increasing issues with migrants attempting to stowaway in vehicles, especially if you're travelling through Calais. The UK government have issued the following instructions to prevent people entering the UK illegally:

- Where possible all access to vehicles or storage compartments should be fitted with locks.
- All locks must be engaged when the vehicle is stationary or unattended.
- Immediately before boarding your ferry or train check that the locks on your vehicle haven't been compromised.
- If you have any reason to suspect someone may have accessed your outfit speak to border control staff or call the police. Do not board the ferry or train or you may be liable for a fine of up to £2000.

Overnight Stops

Overnight stops should always be at campsites and not at motorway service areas, ferry terminal car parks, petrol station forecourts or isolated 'aires de services' on motorways where robberies are occasionally reported. If you decide to use these areas for a rest then take appropriate precautions, for example, shutting all windows, securing locks and making a thorough external check of your vehicle(s) before departing. Safeguard your property, e.g. handbags, while out of the caravan and beware of approaches by strangers.

For a safer place to take a break, there is a wide network of 'Aires de Services' in cities, towns and villages across Europe, many specifically for motorhomes with good security and overnight facilities. They are often less isolated and therefore safer than the motorway aires. It is rare that you will be the only vehicle staying on such areas, but take sensible precautions and trust your instincts.

Personal Security

Petty crime happens all over the world, including in the UK; however as a tourist you are more vulnerable to it. This shouldn't stop you from exploring new horizons, but there are a few sensible precautions you can take to minimise the risk.

- Leave valuables and jewellery at home. If you do take them, fit a small safe in your caravan or lock them in the boot of your car. Don't leave money or valuables in a car glovebox or on view. Don't leave bags in full view when sitting outside at cafés or restaurants, or leave valuables unattended on the beach.
- When walking be security-conscious. Avoid unlit streets at night, walk away from the kerb edge and carry handbags or shoulder bags on the side away from the kerb. The less of a tourist you appear, the less of a target you are.
- Keep a note of your holiday insurance details and emergency telephone numbers in more than one place, in case the bag or vehicle containing them is stolen.
- Beware of pickpockets in crowded areas, at tourist attractions and in cities. Be especially aware when using public transport in cities.
- Be cautious of bogus plain-clothes policemen who may ask to see your foreign currency or credit cards and passport. If approached, decline to show your money or to hand over your passport but ask for credentials and offer instead to go to the nearest police station.
- Laws and punishment vary from country to country so make yourself aware of anything which may affect you before you travel. Be especially careful on laws involving alcohol consumption (such as drinking in public areas), and never buy or use illegal drugs abroad.
- Respect customs regulations - smuggling is a serious offence and can carry heavy penalties. Do not carry parcels or luggage through customs for other people and never cross borders with people you do not know in your vehicle, such as hitchhikers.

The Foreign & Commonwealth Office produces a range of material to advise and inform British citizens travelling abroad about issues affecting their safety - www.gov.uk/foreign-travel-advice has country specific guides.

Money Security

We would rarely walk around at home carrying large amounts of cash, but as you may not have the usual access to bank accounts and credit cards you are more likely to do so on holiday. You are also less likely to have the same degree of security when online banking as you would in your own home. Take the following precautions to keep your money safe:

- Carry only the minimum amount of cash and don't rely on one person to carry everything. Never carry a wallet in your back pocket. Concealed money belts are the most secure way to carry cash and passports.
- Keep a separate note of bank account and credit/debit card numbers. Carry your credit card issuer/bank's 24-hour UK contact number with you.
- Be careful when using cash machines (ATMs) – try to use a machine in an area with high footfall and don't allow yourself to be distracted. Put your cash away before moving away from the cash machine.
- Always guard your PIN number, both at cash machines and when using your card to pay in shops and restaurants. Never let your card out of your sight while paying.
- If using internet banking do not leave the PC or mobile device unattended and make sure you log out fully at the end of the session.

Winter Sports

If you are planning a skiing or snowboarding holiday you should research the safety advice for your destination before you travel. A good starting point may be the relevant embassy

for the country you're visitng. All safety instructions should be followed meticulously given the dangers of avalanches in some areas.

The Ski Club of Great Britain offer a lot of advice for anyone taking to the mountains, visit their website www.skiclub.co.uk to pick up some useful safety tips and advice on which resorts are suitable for different skill levels.

British Consular Services Abroad

British Embassy and Consular staff offer practical advice, assistance and support to British travellers abroad. They can, for example, issue replacement passports, help Britons who have been the victims of crime, contact relatives and friends in the event of an accident, illness or death, provide information about transferring funds and provide details of local lawyers, doctors and interpreters. But there are limits to their powers and a British Consul cannot, for example, give legal advice, intervene in court proceedings, put up bail, pay for legal or medical bills, or for funerals or the repatriation of bodies, or undertake work more properly done by banks, motoring organisations and travel insurers.

If you are charged with a serious offence, insist on the British Consul being informed. You will be contacted as soon as possible by a Consular Officer who can advise on local procedures, provide access to lawyers and insist that you are treated as well as nationals of the country which is holding you. However, they cannot get you released as a matter of course.

British and Irish embassy contact details can be found in the Country Introduction chapters.

Continental Campsites

The quantity and variety of sites across Europe means you're sure to find one that suits your needs – from full facilities and entertainment to quiet rural retreats. If you haven't previously toured outside of the UK you may notice some differences, such as pitches being smaller or closer together. In hot climates hard ground may make putting up awnings difficult.

In the high season all campsite facilities are usually open, however bear in mind that toilet and shower facilities may be busy. Out of season some facilities such as shops and swimming pools may be closed and office opening hours may be reduced. If the site has very low occupancy the sanitary facilities may be reduced to a few unisex toilet and shower cubicles.

Booking a Campsite

To save the hassle of arriving to find a site full it is best to book in advance, especially in high season. If you don't book ahead arrive no later than 4pm (earlier at popular resorts) to secure a pitch, after this time sites fill up quickly. You also need to allow time to find another campsite if your first choice is fully booked.

You can often book directly via a campsite's website using a credit or debit card to pay a deposit if required.

Please be aware that some sites regard the deposit as a booking or admin fee and will not deduct the amount from your final bill.

Overseas Travel Service

The Club's Overseas Travel Service offers members an site booking service on over 250 campsites in Europe. Full details of these sites plus information on ferry special offers and Red Pennant Overseas Holiday Insurance can be found in the Club's Venture Abroad brochure – call 01342 327410 to request a copy or visit camc.com/overseas.

Overseas Site Booking Service sites are marked 'SBS' in the site listings. Many of them can be booked at camc.com. We can't make advance reservations for any other campsites listed in this guide. Only those sites marked SBS have been inspected by Club staff.

Overseas Site Night Vouchers

The Club offers Overseas Site Night Vouchers which can be used at over 300 Club inspected sites in Europe. The vouchers cost £21.95 each (2019 cost) and you'll need one voucher per night to stay on a site in low season and two per night in high season. You'll also be eligable for the Club's special packaged ferry rates when you're buying vouchers. For more information and the view the voucher terms and conditions visit www.camc.com/overseasoffers or call 01342 327 410.

Caravan Storage Abroad

Storing your caravan on a site in Europe can be a great way to avoid a long tow and to save on ferry and fuel costs. Even sites which don't offer a specific long-term storage facility may be willing to negotiate a price to store your caravan for you. Before you leave your caravan in storage abroad always check whether your insurance covers this, as many policies don't.

If you aren't covered then look for a specialist policy - Towergate Insurance (tel: 01242 538431 or www.towergateinsurance.co.uk) or Look Insurance (tel: 0333 777 3035 or www.lookinsuranceservices.co.uk) both offer insurance policies for caravans stored abroad.

Facilities and Site Description

All of the site facilities shown in the site listings of this guide have been taken from member reports, as have the comments at the end of each site entry. Please remember that opinions and expectations can differ significantly from one person to the next.

The year of report is shown at the end of each site listing – sites which haven't been reported on for a few years may have had significant changes to their prices, facilities, opening dates and standards. It is always best to check any specific details you need to know before travelling by contacting the site or looking at their website.

Sanitary Facilities

Facilities normally include toilet and shower blocks with shower cubicles, wash basins and razor sockets. In site listings the abbreviation 'wc' indicates that the site has the kind of toilets we are used to in the UK (pedestal style). Some sites have footplate style toilets and, where this is known, you will see the abbreviation 'cont', i.e. continental. European sites do not always provide sink plugs, toilet paper or soap so take them with you.

Waste Disposal

Site entries show (when known) where a campsite has a chemical disposal and/or a motorhome service point, which is assumed to include a waste (grey) water dump station and toilet cassette-emptying point. You may find fewer waste water disposal facilities as on the continent more people use the site sanitary blocks rather than their own facilities.

Chemical disposal points may be fixed at a high level requiring you to lift cassettes in order to empty them. Disposal may simply be down a toilet. Wastemaster-style emptying points are not very common in Europe. Formaldehyde chemical cleaning products are banned in many countries. In Germany the 'Blue Angel' (Blaue Engel) Standard, and in the Netherlands the 'Milieukeur' Standard, indicates that the product has particularly good environmental credentials.

Finding a Campsite

Directions are given for all campsites listed in this guide and most listings also include GPS co-ordinates. Full street addresses are also given where available. The directions have been supplied by member reports and haven't been checked in detail by The Club.

For information about using satellite navigation to find a site see the Motoring Equipment section.

Overnight Stops

Many towns and villages across Europe provide dedicated overnight or short stay areas specifically for motorhomes, usually with security, electricity, water and waste facilities. These are known as 'Aires de Services', 'Stellplatz' or 'Aree di Sosta' and are usually well signposted with a motorhome icon. Facilities and charges for these overnight stopping areas will vary significantly.

Many campsites in popular tourist areas will also have separate overnight areas of hardstanding with facilities often just outside the main campsite area. There are guidebooks available which list just these overnight stops, Vicarious books publish an English guide to the Aires including directions, GPS co-ordinates and photographs. Please contact 0131 208 3333 or visit their website www.vicarious-shop.co.uk.

For security reasons you shouldn't spend the night on petrol station service areas, ferry terminal car parks or isolated 'Aires de Repos' or 'Aires de Services' along motorways.

Municipal Campsites

Municipal sites are found in towns and villages all over Europe, in particular in France. Once very basic, many have been improved in recent years and now offer a wider range of facilities. They can usually be booked in advance through the local town hall or tourism office. When approaching a town you may find that municipal sites are not always named and signposts may simply state 'Camping' or show a tent or caravan symbol. Most municipal sites are clean, well-run and very reasonably priced but security may be basic.

These sites may be used by seasonal workers, market traders and travellers in low season and as a result there may be restrictions or very high charges for some types of outfits (such as twin axles) in order to discourage this. If you may be affected check for any restrictions when you book.

Naturist Campsites

Some naturist sites are included in this guide and are shown with the word 'naturist' after their site name. Those marked 'part naturist' have separate areas for naturists. Visitors to naturist sites aged 16 and over usually require an INF card or Naturist Licence - covered by membership of British Naturism (tel 01604 620361, visit www.british-naturism.org.uk or email headoffice@british-naturism.org.uk) or you can apply for a licence on arrival at any recognised naturist site (a passport-size photograph is required).

Opening Dates and Times

Opening dates should always be taken with a pinch of salt - including those given in this guide. Sites may close without notice due to refurbishment work, a lack of visitors or bad weather. Outside the high season it is always best to contact campsites in advance, even if the site advertises itself as open all year.

Most sites will close their gates or barriers overnight – if you are planning to arrive late or are delayed on your journey you should call ahead to make sure you will be able to gain access to the site. There may be a late arrivals area outside of the barriers where you can pitch overnight. Motorhomers should also consider barrier closing times if leaving site in your vehicle for the evening.

Check out time is usually between 10am and 12 noon – speak to the site staff if you need to leave very early to make sure you can check out on departure. Sites may also close for an extended lunch break, so if you're planning to arrive or check out around lunchtime check that the office will be open.

Pets on Campsites

Dogs are welcome on many sites, although you may have to prove that all of their vaccinations are up to date before they are allowed onto the site. Certain breeds of dogs are banned in some countries and other breeds will need to be muzzled and kept on a lead at all times.
A list of breeds with restrictions by country can be found at camc.com/pets.

Sites usually charge for dogs and may limit the number allowed per pitch. On arrival make yourself aware of site rules regarding dogs, such as keeping them on a lead, muzzling them or not leaving them unattended in your outfit.

In popular tourist areas local regulations may ban dogs from beaches during the summer. Some dogs may find it difficult to cope with changes in climate. Also watch out for diseases transmitted by ticks, caterpillars, mosquitoes or sandflies - dogs from the UK will have no natural resistance. Consult your vet about preventative treatment before you travel.

Visitors to southern Spain and Portugal, parts of central France and northern Italy should be aware of the danger of Pine Processionary Caterpillars from mid-winter to late spring. Dogs should be kept away from pine trees if possible or fitted with a muzzle that prevents the nose and mouth from touching the ground. This will also protect against poisoned bait sometimes used by farmers and hunters.

In the event that your pet is taken ill abroad a campsite should have information about local vets.

Most European countries require dogs to wear a collar identifying their owners at all times. If your dog goes missing, report the matter to the local police and the local branch of that country's animal welfare organisation.

See the Documents section of this book for more information about the Pet Travel Scheme.

Prices and Payment

Prices per night (for an outfit and two adults) are shown in the site entries. If you stay on site after midday you may be charged for an extra day. Many campsites have a minimum amount for credit card transactions, meaning they can't be used to pay for overnight or short stays. Check which payment methods are accepted when you check in.

Sites with automatic barriers may ask for a deposit for a swipe card or fob to operate it.

Extra charges may apply for the use of facilities such as swimming pools, showers or laundry rooms. You may also be charged extra for dogs, Wi-Fi, tents and extra cars.

A tourist tax, eco tax and/or rubbish tax may be imposed by local authorities in some European countries. VAT may also be added to your campsite fees.

Registering on Arrival

Local authority requirements mean you will usually have to produce an identity document on arrival, which will be retained by the site until you check out. If you don't want to leave your passport with reception then most sites will accept a camping document such as the Camping Key Europe (CKE) or Camping Card International (CCI) - if this is known site entries are marked CKE/CCI.

CKE are available for Club members to purchase by calling 01342 336633 or are free to members if you take out the 'motoring' level of cover from the Club's Red Pennant Overseas Holiday Insurance.

General Advice

If you've visiting a new site ask to take a look around the site and facilities before booking in. Riverside pitches can be very scenic but keep an eye on the water level; in periods of heavy rain this may rise rapidly.

Speed limits on campsites are usually restricted to 10 km/h (6 mph). You may be asked to park your car in a separate area away from your caravan, particularly in the high season.

The use of the term 'statics' in the campsite reports in this guide may to any long-term accommodation on site, such as seasonal pitches, chalets, cottages, fixed tents and cabins, as well as static caravans.

Complaints

If you want to make a complaint about a site issue, take it up with site staff or owners at the time in order to give them the opportunity to rectify the problem during your stay.

The Caravan and Motorhome Club has no control or influence over day to day campsite operations or administration of the sites listed in this guide. Therefore we aren't able to intervene in any dispute you should have with a campsite, unless the booking has been made through our Site Booking Service - see listings marked 'SBS' for sites we are able to book for you.

Campsite Groups

Across Europe there are many campsite 'groups' or 'chains' with sites in various locations.

You will generally find that group sites will be consistent in their format and the quality and variety of facilities they offer. If you liked one site you can be fairly confident that you will like other sites within the same group.

If you're looking for a full facility site, with swimming pools, play areas, bars and restaurants on site you're likely to find these on sites which are part of a group. You might even find organised excursions and activities such as archery on site.

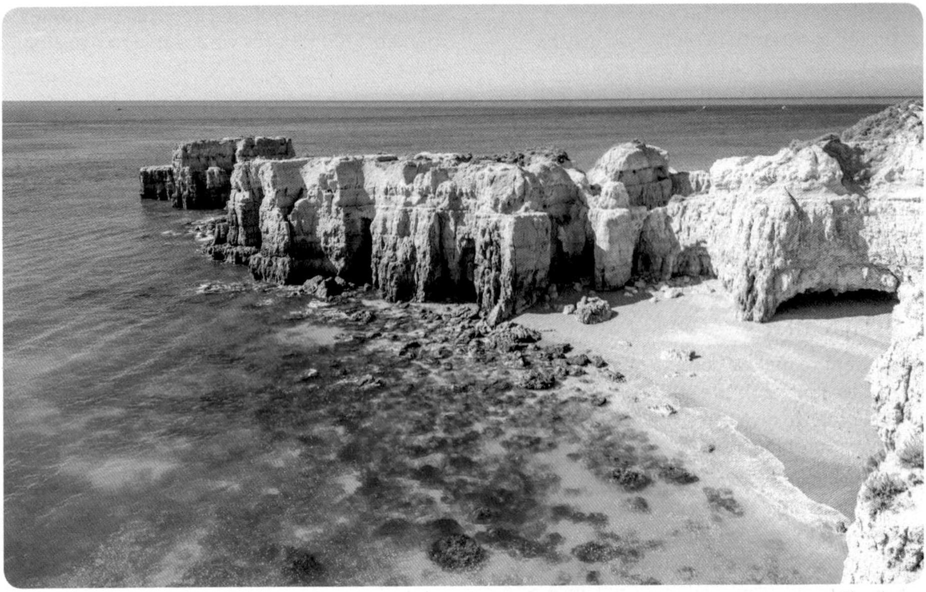

The Algarve

Welcome to Portugal

With around 3000 hours of sunshine a year, 850 kilometres of spectacular beaches and a wonderfully vibrant and varied landscape, Portugal is a visitor's paradise.

The country has been heavily influenced by its nautical tradition and position on the Atlantic. Many local delicacies are fish-based dishes, such as grilled sardines and salt cod, while some of Portugal's most splendid architecture date from when it was a global maritime empire.

Country highlights

Ceramic tiles, or azelujos, are a common element of Portuguese designs. Often depicting aspects of Portuguese culture and history, these tiles are both beautiful and functional, and are a significant part of Portugal's heritage.

Portugal is the birthplace of port, and the Douro region is one of the oldest protected wine regions in the world. Taking its name from the city of Porto, this smooth, fortified wine is exclusively produced in the Duoro Valley of Northern Portugal.

Major towns and cities

- Lisbon – Portugal's capital is known for its museums and café culture.
- Porto – an extravagant city filled with beautiful and colourful sights.
- Braga – an ancient city with filled with churches and Roman ruins.
- Faro – the prefect base to explore the Algarve.

Attractions

- Jerónimos Monastery – this UNESCO heritage site houses two museums.
- Guimarães Castle – this medieval castle is known as the Cradle of Portugal.
- National Palace of Pena – a striking palace filled with wonderful works of art.
- Lisbon Oceanarium – enjoy stunning living ocean exhibits.

Find out more

www.visitportugal.com
E: info@visitportugal.com T: (0)1 21 11 40 200

Country Information

Population (approx): 10.8 million

Capital: Lisbon

Area: 92,100 sq km (inc Azores and Madeira)

Bordered by: Spain

Terrain: Rolling plains in south; mountainous and forested north of River Tagus

Climate: Temperate climate with no extremes of temperature; wet winters in the north influenced by the Gulf Stream; elsewhere Mediterranean with hot, dry summers and short, mild winters

Coastline: 1,794km

Highest Point (mainland Portugal): Monte Torre 1,993m

Language: Portuguese

Local Time: GMT or BST, i.e. the same as the UK all year

Currency: Euros divided into 100 cents; £1 = €1.13, €1 = £0.89 (November 2018)

Emergency numbers: Police 112; Fire brigade 112; Ambulance 112

Public Holidays 2019: Jan 1; Apr 19, 21, 25; May 1; Jun 10 (National Day),20; Aug 15; Oct 5, Nov 1; Dec 1, 8,25.

Other holidays and saints' days are celebrated according to region. School summer holidays run from the end of June to the end of August.

Camping and Caravanning

There are numerous campsites in Portugal, and many of these are situated along the coast. Sites are rated from 1 to 4 stars.

There are 22 privately owned campsites in the Orbitur chain. Caravanners can join the Orbitur Camping Club for discounts of at lease 15% at these sites. The joining fee is €21, with a 50% discount for senior citizens. You can buy membership at Orbitur sites or www.orbitur.com. Casual/wild camping is not permitted.

Motorhomes

A number of local authorities now provide dedicated short stay areas for motorhomes called 'Áreas de Serviçio'. It is rare that yours will be the only motorhome staying on such areas, but take sensible precautions and avoid any that are isolated.

Cycling

In Lisbon there are cycle lanes in Campo Grande gardens, also from Torre de Belém to Cais do Sodré (7km) along the River Tagus, and between Cascais and Guincho. Elsewhere in the country there are few cycle lanes.

Transportation of Bicycles

Legislation stipulates that the exterior dimensions of a vehicle should not be exceeded and, in practice, this means that only caravans or motorhomes are allowed to carry bicycles/motorbikes at the rear of the vehicle. Bicycles may not extend beyond the width of the vehicle or more than 45cms from the back. However, bicycles may be transported on the roof of cars provided that an overall height of 4 metres is not exceeded. Cars carrying bicycles/motorbikes on the back may be subject to a fine.

If you are planning to travel from Spain to Portugal please note that slightly different regulations apply and these are set out in the Spain Country Introduction.

Electricity and Gas

Usually current on campsites varies between 6 and 15 amps. Plugs have two round pins. CEE connections are commonplace.

The full range of Campingaz cylinders is available.

Entry Formalities

Holders of British and Irish passports may visit Portugal for up to three months without a visa. For stays of over three months you will need to apply for a Registration Certificate from the nearest office of Servico de Estrangeiros e Fronteiras (immigration authority) or go to www.sef.pt.

Medical Services

For treatment of minor conditions go to a pharmacy (farmacia). Staff are generally well trained and are qualified to dispense drugs, which may only be available on prescription in Britain. In large towns there is usually at least one pharmacy whose staff speak English, and all have information posted on the door indicating the nearest pharmacy open at night.

All municipalities have a health centre. State emergency health care and hospital treatment is free on production of a European Health

Insurance Card (EHIC). You will have to pay for items such as X-rays, laboratory tests and prescribed medicines as well as dental treatment. Refunds can be claimed from local offices of the Administracão Regional de Saúde (regional health service).

For serious illness you can obtain the name of an English speaking doctor from the local police station or tourist office or from a British or American consulate.

Normal precautions should be taken to avoid mosquito bites, including the use of insect repellents, especially at night.

Opening Hours

Banks: Mon-Fri 8.30am-3pm; some banks in city centres are open until 6pm.

Museums: Tue-Sun 10am-5pm/6pm; closed Mon and may close 12.30pm-2pm.

Post Offices: Mon-Fri 9am-6pm; may close for an hour at lunch.

Shops: Mon-Fri 9am-1pm & 3pm-7pm, Sat 9am-1pm; large supermarkets open Mon-Sun 9am/9.30am-10pm/11pm.

Safety and Security

The crime rate is low but pickpocketing, bag snatching and thefts from cars can occur in major tourist areas. Be vigilant on public transport, at crowded tourist sites and in public parks where it is wise to go in pairs. Keep car windows closed and doors locked while driving in urban areas at night. There has been an increase in reported cases of items stolen from vehicles in car parks. Thieves distract drivers by asking for directions, for example, or other information. Be cautious if you are approached in this way in a car park.

Take care of your belongings at all times. Do not leave your bag on the chair beside you, under the table or hanging on your chair while you eat in a restaurant or café.

Death by drowning occurs every year on Portuguese beaches. Warning flags should be taken very seriously. A red flag indicates danger and you should not enter the water when it is flying. If a yellow flag is flying you may paddle at the water's edge, but you may not swim. A green flag indicates that it is safe to swim, and a chequered flag means that the lifeguard is temporarily absent.

Do not swim from beaches which are not manned by lifeguards. The police are entitled to fine bathers who disobey warning flags.

During long, hot, dry periods forest fires can occur, especially in northern and central parts of the country. Take care when visiting or driving through woodland areas: ensure that cigarettes are extinguished properly, do not light barbecues, and do not leave empty bottles behind.

Portugal shares with the rest of Europe an underlying threat from terrorism. Attacks could be indiscriminate and against civilian targets in public places including tourist sites.

British Embassy

RUA DE SÃO BERNARDO 33,
1249-082 LISBOA
Tel: 21 392 4000
www.ukinportugal.fco.gov.uk
There is also a British Consulate in Portimão.

Irish Embassy

VENIDA DA LIBERDADE No 200, 4th FLOOR
1250-147 LISBON
Tel: 213 308 200
www.embassyofireland.pt

Documents

Driving Licence

All valid UK driving licences should be accepted in Portugal but holders of an older all green style licence are advised to update it to a photocard licence before travelling in order to avoid any local difficulties. Alternatively carry an International Driving Permit, available from the AA, the RAC or selected Post Offices.

Passport

You must carry proof of identity which includes a photograph and signature, e.g. a passport or photocard licence, at all times.

Vehicle(s)

When driving you must carry your vehicle registration certificate (V5C), proof of insurance and MOT certificate (if applicable). There are heavy on the spot fines for those who fail to do so.

Money

The major credit cards are widely accepted and there are cash machines (Multibanco) throughout the country. A tax of €0.50 may be added to credit card transactions, especially at petrol stations. Carry your credit card issuers'/banks' 24 hour UK contact numbers in case of loss or theft.

Motoring in Portugal

Many Portuguese drive erratically and vigilance is advised. By comparison with the UK, the accident rate is high. Particular blackspots are the N125 along the south coast, especially in the busy holiday season, and the coast road between Lisbon and Cascais. In rural areas you may encounter horse drawn carts and flocks of sheep or goats. Otherwise there are no special difficulties in driving except in Lisbon and Porto, which are unlimited 'free-for-alls'.

Accident Procedures

The police must be called in the case of injury or significant material damage.

Alcohol

The maximum permitted level of alcohol is 50 milligrams in 100 millilitres of blood, i.e. lower than permitted in the UK (80 milligrams). For newly qualified drivers (those with under 3 year's experience), the legal limit is 20 milligrams per 100 millilitres of blood. It is advisable to adopt the 'no drink-driving' rule at all times.

Breakdown Service

The Automovel Club de Portugal (ACP) operates a 24 hour breakdown service covering all roads in mainland Portugal. Its vehicles are coloured red and white. Emergency telephones are located at 2km intervals on main roads and motorways. To contact the ACP breakdown service call +351 219 429113 from a mobile or 707 509510 from a landline.

The breakdown service comprises on the spot repairs taking up to a maximum of 45 minutes and, if necessary, the towing of vehicles. The charges for breakdown assistance and towing vary according to distance, time of day and day of the week, plus motorway tolls if applicable. Payment by credit card is accepted.

Alternatively, on motorways breakdown vehicles belonging to the motorway companies (their emergency numbers are displayed on boards along the motorways) and police patrols (GNR/Brigada de Trânsito) can assist motorists.

Essential Equipment

Reflective Jackets/ Waistcoats

If your vehicle is immobilised on the carriageway you should wear a reflective jacket or waistcoat when getting out of your vehicle. This is a legal requirement for residents of Portugal and is recommended for visitors. Passengers who leave a vehicle, for example, to assist with a repair, should also wear one. Keep the jackets within easy reach inside your vehicle, not in the boot.

Warning Triangles

Use a warning triangle if, for any reason, a stationary vehicle is not visible for at least 100 metres. In addition, hazard warning lights must be used if a vehicle is causing an obstruction or danger to other road users.

Child Restraint System

Children under 12 years of age and less than 1.35m in height are not allowed to travel in the front passenger seat. They must be seated in a child restraint system adapted to their size and weight in the rear of the vehicle, unless the vehicle only has two seats, or if the vehicle is not fitted with seat belts.

Children under the age of 3 years old can be seated in the front passenger seat as long as they are in a suitable rear facing child restraint system and the airbag has been deactivated.

Fuel

Credit cards are accepted at most filling stations but a small extra charge may be added and a tax of €0.50 is added to credit card transactions. There are no automatic petrol pumps. LPG (gáz liquido) is widely available.

Low Emission Zone

There is a Low Emission Zone in operation in Lisbon. There are 2 different zones within the city. In zone 1 vehicles must meet European Emission Standard 2 (EURO 3) and in zone 2 vehicles must meet EURO 2 standard. For more information visit www.lowemissionzones.eu.

Mountain Roads and Passes

There are no mountain passes or tunnels in Portugal. Roads through the Serra da Estrela near Guarda and Covilha may be temporarily obstructed for short periods after heavy snow.

Parking

In most cases vehicles must be parked facing in the same direction as moving traffic. Parking is very limited in the centre of main towns and cities and 'blue zone' parking schemes operate. Illegally parked vehicles may be towed away or clamped. Parking in Portuguese is 'estacionamento'.

Priority

In general at intersections and road junctions, road users must give way to vehicles approaching from the right, unless signs indicate otherwise. At roundabouts vehicles already on the roundabout, i.e. on the left, have right of way.

Do not pass stationary trams at a tram stop until you are certain that all passengers have finished entering or leaving the tram.

Roads

Roads are surfaced with asphalt, concrete or stone setts. Main roads generally are well surfaced and may be three lanes wide, the middle lane being used for overtaking in either direction.

Roads in the south of the country are generally in good condition, but some sections in the north are in a poor state. Roads in many towns and villages are often cobbled and rough.

Drivers entering Portugal from Zamora in Spain will notice an apparently shorter route on the CL527/N221 road via Mogadouro. Although this is actually the signposted route, the road surface is poor in places and this route is not recommended for trailer caravans. The recommended route is via the N122/IP4 to Bragança.

Road Signs and Markings

Road signs conform to international standards. Road markings are white or yellow. Signs on motorways (auto-estrada) are blue and on regional roads they are white with black lettering. Roads are classified as follows:

Code	Road Type
AE	Motorways
IP	Principal routes
IC	Complementary routes
EN	National roads
EM	Municipal roads
CM	Other municipal roads

Signs you might encounter are as follows:

Portuguese	English Translation
Atalho	Detour
Entrada	Entrance
Estacão de gasolina	Petrol station
Estacão de policia	Police station
Estacionamento	Parking
Estrada con portagem	Toll road
Saida	Exit

Speed Limits

	Open Road (km/h)	Motorway (km/h)
Car Solo	90-100	120
Car towing caravan/ trailer	70-80	100
Motorhome under 3500kg	90-100	120
Motorhome 3500-7500kg	70-90	100

Drivers must maintain a speed between 40 km/h (25 mph) and 60 km/h (37 mph) on the 25th April Bridge over the River Tagus in Lisbon. Speed limits are electronically controlled.

Visitors who have held a driving licence for less than one year must not exceed 90 km/h (56 mph) on any road subject to higher limits.

In built-up areas there is a speed limit of 50 km/h.

It is prohibited to use a radar detector or to have one installed in a vehicle.

Towing

Motorhomes are permitted to tow a car on a four wheel trailer, i.e. with all four wheels of the car off the ground. Towing a car on an A-frame (two back wheels on the ground) is not permitted.

Traffic Jams

Traffic jams are most likely to be encountered around Lisbon and Porto and on roads to the coast, such as the A1 Lisbon-Porto and the A2 Lisbon-Setúbal, which are very busy on Friday evenings and Saturday mornings. The peak times for holiday traffic are the last weekend in June and the first and last weekends in July and August.

Around Lisbon bottlenecks occur on the bridges across the River Tagus, the N6 to Cascais, the A1 to Vila Franca de Xira, the N8 to Loures and on the N10 from Setúbal via Almada.

Around Porto you may find traffic jams on the IC1 on the Arribada Bridge and at Vila Nova de Gaia, the A28/IC1 from Póvoa de Varzim and near Vila de Conde, and on the N13, N14 and the N15.

Major motorways are equipped with suspended signs which indicate the recommended route to take when there is traffic congestion.

Traffic Lights

There is no amber signal after the red. A flashing amber light indicates 'caution' and a flashing or constant red light indicates 'stop'. In Lisbon there are separate traffic lights in bus lanes.

Violation of Traffic Regulations

Speeding, illegal parking and other infringements of traffic regulations are heavily penalised.

You may incur a fine for crossing a continuous single or double white or yellow line in the centre of the road when overtaking or when executing a left turn into or off a main road, despite the lack of any other 'no left turn' signs. If necessary, drive on to a roundabout or junction to turn, or turn right as directed by arrows.

The police are authorised to impose on the spot fines and a receipt must be given. Most police vehicles are now equipped with portable credit card machines to facilitate immediate payment of fines.

Motorways

Motorway Tolls

Portugal has more than 2,600km of motorways (auto-estradas), with tolls (portagem) payable on most sections. Take care not to use the 'Via Verde' green lanes reserved for motorists who subscribe to the automatic payment system – be sure to go through a ticket booth lane where applicable, or one equipped with the new electronic toll system.

Dual carriageways (auto vias) are toll free and look similar to motoways, but speed limits are lower.

It is permitted to spend the night on a motorway rest or service area with a caravan, although The Caravan and Motorhome Club does not recommend this practice for security reasons. Toll tickets are only valid for 12 hours and fines are incurred if this is exceeded.

Vehicle are classified for tolls as follows:

Class 1	Vehicle with or without trailer with height from front axle less than 1.10m.
Class 2	Vehicle with 2 axles, with or without trailer, with height from front axle over 1.10m.
Class 3	Vehicle or vehicle combination with 3 axles, with height from front axle over 1.10m.
Class 4*	Vehicle or vehicle combination with 4 or more axles with height from front axle over 1.10m.

* Drivers of high vehicles of the Range Rover/ Jeep variety, together with some MPVs, and towing a twin axle caravan pay Class 4 tolls.

Electronic Tolls

An electronic toll collecting system was introduced in Portugal during 2010. The following motorways have tolls but no toll booths: A27, A28, A24, A41, A42, A25, A29, A23, A13, A8, A19, A33, A22 and parts of the A17 and A4. Tolls for these motorways can be paid by one of the following options:

If you are crossing the border from Spain on the A24, A25 or A22 or the A28 (via the EN13) then you can use the EASYToll welcome points. You can input your credit card details and the machine reads and then matches your credit/debit card to your number-plate, tolls are deducted automatically from your credit card, and the EASYToll machine will issue you a 30 day receipt as proof that you have paid.

If you are entering Portugal on a road that does not have an EASYToll machine you can register on-line or at a CTT post office and purchase either €5, €10, €20 or €40 worth of tolls. You can purchase a virtual prepaid ticket up to 6 times a year. For more information visit www.ctt.pt – you can select 'ENG' at the top left of the screen to see the site in English.

Alternatively you can get a temporary device (DT) available from some motorway service stations, post offices and Via Verde offices.

A deposit of €27.50 is payable when you hire the DT and this is refundable when you return it to any of the outlets mentioned above. If you use a debit card, toll costs will automatically be debited from your card. If you pay cash you will be required to preload the DT. For further information see www.visitportugal.com and see the heading 'All about Portugal' then 'Useful Information'.

On motorways where this system applies you will see a sign: 'Lanço Com Portagem' or 'Electronic Toll Only', together with details of the tolls charged. Drivers caught using these roads without a DT will incur a minimum fine of €25.

The toll roads A1 to A15 and A21 continue to have manned toll booths. Most, but not all, accept credit cards or cash.

Toll Bridges

The 2km long 25th April Bridge in Lisbon crosses the River Tagus. Tolls are charged for vehicles travelling in a south-north direction only. Tolls also apply on the Vasco da Gama Bridge, north of Lisbon, but again only to vehicles travelling in a south-north direction. Overhead panels indicate the maximum permitted speed in each lane and, when in use, override other speed limit signs.

In case of breakdown, or if you need assistance, you should try to stop in the emergency hard shoulder areas, wait inside your vehicle and switch on your hazard warning lights until a patrol arrives. Emergency telephones are placed at frequent intervals. It is prohibited to carry out repairs, to push vehicles physically or to walk on the bridges.

Touring

Some English is spoken in large cities and tourist areas. Elsewhere a knowledge of French could be useful.

A Lisboa Card valid for 24, 48 or 72 hours, entitles the holder to free unrestricted access to public transport, including trains to Cascais and Sintra, free entry to a number of museums, monuments and other places of interest in Lisbon and surrounding areas, as well as discounts in shops and places offering services to tourists. These cards are obtainable from tourist information offices, travel agents, some hotels and Carris ticket booths, or by visiting www.welovecitycards.com.

Do ensure when eating out that you understand exactly what you are paying for; appetisers put on the table are not free. Service is included in the bill, but it is customary to tip 5 to 10% of the total if you have received good service. Rules on smoking in restaurants and bars vary according to the size of the premises. The areas where clients are allowed to smoke are indicated by signs and there must be adequate ventilation. Each town in Portugal devotes several days in the year to local celebrations which are invariably lively and colourful. Carnivals and festivals during the period before Lent, during Holy Week and during the grape harvest can be particularly spectacular.

Public Transport

A passenger and vehicle ferry crosses the River Sado estuary from Setúbal to Tróia and there are frequent ferry and catamaran services for cars and passengers across the River Tagus from various points in Lisbon including Belém and Cais do Sodré.

Both Lisbon and Porto have metro systems operating from 6am to 1am. For routes and fares information see www.metrolisboa.pt and www.metrodoporto.pt (English versions).

Throughout the country buses are cheap, regular and mostly on time, with every town connected. In Lisbon the extensive bus and tram network is operated by Carris, together with one lift and three funiculars which tackle the city's steepest hills. Buy single journey tickets on board from bus drivers or buy a rechargeable 'Sete Colinas' or Via Viagem card for use on buses and the metro.

In Porto buy a 'Euro' bus ticket, which can be charged with various amounts, from metro stations and transport offices. Validate tickets for each journey at machines on the buses. Also available is an 'Andante' ticket which is valid on the metro and on buses. Porto also has a passenger lift and a funicular so that you can avoid the steep walk to and from the riverside.

Taxis are usually cream in colour. In cities they charge a standard, metered fare; outside they may run on the meter or charge a flat rate and are entitled to charge for the return fare. Agree a price for your journey before setting off.

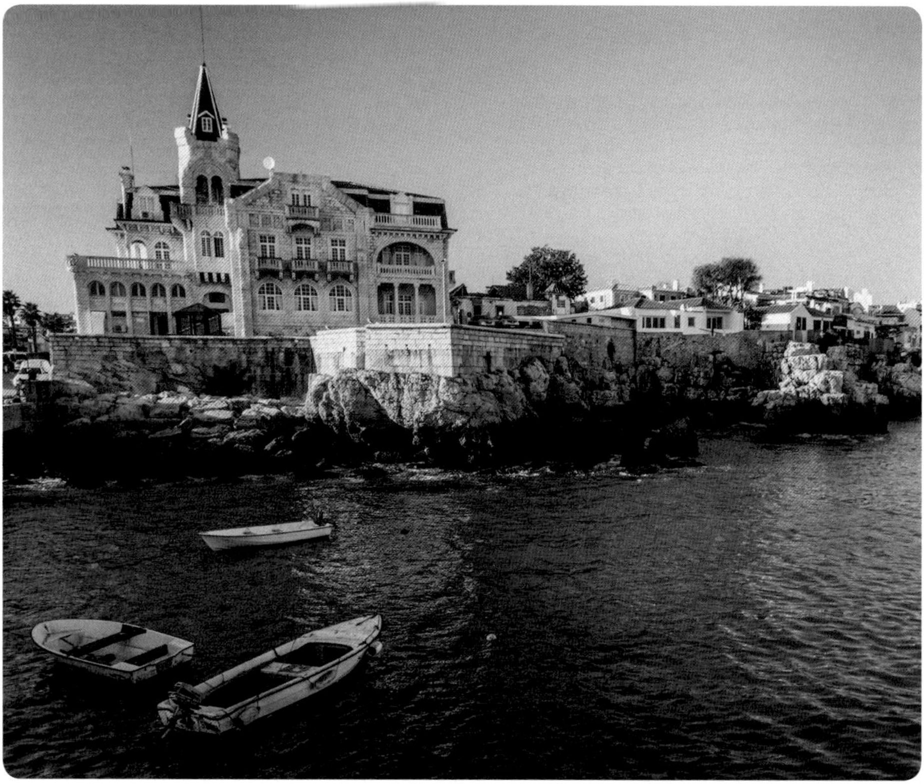

Cascais

Sites in Portugal

ALANDROAL *C3* (13km S Rural) *38.60645, -7.34674* **Camping Rosário, Monte das Mimosas, Rosário, 7250-999 Alandroal [268 459566; info@campingrosario.com; www.campingrosario.com]** Fr E exit IP7/A6 at Elvas W junc 9; at 3rd rndabt take exit sp Espanha, immed 1st R dir Juromenha & Redondo. Onto N373 until exit Rosário. Fr W exit IP7/A6 junc 8 at Borba onto N255 to Alandroal, then N373 E sp Elvas. After 1.5km turn R to Rosário & foll sp to site. Sm, hdstg, pt shd, pt sl, wc; chem disp; shwrs inc; EHU (6A) €2.35; lndry; shop nr; rest; bar; playgrnd; pool; sw nr; red long stay; wifi; TV; dogs €1 (not acc Jul/Aug); Eng spkn; adv bkg acc; quiet; boating; fishing; CKE/CCI. "Remote site beside Alqueva Dam; excel touring base; ltd to 50 people max; excel site; idyllic; peaceful; clean & well maintained; v helpful owner." 1 Mar-1 Oct. € 24.50 2014*

⊞ **ALBUFEIRA** *B4* (3km N Urban) *37.10617, -8.25395* **Camping Albufeira, Estrada de Ferreiras, 8200-555 Albufeira [289 587629 or 289 587630; fax 289 587633; geral@campingalbufeira.net or info@campingalbufeira.net; www.campingalbufeira.net]** Exit IP1/E1 sp Albufeira onto N125/N395 dir Albufeira; camp on L, sp. V lge, mkd, pt shd, pt sl, wc; chem disp; mv service pnt; shwrs inc; EHU (10-12A) €3; gas; lndry; shop; rest; snacks; bar; playgrnd; pool; beach sand 1.5km; red long stay; entmnt; wifi; TV; 20% statics; dogs; phone; bus adj; Eng spkn; quiet; ccard acc; games area; games rm; bike hire; tennis; CKE/CCI. "Friendly, secure site; excel pool area/bar; some pitches lge enough for US RVs; car wash; cash machine; security patrols; disco (soundproofed); sports park; pitches on lower pt of site prone to flooding in heavy rain; conv beach & town; poss lge rallies during Jan-Apr; camp bus to town high ssn." ♦ € 27.60 2016*

⊞ **ALBUFEIRA** *B4* (15km W Urban/Coastal) *37.10916, -8.35333* **Camping Armação de Pêra, 8365-184 Armação de Pêra [282 312260; fax 282 315379; geral@camping-armacao-pera.com; www.camping-armacao-pera.com]** Fr Lagos take N125 coast rd E. At Alcantarilha turn S onto N269-1 sp Armação de Pêra & Campismo. Site at 3rd rndabt in 2km on L. V lge, hdg, shd, pt sl, wc; chem disp; mv service pnt; shwrs inc; EHU (6-10A) €3-4; gas; lndry; shop; rest; snacks; bar; playgrnd; pool; paddling pool; beach sand 500m; entmnt; wifi; TV; 25% statics; phone; bus adj; Eng spkn; quiet; car wash; games area; tennis; bike hire; games rm; CKE/CCI. "Friendly, popular & attractive site; gd pool; min stay 3 days Oct-May; easy walk to town; interesting chapel of skulls at Alcantarilha; birdwatching in local lagoon; vg." ♦ € 20.50 2012*

ALCACER DO SAL *B3* (1km NW Rural) *38.38027, -8.51583* **Parque de Campismo Municipal de Alcácer do Sal, Olival do Outeiro, 7580-125 Alcácer do Sal [265 612303; fax 265 610079; cmalcacer@mail.telepac.pt]** Heading S on A2/IP1 turn L twd Alcácer do Sal on N5. Site on R 1km fr Alcácer do Sal. Sp at rndabt. Site behind supmkt. Sm, hdg, mkd, pt shd, pt sl, wc; chem disp; mv service pnt; shwrs inc; EHU (6-12A) €1.50 (rev pol); lndry; shop nr; rest nr; bar nr; bbq; playgrnd; wifi; dogs; phone; bus 50m; Eng spkn; quiet; ccard acc; games area; clsd mid-Dec to mid-Jan; CKE/CCI. "In rice growing area - major mosquito prob; historic town; spacious pitches; pool, paddling pool adj; poss full in winter - rec phone ahead; pleasant site behing supmkt; wifi at recep; san facs tired & not clean." ♦ 15 Jan-15 Dec. € 13.70 2017*

AMARANTE *C1* (3km NE Rural) *41.27805, -8.07027* **Camping Penedo da Rainha, Rua Pedro Alveollos, Gatão, 4600-099 Amarante [255 437630; fax 255 437353; ccporto@sapo.pt]** Fr IP4 Vila Real to Porto foll sp to Amarante & N15. On N15 cross bdge for Porto & immed take R slip rd. Foll sp thro junc & up rv to site. Lge, hdstg, shd, pt sl, terr, wc; chem disp; mv service pnt; shwrs inc; EHU (6A) €3.25; lndry; shop; rest; bar; playgrnd; pool; entmnt; TV; dogs; phone; bus to Porto fr Amarante; Eng spkn; adv bkg acc; quiet; games rm; rv fishing adj; canoeing; cycling; CKE/CCI. "Well-run site in steep woodland/parkland - take advice or survey rte bef driving to pitch; excel facs but some pitches far fr facs; few touring pitches; friendly, helpful recep; plenty of shd; conv Amarante old town & Douro Valley; Sat mkt; not suitable for long o'fits, better for MH's." ♦ 1 Jan-30 Nov. € 14.00 2017*

"There aren't many sites open at this time of year"

If you're travelling outside peak season remember to call ahead to check site opening dates – even if the entry says 'open all year'.

ARCO DE BAULHE *C1* (0km NE Rural) *41.48659, -7.95845* **Arco Unipessoal, Lugar das Cruzes, 4860-067 Arco de Baúlhe (Costa Verde) [(351) 968176246; campismoarco@hotmail.com]** Dir A7 exit 12 Mondm/Cabeceiras, 2nd R at rndabt dir Arco de Baulhe. Call and they will lead you in. V narr rd access, no mv's over 7m. Med, pt shd, terr, wc; chem disp; shwrs; EHU (6A); lndry; rest; bar; bbq; pool (htd); paddling pool; TV; dogs; Eng spkn; adv bkg rec; rv. "New site run by couple with 20 yrs experience; quiet; centrally located for historic towns & nature parks; gd rest; lovely well maintained site with view; excel facs; 100m fr vill cent; beautiful mountain area." 1 Apr-10 Oct. € 26.00 2014*

ARGANIL *C2* (3km N Rural) *40.2418, -8.06746* **Camp Municipal de Arganil, 3300-432 Sarzedo [235 205706; fax 235 200134; camping@cm-arganil.pt; www.cm-arganil.pt]** Fr Coimbra on N17 twd Guarda; after 50km turn S sp Arganil on N342-4; site on L in 4km in o'skts of Sarzedo bef rv bdge; avoid Góis to Arganil rd fr SW. Med, pt shd, pt sl, terr, wc; chem disp; mv service pnt; fam bthrm; shwrs inc; EHU (5-15A) €2.40; gas; lndry; shop nr; rest; snacks; bar; bbq; playgrnd; pool; red long stay; wifi; TV; dogs; phone; Eng spkn; quiet; ccard acc; canoeing adj; games area; fishing adj; CKE/CCI. "Vg, well-run site; friendly owner; fine views; gd cent for touring; gd walks; ski in Serra da Estrela 50km Dec/Jan; interesting town; gd mkt (Thu)." 1 Mar-31 Oct. € 12.50 2016*

AVEIRO *B2* (14km SW Coastal) *40.59960, -8.74981* **Camping Costa Nova, Estrada da Vagueira, Quinta dos Patos, 3830-453 Ílhavo [234 393220; fax 234 394721; info@campingcostanova.com; www.campingcostanova.com]** Site on Barra-Vagueira coast rd 1km on R after Costa Nova. V lge, mkd, unshd, wc (htd); chem disp; mv service pnt; shwrs inc; EHU (2-6A) €2.40; gas; lndry; shop; rest; snacks; bar; bbq; playgrnd; beach sand; red long stay; entmnt; wifi; TV; 10% statics; dogs €1.40; phone; Eng spkn; adv bkg acc; quiet; ccard acc; games rm; site clsd Jan; fishing; bike hire; games area; CKE/CCI. "Superb, peaceful site adj nature reserve; helpful staff; gd, modern facs; hot water to shwrs only; sm pitches; sep car park high ssn; pool 4km; vg." ♦ 21 Mar-1 Oct. € 22.50 2017*

⊞ **AVEIRO** *B2* (10km W Coastal) *40.63861, -8.74500* **Parque de Campismo Praia da Barra, Rua Diogo Cão 125, Praia da Barra, 3830-772 Gafanha da Nazaré [(234) 369425; barra@cacampings.com; www.cacampings.com]** Fr Aveiro foll sp to Barra on A25/IP5; foll sp to site. Lge, mkd, shd, wc; chem disp; mv service pnt; fam bthrm; shwrs inc; EHU (6-10A) €2.50; gas; lndry; shop; rest; bar; bbq; playgrnd; beach sand 200m; entmnt; wifi; TV; 90% statics; dogs €1.80; phone; bus adj; Eng spkn; adv bkg acc; quiet; bike hire; games area; games rm; CKE/CCI. "Well-situated site with pitches in pine trees; recep open 0900-2200; pool 400m; old san facs." ♦ € 22.60 2013*

AVEIRO *B2* (8km NW Rural/Coastal) *40.70277, -8.7175* **Camping ORBITUR, N327, Km 20, 3800-901 São Jacinto [234 838284; fax 234 838122; infosjacinto@orbitur.pt; www.orbitur.pt]** Fr Porto take A29/IC1 S & exit sp Ovar onto N327. (Note long detour fr Aveiro itself by rd - 30+ km.) Site in trees to N of São Jacinto. Lge, mkd, hdstg, shd, terr, wc; chem disp; mv service pnt; fam bthrm; shwrs inc; EHU (5-15A) €3-4 (poss rev pol); gas; lndry; shop; rest; snacks; bar; bbq; playgrnd; beach sand 2.5km; red long stay; TV; 10% statics; dogs €1.50; phone; bus; Eng spkn; adv bkg acc; quiet; ccard acc; fishing; car wash; CKE/CCI. "Excel site; best in area; pool 5km; gd children's park; 15 min to (car) ferry; gd, clean san facs." ♦ 1 Jun-30 Sep. € 27.00 2014*

⊞ **BEJA** *C4* (2km S Urban) *38.00777, -7.86222* **Parque de Campismo Municipal de Beja, Avda Vasco da Gama, 7800-397 Beja [284 311911; cmb.dcd@iol.pt; www.cm-beja.pt]** Fr S (N122) take 1st exit strt into Beja. In 600m turn R at island then L in 100m into Avda Vasco da Gama & foll sp for site on R in 300m - narr ent. Fr N on N122 take by-pass round town then 1st L after Intermarche supmkt, then as above. Lge, hdstg, shd, wc; chem disp; shwrs inc; EHU (6A) €1.85; shop nr; rest nr; bar nr; bus 300m, train 1.5km; tennis adj; CKE/CCI. "C'van storage facs; helpful staff; NH only; pool adj; gravel pitches; football stadium adj." ♦ € 10.40 2013*

BRAGANCA *D1* (6km N Rural) *41.84361, -6.74722* **Inatel Parque Campismo Bragança, Estrada de Rabal, 5300-671 Meixedo [351 273 326 080 or 351 96 420 66 22; campismobraganca@gmail.com; www.campismobraganca.com]** Fr Bragança N for 6km on N103.7 twd Spanish border. Site on R, sp Inatel. Med, hdstg, pt shd, pt sl, terr, wc; chem disp; shwrs inc; EHU (6A) inc; gas; lndry; shop; rest; snacks; bar; playgrnd; dogs; bus; Eng spkn; quiet; bike hire; fishing. "On S boundary of National Park; rv runs thro site; friendly staff; gd rest; vg facs; lovely location by rv." 1 Jun-15 Sep. € 12.50 2016*

⊞ **BRAGANCA** *D1* (12km W Rural) *41.84879, -6.86120* **Cepo Verde Camping, Gondesende, 5300-561 Bragança [273 999371; fax 273 323577; cepoverde@montesinho.com; www.bragancanet.pt/cepoverde]** Fr IP4 fr W take N103 fr Bragança for 8km. Site sp fr IP4 ring rd. R off N103, foll lane & turn R at sp. NB Camping sp to rd 103-7 leads to different site (Sabor) N of city. Med, hdstg, mkd, pt shd, terr, wc; chem disp; shwrs inc; EHU (6A) €2 (poss rev pol & long lead poss req); shop; rest; snacks; bar; playgrnd; pool; wifi; dogs €1; phone; bus 1km; Eng spkn; adv bkg acc; quiet; CKE/CCI. "Remote, friendly, v pleasant, scenic site adj Montesinho National Park; clean but poorly maintained facs; vg value." ♦ € 20.00 2014*

⊞ **CAMINHA** *B1* (3km SW Coastal) *41.86611, -8.85888* **Camping ORBITUR-Caminha, Mata do Camarido, N13, Km 90, 4910-180 Caminha [258 921295; fax 258 921473; infocaminha@orbitur.pt; www.orbitur.pt]** Foll seafront rd N13/E1 fr Caminha dir Viana/Porto, at sp Foz do Minho turn R, site in approx 1km. Long o'fits take care at ent. Med, shd, terr, wc; chem disp; mv service pnt; shwrs inc; EHU (5-15A) €3-4; gas; lndry; shop; rest; snacks; bar; playgrnd; beach sand 150m; red long stay; wifi; TV; 5% statics; dogs €1.50; Eng spkn; adv bkg acc; quiet; ccard acc; fishing; bike hire; CKE/CCI. "Pleasant, woodland site; pool 2.5km; care in shwrs - turn cold water on 1st as hot poss scalding; Gerês National Park & Viana do Castelo worth visit; poss to cycle to Caminha; vg site, nr attractive beach and sh walk to pleasant town." ♦ € 38.60 2017*

⊞ **CAMPO MAIOR** *C3* (2km SE Rural) *39.00833, -7.04833* **Camping Rural Os Anjos, Estrada da Senhora da Saúde, 7371-909 Campo Maior [268 688138 or 965 236625 (mob); info@campingosanjos.com; www.campingosanjos.com]** Fr Elvas foll rd N373 to Campo Maior. Foll sm sp thro vill. Site on L down country lane. Sm, hdstg, pt shd, terr, wc; chem disp; fam bthrm; shwrs inc; EHU (6A) €2.60; lndry; shop nr; bar; bbq; pool; wifi; TV; dogs €1 (max 1); phone; Eng spkn; adv bkg rec; quiet; games rm; games area; CKE/CCI. "Excel, lovely, peaceful site; v helpful, friendly, Dutch owners; gd touring base for unspoiled, diverse area; conv Spanish border, Badajoz & Elvas; 15 Nov-15 Feb open with adv bkg only; Campo Maior beautiful, white town; LS call or email bef arr; gd walks & bike rides; v clean shwrs; lake sw 8km; fishing 8km; watersports 8km; modern facs; fantastic views; excel." ♦ € 17.00 2015*

"That's changed – Should I let the Club know?"

If you find something on site that's different from the site entry, fill in a report and let us know. See camc.com/europereport.

⊞ **CASCAIS** *A3* (7km NW Urban/Coastal) *38.72166, -9.46666* **Camping ORBITUR-Guincho, Lugar de Areia, EN 247-6, Guincho 2750-053 Cascais [(214) 870450; fax (214) 857413; infoguincho@orbitur.pt; www.orbitur.pt/camping-orbitur-guincho]** Fr Lisbon take A5 W, at end m'way foll sp twd Cascais. At 1st rndabt turn R sp Birre & Campismo. Foll sp for 2.5km. Steep traff calming hump - care needed. V lge, mkd, hdg, shd, terr, wc; chem disp; mv service pnt; fam bthrm; shwrs inc; EHU (6A) inc; gas; lndry; shop; rest; snacks; bar; bbq; playgrnd; pool; beach sand 800m; twin axles; red long stay; entmnt; wifi; TV (pitch); 50% statics; dogs on lead €3; phone; Eng spkn; adv bkg acc; quiet; ccard acc; games rm; horseriding 500m; tennis; watersports 1km; car wash; fishing 1km; golf 3km; bike hire; CKE/CCI. "Sandy, wooded site behind dunes; poss stretched & v busy high ssn; poss diff lge o'fits due trees; steep rd to beach; gd san facs, gd value rest; vg LS; buses to Cascais for train to Lisbon; beautiful coastline within 20 min walk; rec lge o'fits prebook to ensure suitable pitch." ♦ € 37.00 2017*

CASTELO BRANCO *C2* (6km N Rural) *39.85777, -7.49361* **Camp Municipal Castelo Branco, Estrada Nacional 18, 6000-113 Castelo Branco [(272) 322577; fax (272) 322578; albigec@sm-castelobranco.pt; www.cm-castelobranco.pt]** Fr IP2 take Castelo Branco Norte, exit R on slip rd, L at 1st rndabt, site sp at 2nd rndabt. Turn L at T junc just bef Modelo supmkt, site 2km on L, well sp. Lge, shd, pt sl, wc; chem disp; mv service pnt; shwrs; EHU (12A) €2.25; gas; lndry; shop nr; rest nr; bar nr; playgrnd; bus 100m; Eng spkn; quiet; CKE/CCI. "Useful NH on little used x-ing to Portugal; pool 4km; gd site but rds to it poor." 2 Jan-15 Nov. € 9.00 2013*

⊞ **CASTELO DE VIDE** *C3* (7km SW Rural) *39.39805, -7.48722* **Camping Quinta do Pomarinho, N246, Km 16.5, Castelo de Vide [(00351) 965 755 341; info@pomarinho.com; www.pomarinho.com]** On N246 at km 16.5 by bus stop, turn into dirt track. Site in 500m. Sm, mkd, hdstg, unshd, wc; chem disp; shwrs inc; EHU (6A) €2.50-3.50; lndry; shop nr; pool; wifi; dogs; bus adj; Eng spkn; adv bkg acc; quiet; bike hire. "On edge of Serra de São Mamede National Park; gd walking, fishing, birdwatching, cycling; vg; gd views twd Serra de Sao Marmede." € 18.50 2016*

⊞ **CELORICO DE BASTO** *C1* (1km NW Rural) *41.39026, -8.00581* **Parque de Campismo de Celorico de Basto, Adaufe-Gemeos, 4890-361 Celorico de Basto [(255) 323340 or 964-064436 (mob); fax (255) 323341; geral@celoricodebastocamping.com; www.celoricodebastocamping.com]** E fr Guimarães exit A7/IC5 S sp Vila Nune (bef x-ing rv). Foll sp Fermil & Celorico de Basto, site sp. Rte narr and winding or take the N210 fr Amarente and foll sp to site. Med, hdstg, mkd, shd, wc; chem disp; mv service pnt; shwrs inc; EHU (6-16A) €2-3.20; gas; lndry; shop; rest; bar; bbq; playgrnd; sw nr; red long stay; entmnt; wifi; TV; 10% statics; dogs €1.80; phone; quiet; ccard acc; games area; fishing adj; CKE/CCI. "Peaceful, well-run site; gd facs; pool 500m; gd cycling & walking; vg." ♦ € 16.00 2014*

"I like to fill in the reports as I travel from site to site"

You'll find report forms at the back of this guide, or you can fill them in online at camc.com/europereport.

CHAVES *C1* (6km S Rural) *41.70166, -7.50055* **Camp Municipal Quinta do Rebentão, Vila Nova de Veiga, 5400-764 Chaves [276 322733; parquedecampismo@chaves.pt; www.cccchaves.com]** Fr o'skts Chaves take N2 S. After about 3km in vill of Vila Nova de Veiga turn E at sp thro new estate, site in about 500m. Med, hdstg, pt shd, terr, wc; chem disp; mv service pnt; shwrs inc; EHU (6A) inc; lndry; shop nr; snacks; bar; bbq; playgrnd; sw nr; wifi; dogs; phone; bus 800m; adv bkg acc; bike hire; fishing 4km; games rm; CKE/CCI. "Gd site in lovely valley but remote; excel helpful, friendly staff; facs block quite a hike fr some terr pitches, old but clean facs; Chaves interesting, historical Roman town; baker visits every morn; vg rest & bar; site clsd Dec; pool adj; easy access; wifi at recep." ♦ 1 Feb-30 Nov. € 15.00 2017*

⊞ **COIMBRA** *B2* (6km SE Urban) *40.18888, -8.39944* **Coimbra Camping, Rua de Escola, Alto do Areeiro, Santo António dos Olivais, 3030-011 Coimbra [239 086902; coimbra@cacampings.com; www.coimbracamping.com]** Fr S on AP1/IP1 at junc 11 turn twd Lousa & in 1km turn twd Coimbra on IC2. In 9.5km turn R at rndabt onto Ponte Rainha, strt on at 3 rndabts along Avda Mendes Silva. Then turn R along Estrada des Beiras & cross rndabt to Rua de Escola. Or fr N17 dir Beira foll sp sports stadium/campismo. Fr N ent Coimbra on IC2, turn L onto ring rd & foll Campismo sps. V lge, hdstg, pt shd, pt sl, terr, wc (htd); chem disp; mv service pnt; fam bthrm; shwrs inc; EHU (6A) €3.90 (rev pol); gas; lndry; shop; rest; snacks; bar; bbq; playgrnd; sw nr; red long stay; wifi; TV; 10% statics; dogs €2.60; bus 100m; Eng spkn; adv bkg acc; quiet; ccard acc; bike hire; games area; tennis; games rm; sauna; CKE/CCI. "Vg site & facs; health club; pool adj; v interesting, lively university town." ♦ € 23.70 2017*

⊞ **COIMBRAO** *B2* (1km NW Urban) *39.90027, -8.88805* **Camping Coimbrão, 185 Travessa do Gomes, 2425-452 Coimbrão [244 606007; campingcoimbrao@web.de]** Site down lane in vill cent. Care needed lge o'fits, but site worth the effort. Sm, unshd, wc; chem disp; mv service pnt; shwrs inc; EHU (6-10A) €2.20-3.30; lndry; shop nr; bbq; playgrnd; pool; sw nr; red long stay; wifi; TV; bus 200m; Eng spkn; quiet; fishing 4km. "Excel site; helpful & friendly staff; gd touring base; gd loc for exploring; office open 0800-1200 & 1500-2200; canoeing 4km; German owners." ♦ € 16.70 2017*

⊞ **COVILHA** *C2* (48km NW Rural) *40.40406, -7.58770* **Vale do Rossim, Penhas Douradas, 6270 Gouveia [275 981 029; vrecoresort@gmail.com; valedorossimecoresort.com]** 20km W on N232 fr Manteigas. Turn L at sp. Site 1km at end of rd. Sm, pt shd, pt sl, wc (htd); chem disp; mv service pnt; fam bthrm; shwrs; EHU (6A) €3.50; gas; lndry; rest; snacks; bar; bbq; sw nr; twin axles; entmnt; wifi; dogs; phone; Eng spkn; adv bkg acc; quiet; ccard acc. "High in mountains by lake wth beaches; sw, kayaking, biking, climbing, walking; rest & bar by lake; peaceful, rural site in beautiful surroundings; friendly, helpful staff; vg." ♦ € 11.00 2016*

ELVAS *C3* (2km SW Urban) *38.87305, -7.1800* **Parque de Campismo da Piedade, 7350-901 Elvas [268 628997 or 268 622877; fax 268 620729]** Exit IP7/E90 junc 9 or 12 & foll site sp dir Estremoz. Med, mkd, hdstg, pt shd, sl, wc; chem disp; shwrs inc; EHU (16A) inc; gas; lndry; shop nr; rest; snacks; bar; bbq; playgrnd; dogs; phone; bus 500m; CKE/CCI. "Attractive aqueduct & walls; Piedade church & relics adj; traditional shops; pleasant walk to town; v quiet site, even high ssn; adequate san facs; conv NH en rte Algarve; c'vans parked to cls; noisy; facs need updating." 1 Apr-15 Sep. € 20.00 2018*

⊞ **ERMIDAS SADO** *B4* (11km W Rural) *38.01805, -8.48500* **Camping Monte Naturista O Barão (Naturist), Foros do Barão, 7566-909 Ermidas-Sado [936710623 (mob); info@montenaturista.com; www.montenaturista.com]** Fr A2 turn W onto N121 thro Ermidas-Sado twd Santiago do Cacém. At x-rds nr Arelãos turn R at bus stop (km 17.5) dir Barão. Site in 1km along unmade rd. Sm, mkd, pt shd, pt sl, wc; chem disp; mv service pnt; fam bthrm; shwrs inc; EHU (6A) €3.20; lndry; bar; bbq; pool; red long stay; wifi; TV; 10% statics; dogs €1; Eng spkn; adv bkg acc; quiet; ccard acc; games area; CKE/CCI. "Gd, peaceful 'retreat-type' site in beautiful wooded area; meals Tue, Thurs, Sat & Sun; friendly atmosphere; spacious pitches - sun or shd; rec." ♦ € 22.50 2016*

"We must tell the Club about that great site we found"

Get your site reports in by mid-August and we'll do our best to get your updates into the next edition.

⊞ **EVORA** *C3* (3km SW Urban) *38.55722, -7.92583* **Camping Orbitur Evora, Estrada das Alcaçovas, Herdade Esparragosa, 7005-206 Évora [(266) 705190; fax (266) 709830; evora@orbitur.pt; www.orbitur.pt]** Fr N foll N18 & by-pass, then foll sps for Lisbon rd at each rndabt or traff lts. Fr town cent take N380 SW sp Alcaçovas, foll site sp, site in 2km. NB Narr gate to site. Med, mkd, hdstg, pt shd, pt sl, wc; chem disp; mv service pnt; shwrs inc; EHU (6A) €3.50-4.60 (long lead poss req, rev pol); gas; lndry; shop; snacks; bar; playgrnd; pool; paddling pool; red long stay; wifi; TV; dogs €3; phone; bus; Eng spkn; adv bkg acc; quiet; ccard acc; games area; car wash; tennis; CKE/CCI. "Conv town cent, Évora World Heritage site with wealth of monuments & prehistoric sites nrby; cycle path to town; free car parks just outside town walls; poss flooding some pitches after heavy rain; beautiful site." ♦ € 29.00 2017*

⊞ **EVORAMONTE** *C3* (6km NE Rural) *38.79276, -7.68701* **Camping Alentejo, Novo Horizonte, 7100-300 Evoramonte [268 959283 or 936 799249 (mob); info@campingalentejo.com; www.campingalentejo.com]** Fr E exit A6/E90 junc 7 Estremoz onto N18 dir Evora. Site in 8km at km 236. Sm, hdstg, pt shd, terr, wc; chem disp; mv service pnt; shwrs inc; EHU (16A) inc; lndry; bbq; pool; wifi; dogs €1; bus adj; Eng spkn; adv bkg acc; quiet; horseriding; CKE/CCI. "Excel site; gd birdwatching, v friendly and helpful owner; conv NH on the way to S; easy parking; excel modern clean facs; gd library in off; bus stop outside gate to Evora; care req ent site; MH stay at lower price if no hookups; improved site; gd; secure." ♦ € 15.00 2018*

⊞ **FIGUEIRA DA FOZ** *B2* (7km S Urban/Coastal) *40.11861, -8.85666* **Camping ORBITUR-Gala, N109, Km 4, Gala, 3090-458 Figueira da Foz [233 431231; infogala@orbitur.pt; www.orbitur.pt]** Fr Figueira da Foz on N109 dir Leiria for 3.5km. After Gala site on R in approx 400m. Ignore sp on R 'Campismo' after long bdge. Lge, hdstg, shd, terr, wc; chem disp; mv service pnt; shwrs; EHU (6-10A) €3-4; gas; lndry; shop; rest; snacks; bar; bbq; playgrnd; pool (htd); paddling pool; beach sand 400m; red long stay; entmnt; wifi; TV; 80% statics; dogs €1.50; phone; Eng spkn; adv bkg acc; ccard acc; fishing 1km; car wash; tennis; games rm; CKE/CCI. "Gd, renovated site adj busy rd; luxury san facs (furthest fr recep); excel pool; busy site; in pine woods." ♦ € 29.00 2016*

GERES *C1* (2km N Rural) *41.73777, -8.15805* **Vidoeiro Camping, Lugar do Vidoeiro, 4845-081 Gerês [253 391289; aderepg@mail.telepac.pt; www.adere-pg.pt]** NE fr Braga take N103 twds Chaves for 25km. 1km past Cerdeirinhas turn L twds Gerês onto N308. Site on L 2km after Caldos do Gerês. Steep rds with hairpins. Cross bdge & reservoir, foll camp sps. Lge, hdstg, mkd, pt shd, terr, wc; chem disp; shwrs inc; EHU (12A) €1.20; lndry; rest nr; bar nr; bbq; sw nr; dogs €0.60; phone; quiet. "Attractive, wooded site in National Park; gd, clean facs; thermal spa in Gerês; pool 500m; diff access for lge o'fits, mountain rd." ♦ 15 May-15 Oct. € 17.00 2013*

⊞ **GERES** *C1* (14km NW Rural) *41.76305, -8.19111* **Parque de Campismo de Cerdeira, Rue de Cerdeira 400, Campo do Gerês, 4840-030 Terras do Bouro [(253) 351005; fax (253) 353315; info@parquecerdeira.com; www.parquecerdeira.com]** Fr N103 Braga-Chaves rd, 28km E of Braga turn N onto N304 at sp to Poussada. Cont N for 18km to Campo de Gerês. Site in 1km; well sp. V lge, shd, wc; chem disp; shwrs inc; EHU (5-10A) €4; gas; lndry; shop; rest; bar; playgrnd; sw; entmnt; TV; 10% statics; dogs €3; bus 500m; Eng spkn; quiet; ccard acc; canoeing; bike hire; fishing 2km; CKE/CCI. "Beautiful scenery; unspoilt area; fascinating old vills nrby & gd walking; ltd facs LS; Nat pk campsite; wooded site; san facs modern & clean." ♦ € 30.00 2018*

⊞ **GOUVEIA** *C2* (7km NE Rural) *40.52083, -7.54149* **Camping Quinta das Cegonhas, Nabaínhos, 6290-122 Melo [238 745886; cegonhas@cegonhas.com; www.cegonhas.com]** Turn S at 114km post on N17 Seia-Celorico da Beira. Site sp thro Melo vill. Sm, pt shd, wc; chem disp; mv service pnt; shwrs; EHU (6-10A) €3; lndry; shop nr; rest; snacks; bar; playgrnd; pool; red long stay; entmnt; TV; dogs €1.10; bus 400m; Eng spkn; adv bkg acc; quiet; games rm; CKE/CCI. "Vg, well-run, busy site in grnds of vill manor house; friendly Dutch owners; beautiful location conv Torre & Serra da Estrella; guided walks; gd walks; highly rec." ♦ € 18.00 2013*

⊞ **GUARDA** *C2* (2km W Urban) *40.53861, -7.27944* **Camp Municipal da Guarda, Avda do Estádio Municipal, 6300-705 Guarda [271 221200; fax 271 210025]** Exit A23 junc 35 onto N18 to Guarda. Foll sp cent & sports cent. Site adj sports cent off rndabt. Med, hdstg, shd, sl, wc (cont); own san rec; chem disp; shwrs inc; EHU (15A) €1.45; gas; lndry; shop; rest; snacks; bar; bbq; playgrnd; TV; phone; bus adj; Eng spkn; CKE/CCI. "Access to some pitches diff for c'vans, OK for m'vans; poss run down facs & site poss neglected low/mid ssn; walking dist to interesting town - highest in Portugal & poss v cold at night; music festival 1st week Sep." ♦ € 12.00 2014*

GUIMARAES *B1* (32km E Rural) *41.46150, -8.01120* **Quinta Valbom, Quintã 4890-505 Ribas [351 253 653 048; info@quintavalbom.nl; www.quintavalbom.nl]** Fr Guimaraes take A7 SE. Exit 11 onto N206 Fafe/Gandarela. Turn R bef tunnel twds Ribas. Foll blue & red signs of campsite. Med, mkd, pt shd, terr, wc; shwrs inc; EHU (10A); lndry; bar; bbq; pool; twin axles; wifi; dogs; Eng spkn; adv bkg acc; quiet; CCI. "Very nice site; friendly, extremely helpful Dutch owners; quiet surroundings; lots of space in beautiful setting; if driving c'van, park at white chapel and call campsite for their 4WD assistance up last bit of steep hill; owner won't acc c'vans over 6mtrs." ♦ 1 Apr-1 Oct. € 26.70 2017*

GUIMARAES *B1* (6km SE Rural) *41.42833, -8.26861* **Camping Parque da Penha, Penha-Costa, 4800-026 Guimarães [253 515912 or 253 515085; geral@turipenha.pt; www.turipenha.pt]** Take N101 SE fr Guimarães sp Felgueiras. Turn R at sp for Nascente/Penha. Site sp. Lge, hdstg, shd, pt sl, terr, wc; shwrs inc; EHU (6A); gas; rest nr; snacks; bar; playgrnd; pool; wifi; dogs; phone; bus; Eng spkn; adv bkg acc; fishing; car wash; CKE/CCI. "Excel staff; gd san facs; lower terrs not suitable lge o'fits; densely wooded hilltop site; conv Guimarães World Heritage site European City of Culture 2012; cable car down to Guimaraes costs €5 return; excel rest." ♦ 1 May-15 Sep. € 19.00 2017*

⊞ **IDANHA A NOVA** *C2* (10km NE Rural) *39.95027, -7.18777* **Camping ORBITUR-Barragem de Idanha-a-Nova, N354-1, Km 8, Barragem de Idanha-a-Nova, 6060 Idanha-a-Nova [(277) 202793; fax (277) 202945; infoidanha@orbitur.pt; www.orbitur.pt]** Exit IP2 at junc 25 sp Lardosa & foll sp Idanha-a-Nova on N18, then N233, N353. Thro Idanha & cross Rv Ponsul onto N354 to site. Avoid rte fr Castelo Branco via Ladoeiro as rd narr, steep & winding in places. Lge, mkd, hdstg, shd, terr, wc; chem disp; mv service pnt; fam bthrm; shwrs inc; EHU (6A) €3-4; gas; lndry; shop; rest; snacks; bar; bbq; playgrnd; pool (htd); paddling pool; sw; red long stay; entmnt; wifi; TV (pitch); 10% statics; dogs €1.50; phone; Eng spkn; adv bkg acc; quiet; car wash; tennis; fishing; watersports 150m; games rm. "Uphill to town & supmkt; hot water to shwrs only; pitches poss diff - a mover req; level pitches at top of site; excel." ♦ € 29.40 2013*

⊞ **LAGOS** *B4* (7km W Rural/Coastal) *37.10095, -8.73220* **Camping Turiscampo, N125 Espiche, 8600-109 Luz-Lagos [282 789265; fax 282 788578; info@turiscampo.com; www.turiscampo.com]** Exit A22/IC4 junc 1 to Lagos then N125 fr Lagos dir Sagres, site 3km on R. Lge, hdg, hdstg, mkd, shd, pt sl, terr, wc (htd); chem disp; mv service pnt; fam bthrm; shwrs inc; EHU (6A) inc - extra for 10A; gas; lndry; shop; rest; snacks; bar; bbq; playgrnd; pool; paddling pool; beach sand 2km; red long stay; entmnt; wifi; TV; 25% statics; dogs €1.50; phone; bus to Lagos 100m; Eng spkn; adv bkg acc; quiet; ccard acc; games area; tennis 2km; bike hire; games rm; fishing 2.5km; CKE/CCI. "Superb, well-run, busy site; fitness cent; v popular for winter stays & rallies; all facs (inc excel pool) open all yr; gd san facs; helpful staff; lovely vill, beach & views; varied & interesting area, Luz worth visit; vg." ♦ € 39.00 2014*

See advertisement on next page

⊞ **LAMAS DE MOURO** *C1* (2km S Rural) *42.03587, -8.19644* **Camping Lamas de Mouro, 4960-170 Lamas de Mouro [(251) 466041; geral@camping-lamas.com; www.camping-lamas.com]** Fr N202 at Melgaco foll sp Peneda- Gerês National Park, cont R to rd sp Porta de Lamas de Mouro. Cont 1km past park info office, site on L in pine woods. Med, pt shd, wc; chem disp; mv service pnt; shwrs inc; EHU (10A) €3; lndry; shop; rest; snacks; bar; cooking facs; playgrnd; pool; dogs €2; phone; bus 1km; quiet; CKE/CCI. "Ideal for walking in National Park; natural pool." € 15.00 2017*

⊞ **LAMEGO** *C1* (1km NW Urban) *41.09017, -7.82212* **Camping Lamego, EN2 Lugar da Raposeira, 5101-909 Lamego [351 969 021 408; fax 351 254 619 305; campinglamego@gmail.com; campinglamego.wix.com]** Foll N225. Turn L onto N2. Sm, hdstg, pt shd, wc (htd); chem disp; mv service pnt; fam bthrm; shwrs; EHU (6A) €4; lndry; rest; snacks; bar; wifi; Eng spkn; quiet. "Easy walk to Bom Jesus do Monte; site only suitable for MH's; san facs new & excel (2016); v friendly owners; excel." € 15.00 2016*

⊞ **LISBOA** *B3* (17km SW Coastal) *38.65111, -9.23777* **Camping ORBITUR-Costa de Caparica, Ave Afonso de Albuquerque, Quinta de S. António, 2825-450 Costa de Caparica [212 901366 or 903894; fax 212 900661; caparica@orbitur.pt; www.orbitur.pt]** Take A2/IP7 S fr Lisbon; after Rv Tagus bdge turn W to Costa de Caparica. At end of rd turn N twd Trafaria, & site on L. Well sp fr a'strada. Lge, hdg, mkd, shd, terr, wc; chem disp; mv service pnt; shwrs inc; EHU (6A) €3; gas; lndry; shop; rest; snacks; bar; bbq; playgrnd; beach sand 1km; red long stay; entmnt; wifi; TV; 75% statics; dogs €3; phone; bus to Lisbon; Eng spkn; adv bkg acc; ccard acc; car wash; games rm; tennis; fishing; CKE/CCI. "Gd, clean, well run site; heavy traff into city; rec use free parking at Monument to the Discoveries & tram to city cent; ferry to Belém; ltd facs LS; pleasant, helpful staff; pool 800m; aircraft noise early am & late pm." ♦ € 34.50 2017*

See advertisement above

⊞ **LISBOA** *B3* (9km W Urban) *38.72472, -9.20805* **Parque Municipal de Campismo de Monsanto, Estrada da Circunvalação, 1400-061 Lisboa [(217) 628200; fax (217) 628299; info@lisboacamping.com; www.lisboacamping.com]** Fr W on A5 foll sp Parque Florestal de Monsanto/Buraca. Fr S on A2, cross toll bdge & foll sp for Sintra; join expressway, foll up hill; site well sp; stay in RH lane. Fr N on A1 pass airport, take Benfica exit & foll sp under m'way to site. Site sp fr all major rds. Avoid rush hrs! V lge, hdstg, mkd, pt shd, pt sl, terr, serviced pitches; wc (htd); chem disp; mv service pnt; fam bthrm; shwrs inc; EHU (6-16A) inc; gas; lndry; shop; rest; snacks; bar; playgrnd; pool; entmnt; TV; 5% statics; dogs free; bus to city; Eng spkn; adv bkg acc; ccard acc; tennis; CKE/CCI. "Well laid-out, spacious, guarded site in trees; bank; PO; car wash; ltd mv service pnt; take care hygiene at chem disp/clean water tap; facs poss badly maintained & stretched when site full; friendly, helpful staff; in high ssn some o'fits placed on sl forest area (quiet); few pitches take awning; excel excursions booked at TO on site." ♦ € 32.00 2013*

⊞ **LOURICAL** *B2* (5km SW Rural) *39.99149, -8.78880* **Campismo O Tamanco, Rua do Louriçal 11, Casas Brancas, 3105-158 Louriçal [236 952551; tamanco@me.com; www.campismo-o-tamanco.com]**
S on N109 fr Figuera da Foz S twds Leiria foll sp at rndabt Matos do Corrico onto N342 to Louriçal. Site 800m on L. Med, hdstg, hdg, mkd, pt shd, wc; chem disp; mv service pnt; fam bthrm; shwrs inc; EHU inc (6A) €2.25-3.50; gas; lndry; rest; snacks; bar; bbq; pool (htd); paddling pool; beach sand 5km; sw nr; red long stay; twin axles; wifi; 5% statics; dogs; bus 500m; Eng spkn; adv bkg acc; quiet; CKE/CCI. "Excel; friendly Dutch owners; chickens & ducks roaming site; superb mkt on Sun at Louriçal; a bit of real Portugal; gd touring base; mkd walks thro pine woods; v clean; relaxed; sm farm animal area." ♦ € 25.00 2016*

⊞ **MEDA** *C2* (1km N Urban) *40.96972, -7.25916* **Parque de Campismo Municipal, Av. Professor Adriano Vasco Rodrigues, 6430 Mêda [(351) 925 480 500 or (351) 279 883 270; campismo@cm-meda.pt; www.cm-meda.pt/turismo/Paginas/Parque_Camsimo.aspx]** Head N fr cent of town, take 1st R, take 1st L & site on L within the Meda Sports Complex. Sm, hdstg, pt shd, pt sl, wc; chem disp; mv service pnt; fam bthrm; shwrs; shop; rest; snacks; bar; pool; wifi. "Pt of the Municipal Sports Complex with facs avail; conv for town cent & historic ctr; lovely sm site; lge pitches; some pull thro; blocks esential; clean modern san facs." ♦ € 16.00 2017*

MOGADOURO *D1* (2km SW Rural) *41.33527, -6.71861* **Parque de Campismo da Quinta da Agueira, Complexo Desportivo, 5200-244 Mogadouro [279 340231 or 936-989202 (mob); fax 279 341874; campismo@mogadouro.pt; www.mogadouro.pt]** Fr Miranda do Douro on N221 or fr Bragança on IP2 to Macedo then N216 to Mogadouro. Site sp adj sports complex. Lge, shd, wc; chem disp; mv service pnt; shwrs inc; EHU (15A) €2; gas; lndry; shop nr; rest nr; snacks; bar; bbq; playgrnd; entmnt; wifi; TV; dogs €1.50; phone; bus 300m; Eng spkn; adv bkg acc; quiet; tennis; car wash; waterslide. "Brilliant site in lovely area; value for money; steep hill to town; pool adj; gd touring base." ♦ 1 Apr-30 Sep. € 11.50 2013*

⊞ **MONTARGIL** *B3* (4km N Rural) *39.10083, -8.14472* **Camping ORBITUR, Baragem de Montargil, N2, 7425-017 Montargil [242 901207; fax 242 901220; infomontargil@orbitur.pt; www.orbitur.pt]**
Fr N251 Coruche to Vimiero rd, turn N on N2, over dam at Barragem de Montargil. Fr Ponte de Sor S on N2 until 3km fr Montargil. Site clearly sp bet rd & lake. Med, hdstg, mkd, pt shd, terr, wc; chem disp; mv service pnt; shwrs inc; EHU (6-10A) €3-4; gas; lndry; shop; rest; snacks; bar; bbq; playgrnd; pool; paddling pool; red long stay; entmnt; wifi; TV (pitch), 60% statics; dogs €1.50; phone; Eng spkn; adv bkg acc; ccard acc; fishing; car wash; tennis; games rm; boating; watersports; CKE/CCI. "Friendly site in beautiful area." ♦ € 26.70 2016*

See advertisement

⊞ **NAZARE** *B2* (2km N Rural) *39.62036, -9.05630* **Camping Vale Paraíso, N242, 2450-138 Nazaré [262 561800; fax 262 561900; info@valeparaiso-naturpark.com; www.valeparaiso-naturpark.com]** Site thro pine reserve on N242 fr Nazaré to Leiria. V lge, hdstg, mkd, shd, terr, wc; chem disp; mv service pnt; fam bthrm; shwrs inc; EHU (4-10A) €3; gas; lndry; shop; rest; snacks; bar; playgrnd; pool; paddling pool; beach sand 2km; red long stay; wifi; TV; 20% statics; dogs €2.10; bus; Eng spkn; adv bkg acc; quiet; ccard acc; games area; games rm; bike hire; site clsd 19-26 Dec; fishing; CKE/CCI. "Gd, clean site; well run; gd security; pitches vary in size & price, & divided by concrete walls, poss not suitable lge o'fits, bus outside gates to Nazare, exit down steep hill." ♦ € 23.60 2017*

⊞ **ODEMIRA** *B4* (13km W Rural) *37.60565, -8.73786* **Zmar Eco Camping Resort & Spa, Herdade A-de-Mateus, N393/1, San Salvador, 7630-011 Odemira [(707) 200626 or (283) 690010; fax (283) 690014; info@zmar.eu; www.zmar.eu]** Fr N on A2 take IC33 dir Sines. Just bef ent Sines take IC4 to Cercal (sp Sul Algarve) then foll N390/393 & turn R dir Zambujeira do Mar, site sp. Med, mkd, hdstg, pt shd, wc; chem disp; mv service pnt; fam bthrm; shwrs inc; EHU (10A) inc; lndry; shop; rest; snacks; bar; bbq; cooking facs; playgrnd; pool (covrd, htd); paddling pool; twin axles; entmnt; wifi; TV; 17% statics; dogs €2.50; Eng spkn; adv bkg acc; quiet; bike hire; games area; sauna; excursions; tennis. "Superb new site 2009 (eco resort) with excel facs; private bthrms avail; wellness cent; fitness rm; in national park; vg touring base." ♦ € 46.00 2013*

⊞ **ODIVELAS** *B3* (8km NE Rural) *38.18361, -8.10361* **Camping Markádia, Barragem de Odivelas, 7920-999 Alvito [(284) 763141; fax (284) 763102; markadia@hotmail.com; www.markadia.com]** Fr Ferreira do Alentejo on N2 N twd Torrão. After Odivelas turn R onto N257 twd Alvito & turn R twd Barragem de Odivels. Site in 7km, clearly sp. Med, hdstg, pt shd, pt sl, wc; chem disp; mv service pnt; shwrs inc; EHU (16A) inc; gas; lndry; shop; rest; snacks; bar; playgrnd; paddling pool; beach sand 500m; sw nr; dogs (except Jul-Aug); phone; adv bkg acc; quiet; horseriding; boating; fishing; car wash; tennis; CKE/CCI. "Beautiful, secluded site on banks of reservoir; spacious pitches; pool 50m; gd rest; site lighting low but san facs well lit; excel walking, cycling, birdwatching; wonderful." ♦ € 32.00 2013*

⊞ **OLHAO** *C4* (12km NE Rural) *37.09504, -7.77430* **Camping Caravanas Algarve, Sitio da Cabeça Moncarapacho, 8700-618 Moncarapacho [(289) 791669]** Exit IP1/A22 sp Moncarapacho. In 2km turn L sp Fuzeta. At traff lts turn L & immed L opp supmkt in 1km. Turn R at site sp. Site on L. Sm, hdstg, unshd, pt sl, wc; chem disp; shwrs inc; EHU (6A) inc; lndry; shop nr; rest nr; bar nr; beach sand 4km; 10% statics; dogs; Eng spkn; adv bkg acc; quiet; CKE/CCI. "Situated on a farm in orange groves; pitches ltd in wet conditions; gd, modern san facs; gd security; Spanish border 35km; National Park Ria Formosa 4km; lovely popular site." € 10.00 2013*

⊞ **OLHAO** *C4* (2km NE Rural) *37.03527, -7.82250* **Camping Olhão, Pinheiros do Marim, 8700-912 Olhão [289 700300; fax 289 700390 or 700391; parque.campismo@sbsi.pt; www.sbsi.pt]** Turn S twd coast fr N125 1.5km E of Olhão by filling stn. Clearly sp on S side of N125, adj Ria Formosa National Park. V lge, hdg, mkd, shd, pt sl, wc; chem disp; mv service pnt; shwrs inc; EHU (6A) €1.90; gas; lndry; shop; rest; bar; playgrnd; pool; paddling pool; beach 1.5km; red long stay; wifi; TV; 75% statics; dogs €1.60; phone; bus adj, train 1.5km; Eng spkn; adv bkg acc; ccard acc; tennis; games area; horseriding 1km; bike hire; games rm; CKE/CCI. "Pleasant, helpful staff; sep car park for some pitches; car wash; security guard; excel pool; gd san facs; very popular long stay LS; many sm sandy pitches, some diff access for lge o'fits; gd for cycling, birdwatching; ferry to islands." ♦ € 24.00 2014*

See advertisement opposite

⊞ **OLHAO** *C4* (8km NE Rural) *37.07245, -7.79928* **Campismo Casa Rosa, Apt 209 8700 Moncarapacho [(289) 794400 or (9191) 73132 (mob); fax (289) 792952; casarosa@sapo.pt; www.casarosa.eu.com]** Fr A22 (IP1) E twd Spain, leave at exit 15 Olhão/ Moncarapacho. At rndabt take 2nd exit dir Moncarapacho. Cont past sp Moncarapacho Centro dir Olhão. In 1km at Lagoão, on L is Café Da Lagoão with its orange awning. Just past café is sp for Casa Rosa. Foll sp. Sm, hdstg, unshd, terr, wc (htd); chem disp; shwrs inc; EHU (6A) inc; lndry; rest; pool; wifi; TV (pitch); dogs; Eng spkn; adv bkg acc; CKE/CCI. "Excel CL-type site adj holiday apartments; adults only; helpful, friendly, Norwegian owners; evening meals avail; ideal for touring E Algarve; conv Spanish border; rec; 30% dep req, no refunds if leaving early; insufficient san facs, but still a gd site; drinkable water taps." ♦ € 13.50 2013*

OLIVEIRA DO HOSPITAL *C2* (9km NE Rural) *40.40338, -7.82684* **Camping Toca da Raposa, 3405-351 Meruge [238 601547 or 926 704218 (mob); campingtocadaraposa@gmail.com; toca-da-raposa.com]** Fr N: N170 Oliveira Do Hospital head SW. Drive thro EM540-2, EM503-1, R. Principal, Estr. Principal and EM504-3 to Coimbra. Foll sp to site. Sm, hdg, pt shd, pt sl, terr, wc; chem disp; mv service pnt; shwrs inc; EHU (6A) €2.50; gas; rest; snacks; bar; playgrnd; pool; wifi; dogs; bus; Eng spkn; quiet. "Charming; lovely pool; bar & eve meals; friendly Dutch owner; vg." ♦ 15 Mar-1 Nov. € 18.50 2017*

⊞ **ORTIGA** *C3* (6km SE Rural) *39.48277, -8.00305* **Parque Campismo de Ortiga, Estrada da Barragem, 6120-525 Ortiga [241 573464; fax 241 573482; campismo@cm-macao.pt]** Exit A23/IP6 junc 12 S to Ortiga. Thro Ortiga & foll site sp for 1.5km. Site beside dam. Sm, mkd, hdstg, pt shd, terr, wc; chem disp; shwrs inc; EHU (10A) €1.50; lndry; shop nr; rest nr; bar nr; bbq; playgrnd; sw nr; TV; 50% statics; dogs free; Eng spkn; quiet; watersports; CKE/CCI. "Lovely site in gd position; useful NH; lge o'fits should avoid acc thro town." ♦ € 21.00 2016*

PENACOVA *B2* (3km N Rural) *40.27916, -8.26805* **Camp Municipal de Camp de Penacova (Vila Nova), Rua dos Barqueiros, Vila Nova, 3360-204 Penacova [919 121967; fax 239 470009; penaparque2@iol.pt]** IP3 fr Coimbra, exit junc 11, cross Rv Mondego N of Penacova & foll to sp to Vila Nova & site. Med, pt shd, wc; shwrs inc; EHU (6A) €1; shop nr; rest nr; snacks; bar; bbq; playgrnd; sw nr; wifi; TV; phone; bus 150m; Eng spkn; bike hire; fishing; CKE/CCI. "Open, attractive site." 31 May-30 Sep. € 20.00 2017*

⊞ **PENICHE** *B3* (2km NW Urban/Coastal) *39.36944, -9.39194* **Camping Peniche Praia, Estrada Marginal Norte, 2520 Peniche [262 783460; fax 262 784140; geral@penichepraia.pt; www.penichepraia.pt]** Travel S on IP6 then take N114 sp Peniche; fr Lisbon N on N247 then N114 sp Peniche. Site on R on N114 1km bef Peniche. Med, hdg, hdstg, mkd, unshd, wc; chem disp; mv service pnt; shwrs inc; EHU (6A) inc; lndry; shop nr; rest; snacks; bar; bbq; playgrnd; pool (covrd); paddling pool; beach sand 1.5km; red long stay; entmnt; wifi; TV; 30% statics; dogs €2.10; phone; Eng spkn; adv bkg rec; car wash; games rm; bike hire; CKE/CCI. "Vg site in lovely location; some sm pitches; rec, espec LS." € 18.00 2013*

PONTE DA BARCA *B1* (11km E Rural) *41.82376, -8.31723* **Camping Entre-Ambos-os-Rios, Lugar da Igreja, Entre-Ambos-os-Rios, 4980-613 Ponte da Barca [258 588361; fax 258 452450; aderepg@mail.telepac.pt; www.adere-pg.pt]** N203 E fr Ponte da Barca, pass ent sp for vill. Site sp N twd Rv Lima, after 1st bdge. Lge, shd, pt sl, wc; shwrs inc; EHU (6A) €1.20; gas; lndry; shop nr; rest nr; snacks; bar; playgrnd; entmnt; TV; dogs €0.60; phone; bus 100m; adv bkg acc; fishing; canoeing; CKE/CCI. "Beautiful, clean, well run & maintained site in pine trees; well situated for National Park; vg rest." 15 May-30 Sep. € 18.00 2014*

⊞ **PORTIMAO** *B4* (7km W Rural) *37.13500, -8.59027* **Parque Campismo de Alvor (Formaly da Dourada), R Serpa Pinto 8500-053 Alvor [(282) 459178; info@campingalvor.com; www.campingalvor.com]** Turn S at W end of N125 Portimão by-pass sp Alvor. Site on L in 4km bef ent town. V lge, shd, pt sl, terr, wc; chem disp; shwrs; EHU (6-16A) €3-5; gas; lndry; shop; rest; snacks; bar; playgrnd; pool; paddling pool; beach sand 1km; red long stay; entmnt; TV; dogs €2.50; bus adj; fishing; games area; CKE/CCI. "Friendly & helpful, family-run site; office poss unattended in winter, ltd facs & site untidy; excel rest; lovely town & beaches; site much improved, never untidy (2013); v welcoming; popular with wintering Brits." ♦ € 21.50 2013*

> **"I need an on-site restaurant"**
>
> We do our best to make sure site information is correct, but it is always best to check any must-have facilities are still available or will be open during your visit.

⊞ **PORTO** *B1* (17km N Coastal) *41.2675, -8.71972* **Camping Orbitur-Angeiras, Rua de Angeiras, Matosinhos, 4455-039 Lavra [229 270571; fax 229 271178; infoangeiras@orbitur.pt; www.orbitur.pt]** Fr ICI/A28 take turn-off sp Lavra, site sp at end of slip rd. Site in approx 3km - app rd potholed & cobbled. Lge, shd, pt sl, wc; chem disp; mv service pnt; shwrs inc; EHU (6A) €3-4 (check earth); gas; lndry; shop; rest; snacks; bar; bbq; playgrnd; pool; paddling pool; beach sand 400m; red long stay; entmnt; wifi; TV (pitch); 70% statics; dogs €1.50; phone; bus to Porto at site ent; Eng spkn; adv bkg acc; ccard acc; car wash; games area; tennis; games rm; fishing; CKE/CCI. "Friendly & helpful staff; gd rest; gd pitches in trees at end of site but ltd space lge o'fits; ssnl statics all yr; fish & veg mkt in Matosinhos; excel new san facs (2015); vg pool; poss noisy on Sat nights (beach parties)." ♦ € 40.70 2017*

⊞ **PORTO** *B1* (32km SE Rural) *41.03972, -8.42666* **Campidouro Parque de Medas, Lugar do Gavinho, 4515-397 Medas-Gondomar [224 760162; fax 224 769082; geral@campidouro.pt; www.campidouro.pt]** Take N12 dir Gondomar off A1. Almost immed take R exit sp Entre-os-Rios. At rndabt pick up N108 & in approx 14km. Sp for Medas on R, thro hamlet & forest for 3km & foll sp for site on R. Long, steep app. New concrete access/site rds. Lge, mkd, hdstg, pt shd, terr, serviced pitches; wc; chem disp; mv service pnt; shwrs inc; EHU (6A) inc (poss rev pol); gas; lndry; shop; rest; bar; bbq; playgrnd; pool; paddling pool; sw; entmnt; wifi; TV; 90% statics; dogs; phone; bus to Porto; Eng spkn; adv bkg acc; quiet; ccard acc; games rm; tennis; boating; fishing; CKE/CCI. "Beautiful site on Rv Douro; helpful owners; gd rest; clean facs; sm level area (poss cr by rv & pool) for tourers - poss noisy at night & waterlogged after heavy rain; bus to Porto (just outside site) rec as parking diff (ltd buses at w/end)." ♦ € 26.00 2016*

⊞ **PORTO** *B1* (9km SW Urban/Coastal) *41.10777, -8.65611* **Camping ORBITUR-Madalena, Rua do Cerro 608, Praia da Madalena, 4405-736 Vila Nova de Gaia [(227) 122520; fax (227) 122534; infomadalena@orbitur.pt; www.orbitur.pt/camping-orbitur-madalena]** Fr Porto ring rd IC1/A44 take A29 exit dir Espinho. In 1km take exit slip rd sp Madalena opp Volvo agent. Watch for either 'Campismo' or 'Orbitur' sp to site along winding, cobbled rd (beware campismo sp may take you to another site nrby). Lge, pt shd, pt sl, terr, wc; chem disp; mv service pnt; fam bthrm; shwrs inc; EHU (6A)inc; gas; lndry; shop; rest; snacks; bar; bbq; playgrnd; pool; paddling pool; beach sand 250m; red long stay; entmnt; wifi; TV; 40% statics; dogs €2.20; phone; bus to Porto; Eng spkn; adv bkg acc; ccard acc; car wash; games area; tennis; games rm; CKE/CCI. "Site in forest; restricted area for tourers; slight aircraft noise; some uneven pitches; poss ltd facs LS; excel bus to Porto cent fr site ent - do not take c'van into Porto; facs in need of refurb (2013)." ♦ € 22.00 2017*

⊞ **SAGRES** *B4* (1km N Coastal) *37.02305, -8.94555* **Camping ORBITUR-Sagres, Cerro das Moitas, 8650-998 Vila de Sagres [282 624371; fax 282 624445; infosagres@orbitur.pt; www.orbitur.pt]** On N268 to Cape St Vincent; well sp. Lge, hdg, hdstg, mkd, pt shd, wc; chem disp; mv service pnt; shwrs inc; EHU (6-10A) €3-4; gas; lndry; shop; rest; snacks; bar; bbq; playgrnd; beach sand 2km; red long stay; wifi; TV; dogs €1.50; Eng spkn; adv bkg acc; quiet; ccard acc; bike hire; car wash; games rm. "Vg, clean, tidy site in pine trees; helpful staff; hot water to shwrs only; cliff walks." ♦ € 25.60 2012*

SANTIAGO DO CACEM *B4* (17km NW Coastal) *38.10777, -8.78690* **Camping Lagoa de Santo Andre, Campismo Lago de Santo Andre, Vila Nova de Santo Andre [269 708550; fax 269 708559]** Take N261 sp Melides out of town & foll sps to Lagoa de Santo Andre to site on L of rd. On shore but fenced off fr unsafe banks of lagoon. Med, pt shd, pt sl, wc; shwrs inc; EHU inc (4-6A); shop; rest; bar; sw; boating; fishing. "Ltd facs LS." 18 Jan-18 Dec. € 17.70 2014*

⊞ **SANTO ANTONIO DAS AREIAS** *C3* (5km N Rural) *39.41370, -7.37575* **Quinta Do Maral (Naturist), PO Box 57, Cubecudos, 7330-205 Santo Antonio das Areias [(963) 462169; info@quintadomaral.com; www.quintadomaral.com]** Take N359 twds Santo Antonio Das Areias/Beira. Pass turn off to Santo Antonio in dir to Beira. At Ranginha turn L to Cubecudos. Turn R past vill sp by a school bldg, keep on this rd for 1.7km. Campsite on L, white hse with blue stripe. Do not use sat nav as they lead to unsuitable rds. Sm, hdg, pt shd, wc; chem disp; mv service pnt; shwrs; EHU (6-16A); lndry; snacks; bar; bbq; pool; red long stay; twin axles; wifi; TV; 5% statics; dogs; Eng spkn; adv bkg rec; quiet. "In S Mamede nature park; vill & castle of Marvao an hr's walk; young, friendly owners; excel." ♦ € 18.50 2016*

SANTO ANTONIO DAS AREIAS *C3* (0km S Rural) *39.40992, -7.34075* **Camping Asseiceira, Asseiceira, 7330-204 Santo António das Areias [(245) 992940 or (960) 150352 (mob); gary-campingasseiceira@hotmail.com; www.campingasseiceira.com]** Fr N246-1 turn off sp Marvão/Santo António das Areias. Turn L to Santo António das Areias then 1st R on ent town then immed R again, up sm hill to rndabt. At rndabt turn R then at next rndabt cont strt on. There is a petrol stn on R, cont down hill for 400m. Site on L. Sm, pt shd, pt sl, wc; chem disp; shwrs inc; EHU (10A) €4; shop nr; rest nr; snacks; bar; pool; wifi; dogs free; bus 1km; quiet; CKE/CCI. "Attractive area; peaceful, well-equipped, remote site among olive trees; clean, tidy; gd for walking, birdwatching; helpful, friendly, British owners; excel san facs, maintained to a high standard; nr Spanish border; excel; ideal cent for walking, cycling, visit hilltop castle Marvao." ♦ 1 Jan-31 Oct. € 21.00 2013*

⊞ **SAO MARCOS DA SERRA** *B4* (5km SE Rural) *37.3350, -8.3467* **Campismo Rural Quinta Odelouca, Vale Grande de Baixo, CxP 644-S, 8375-215 São Marcos da Serra [282 361718; info@quintaodelouca.com; www.quintaodelouca.com]** Fr N (Ourique) on IC1 pass São Marcos da Serra & in approx 2.5km turn R & cross blue rlwy bdge. At bottom turn L & at cont until turn R for Vale Grande (paved rd changes to unmade). Foll sp to site. Fr S exit A22 junc 9 onto IC1 dir Ourique. Pass São Bartolomeu de Messines & at km 710.5 turn L & cross blue rlwy bdge, then as above. Sm, pt shd, terr, wc; chem disp; fam bthrm; shwrs inc; EHU (10A) €2.10; lndry; shop nr; rest nr; bar; bbq; pool; sw; wifi; dogs €1; Eng spkn; adv bkg acc; quiet; CKE/CCI. "Helpful, friendly Dutch owners; phone ahead bet Nov & Feb; beautiful views; gd walks; vg; v little shd; access via bad rd." € 18.00 2015*

SAO MARTINHO DO PORTO *B2* (2km NE Coastal) *39.52280, -9.12310* **Parque de Campismo Colina do Sol, Serra dos Mangues, 2460-697 São Martinho do Porto [(262) 989764; fax (262) 989763; parque.colina.sol@clix.pt or geral@colinadosol.net; www.colinadosol.net]** Leave A8/IC1 SW at junc 21 onto N242 W to São Martinho, by-pass town on N242 dir Nazaré. Site on L. Lge, mkd, hdstg, pt shd, terr, wc; chem disp; mv service pnt; shwrs inc; EHU (6A) €2.75; gas; lndry; shop; rest; snacks; bar; bbq; playgrnd; pool; paddling pool; beach sand 2km; TV; dogs €1; phone; Eng spkn; adv bkg acc; quiet; ccard acc; games rm; fishing; site clsd at Xmas; games area; CKE/CCI. "Gd touring base on attractive coastline; mob homes/c'vans for hire; gd walking, cycling; vg san facs; excel site; san facs a bit tired (2013), water v hot." ♦ € 30.50 2013*

SAO PEDRO DE MOEL *B2* (1km E Urban/Coastal) *39.75861, -9.02583* **Camping ORBITUR-São Pedro de Moel, Rua Volta do Sete, São Pedro de Moel, 2430 Marinha Grande [244 599168; fax 244 599148; infospedro@orbitur.pt; www.orbitur.pt]** Site at end of rd fr Marinha Grande to beach; turn R at 1st rndabt on ent vill. Site S of lighthouse. V lge, mkd, hdstg, hdg, shd, terr, wc; chem disp; mv service pnt; shwrs inc; EHU (6A) €3-4 (poss rev pol); gas; lndry; shop; rest; snacks; bar; bbq; playgrnd; pool (htd); paddling pool; beach sand 500m; red long stay; entmnt; wifi; TV (pitch); 10% statics; dogs €1.50; phone; Eng spkn; adv bkg acc; quiet; ccard acc; waterslide; fishing; car wash; tennis; games rm; bike hire; CKE/CCI. "Friendly, well-run, clean site in pine woods; easy walk to shops, rests; heavy surf; gd cycling to beaches; São Pedro smart resort; ltd facs LS site in attractive area and well run." ♦ € 28.60 2016*

"Satellite navigation makes touring much easier"

Remember most sat navs don't know if you're towing or in a larger vehicle – always use yours alongside maps and site directions.

SAO TEOTONIO *B4* (11km W Coastal) *37.49497, -8.78667* **Camping Monte Carvalhal da Rocha, Praia do Carvalhal, 7630-569 S Teotónio [282 947293; fax 282 947294; geral@montecarvalhalr-turismo.com; www.montecarvalhaldarocha.com]** Turn W off N120 dir Brejão & Carvalhal; site in 4.5km. Site sp. Med, shd, wc; shwrs inc; EHU (16A) inc; gas; lndry; shop; rest; snacks; bar; bbq; playgrnd; beach sand 500m; TV; 10% statics; phone; Eng spkn; adv bkg acc; quiet; ccard acc; bike hire; car wash; fishing. "Beautiful area; friendly, helpful staff." € 26.00 2012*

SAO TEOTONIO *B4* (7km W Coastal) *37.52560, -8.77560* **Parque de Campismo da Zambujeira, Praia da Zambujeira, 7630-740 Zambujeira do Mar [(283) 958 407; fax (283) 959 940; info@campingzambujeira; www.campingzambujeira.com]** S on N120 twd Lagos, turn W when level with São Teotónio on unclassified rd to Zambujeira. Site on L in 7km, bef vill. V lge, pt shd, pt sl, wc; chem disp; mv service pnt; shwrs inc; EHU (6-10A) €3.50; gas; shop; rest; snacks; bar; playgrnd; beach sand 1km; red long stay; TV; dogs €4; phone; bus adj; Eng spkn; tennis. "Welcoming, friendly owners; in pleasant rural setting; hot water to shwrs only; sh walk to unspoilt vill with some shops & rest; cliff walks; nice site; facs & pool gd; rest food basic." 1 Apr-31 Oct. € 18.00 2016*

SATAO *C2* (12km N Rural) *40.82280, -7.6961* **Camping Quinta Chave Grande, Rua do Barreiro 462, Casfreires, Ferreira d'Aves, 3560-043 Sátão [232 665552; info@chavegrande.com; www.chavegrande.com]** Leave IP5 Salamanca-Viseu rd onto N229 to Sátão, site sp in Satão - beyond Lamas. Med, pt shd, terr, wc; chem disp; fam bthrm; shwrs inc; EHU (6A) €3.50; gas; lndry; shop nr; rest nr; snacks; bar; playgrnd; pool; paddling pool; red long stay; wifi; TV; dogs leashed €2.50; Eng spkn; quiet; tennis; games area; games rm. "Warm welcome fr friendly Dutch owners; gd facs; well organised BBQ's - friendly atmosphere; gd touring base; gd walks fr site; excel." 15 Mar-31 Oct. € 20.50 2013*

SERPA *C4* (1km SW Urban) *37.94090, -7.60404* **Parque Municipal de Campismo Serpa, Rua da Eira São Pedro, 7830-303 Serpa [284 544290; fax 284 540109]** Fr IP8 take 1st sp for town; site well sp fr most dirs - opp sw pool. Do not ent walled town. Med, pt shd, pt sl, wc; chem disp; shwrs inc; EHU (6A) €1.25; gas; lndry; shop nr; rest nr; bar nr; bbq; sw nr; wifi; 20% statics; dogs; phone; adv bkg acc; CKE/CCI. "Popular gd site; daily mkt 500m; pool adj; simple, high quality facs; interesting, historic town." ♦ € 10.00 2014*

TAVIRA *C4* (5km E Rural/Coastal) *37.14506, -7.60223* **Camping Ria Formosa, Quinta da Gomeira, 8800-591 Cabanas-Tavira [281 328887; info@campingriaformosa.com; www.campingriaformosa.com]** Fr spain onto A22 take exit junc 17 (bef tolls) Fr N125 turn S at Conceição dir 'Cabanas Tavira' & 'Campismo'. Cross rlwy line & turn L to site, sp. V lge, hdstg, mkd, pt shd, terr, wc (htd); chem disp; mv service pnt; fam bthrm; shwrs inc; EHU (16A) €3; gas; lndry; shop; rest; snacks; bar; bbq; playgrnd; pool; paddling pool; beach sand 1.2km; red long stay; wifi; TV; dogs €2; bus 100m, train 100m; Eng spkn; adv bkg acc; quiet; ccard acc; games area; car wash; bike hire; CKE/CCI. "Excel, comfortable site; friendly, welcoming owner & staff; vg, modern san facs; various pitch sizes; cycle path to Tavira, excel facs." ♦ € 23.00 2014*

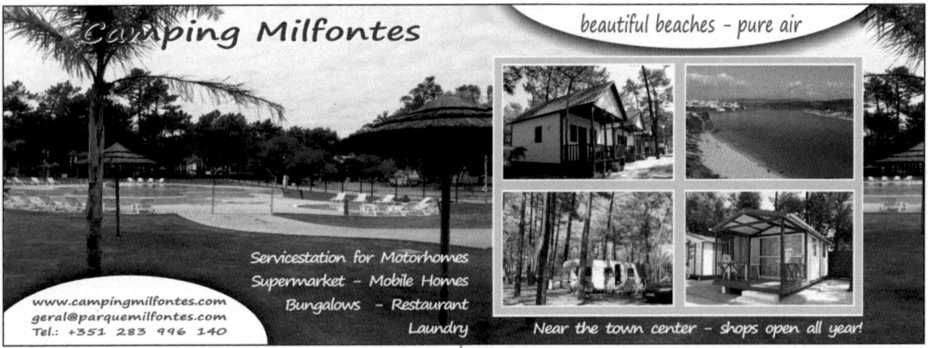

Camping Milfontes
beautiful beaches - pure air
Servicestation for Motorhomes
Supermarket - Mobile Homes
Bungalows - Restaurant
Laundry
www.campingmilfontes.com
geral@parquemilfontes.com
Tel.: +351 283 996 140
Near the town center - shops open all year!

⊞ **TOMAR** *B2* (1km NW Urban) *39.60694, -8.41027* **Campismo Parque Municipal, 2300-000 Tomar [249 329824; fax 249 322608; camping@cm-tomar.pt; www.cm-tomar.pt]** Fr S on N110 foll sp to town cent at far end of stadium. Fr N (Coimbra) on N110 turn R immed bef bdge. Site well sp fr all dirs. Med, mkd, pt shd, wc (htd); chem disp; mv service pnt; fam bthrm; shwrs inc; EHU (10A) inc; lndry; shop; rest; snacks; bar nr; bbq; playgrnd; wifi; TV; dogs; Eng spkn; adv bkg acc; quiet; ccard acc; CKE/CCI. "Useful base for touring Alcobaca, Batalha & historic monuments in Tomar; conv Fatima; Convento de Cristo worth visit; vg, popular, improved site; lovely walk to charming rvside town; easy access for lge vehicle, sh walk thro gdns to Knights Templar castle; free wifi nr recep; camp entry ticket gives free access to adj pool; pool adj; v helpful, friendly staff; great site" € 19.00 2016*

VIANA DO CASTELO *B1* (5km S Urban/Coastal) *41.67888, -8.82583* **Camping ORBITUR-Viana do Castelo, Rua Diogo Álvares, Cabedelo, 4935-161 Darque [258 322167; fax 258 321946; infoviana@orbitur.pt; www.orbitur.pt]** Exit IC1 junc 11 to W sp Darque, Cabedelo, foll sp to site in park. Lge, mkd, shd, pt sl, wc; chem disp; mv service pnt; shwrs inc; EHU (5-15A) inc; gas; lndry; shop; rest; snacks; bar; bbq; playgrnd; pool (htd); beach sand adj; red long stay; entmnt; wifi; TV; dogs €2.20; phone; Eng spkn; adv bkg acc; quiet; ccard acc; surfing; fishing; car wash; CKE/CCI. "Site in pine woods; friendly staff; gd facs; plenty of shd; major festival in Viana 3rd w/end in Aug; lge mkt in town Fri; sm passenger ferry over Rv Lima to town high ssn; Santa Luzia worth visit." ♦ 1 Mar-31 Oct. € 42.00 2013*

⊞ **VILA DO BISPO** *B4* (12km E Rural) *37.08834, -8.80006* **Quinta Manjericao, Sitio do Boieira, Budens, Vila do Bispo 8650-151 Algarve [351 922 239 070; info@quintamanjericao.com; www.quintamanjericao.com]** Turn L off N125 bet Almadena & Budens at the Pottery. Foll unmade rd for 300mtrs to fork by big tree. Turn R, go past fm on R and then L into site. Sm, hdstg, hdg, mkd, unshd, chem disp; mv service pnt; EHU (10A) inc; lndry; bbq; wifi; dogs inc; bus 1km; Eng spkn; adv bkg acc; quiet; CKE/CCI. "Excel site; horseriding 5km; cliff/coastal walks; supmkt/fuel 3km." ♦ € 10.00 2017

⊞ **VILA DO BISPO** *B4* (11km SE Rural/Coastal) *37.07542, -8.83133* **Quinta dos Carriços (Part Naturist), Praia de Salema, 8650-196 Budens [282 695201; fax 282 695122; quintacarrico@oninet.pt; www.quintadoscarricos.com]** Take N125 out of Lagos twd Sagres. In approx 14km at sp Salema, turn L & again immed L twd Salema. Site on R 300m. Lge, pt shd, terr, wc (htd); chem disp; mv service pnt; shwrs; EHU (6-10A) €3.90 (metered for long stay); gas; lndry; shop; rest; snacks; bar; playgrnd; beach sand 1.5km; red long stay; TV; 8% statics; dogs €2.45; phone; bus; Eng spkn; adv bkg req; ccard acc; golf 1km; CKE/CCI. "Naturist section in sep valley; apartments avail on site; ltd pitches for lge o'fits; friendly Dutch owners; tractor avail to tow to terr; noise fr adj quarry; area of wild flowers in spring; beach 30 mins walk; buses pass ent for Lagos, beach & Sagres; excel." ♦ € 28.00 2012*

"There aren't many sites open at this time of year"

If you're travelling outside peak season remember to call ahead to check site opening dates – even if the entry says 'open all year'.

VILA NOVA DE CERVEIRA *B1* (5km E Rural) *41.94362, -8.69365* **Parque de Campismo Convívio, Rua de Badão, 1 Bacelo, 4920-020 Candemil [251 794404; info@campingconvivio.net; www.campingconvivio.net]** Fr Vila Nova de Cerveira dir Candemil on N13/N302, turn L at Bacelo, site sp. Sm, pt shd, terr, wc; chem disp; fam bthrm; shwrs inc; EHU (6A); lndry; rest; snacks; bar; bbq; pool; red long stay; wifi; dogs €1.10; Eng spkn; adv bkg acc; quiet; games rm; CKE/CCI. "V helpful Dutch owners; gd area to visit; vg." ♦ 1 Mar-15 Oct. € 21.00 2017*

⊞ **VILA NOVA DE MILFONTES** *B4* (1km N Coastal) *37.73194, -8.78277* **Camping Milfontes, Apartado 81, 7645-300 Vila Nova de Milfontes [(283) 996140; fax (283) 996104; reservas@campingmilfontes.com; www.campingmilfontes.com]** S fr Sines on N120/IC4 for 22km; turn R at Cercal on N390 SW for Milfontes on banks of Rio Mira; clear sp. V lge, mkd, hdg, pt shd, wc; chem disp; mv service pnt; shwrs inc; EHU (6A) €3 (long lead poss req); gas; lndry; shop; rest; snacks; bar; playgrnd; beach sand 800m; TV; 80% statics; phone; bus 600m; quiet; ccard acc; CKE/CCI. "Pitching poss diff for lge o'fits due trees & statics; supmkt & mkt 5 mins walk; nr fishing vill at mouth Rv Mira with beaches & sailing on rv; pleasant site." ♦ € 27.00 2013*

See advertisement opposite

VILA REAL *C1* (1km NE Urban) *41.30361, -7.73694* **Camping Vila Real, Rua Dr Manuel Cardona, 5000-558 Vila Real [259 324724]** On IP4/E82 take Vila Real N exit & head S into town. Foll 'Centro' sp to Galp g'ge; at Galp g'ge rndabt, turn L & in 30m turn L again. Site at end of rd in 400m. Site sp fr all dirs. Med, pt shd, pt sl, terr, wc; chem disp; mv service pnt; fam bthrm; shwrs inc; EHU (6A); gas; shop nr; rest; snacks; bar; bbq; playgrnd; wifi; 10% statics; dogs; phone; bus 150m; tennis; CKE/CCI. "Conv upper Douro; pool complex adj; gd facs ltd when site full; gd mkt in town (15 min walk); Lamego well worth a visit; excel rest adj." ♦ 1 Feb-31 Dec. € 18.70 2017*

⊞ **VILA REAL DE SANTO ANTONIO** *C4* (14km W Rural) *37.18649, -7.55003* **Camping Caliço Park, Sitio do Caliço, 8900-907 Vila Nova de Cacela [281 951195; geral@calico-park.com; www.calico-park.com]** On N side of N125 Vila Real to Faro rd. Sp on main rd & in Vila Nova de Cacela vill, visible fr rd. Lge, hdstg, shd, pt sl, terr, wc; chem disp; shwrs inc; EHU (6A) €2.80; gas; lndry; shop; rest; snacks; bar; playgrnd; pool; beach sand 4km; red long stay; wifi; 80% statics; dogs €1.60; phone; Eng spkn; adv bkg acc; ccard acc; bike hire; CKE/CCI. "Friendly staff; noisy in ssn & rd noise; not suitable for m'vans or tourers in wet conditions - ltd touring pitches & poss diff access; gd NH." € 17.00 2016*

⊞ **VILAMOURA** *B4* (5km N Rural) *37.112371, -8.106409* **Vilamoura Rustic Motorhome Aire, N125 436A, 8100 Consiguinte, Loule [289 149315 or 918 721948 (mob) or 917 428356 (mob); info@vilamoura-rustic-motorhome-aire.com; www.vilamoura-rustic-motorhome-aire.com]** Fr A22 (tollrd) take exit for Quarteria and drive for 3km for exit Quarteira/Portimao. Turn R at exit to join N125 and drive for approx 6km W direct to site. Sm, hdstg, pt shd, wc; chem disp; mv service pnt; shwrs; EHU (6A); lndry; snacks; bar; bbq; beach 5km; twin axles; wifi; TV; dogs €1; bus 200m; Eng spkn; adv bkg acc; games area; bike hire. "Boutique, adults only (18+), MH & c'van Aire; outdoor cinema; Sky Sports; events throughout the year; vg; run by past CAMC memb." € 9.00 2018*

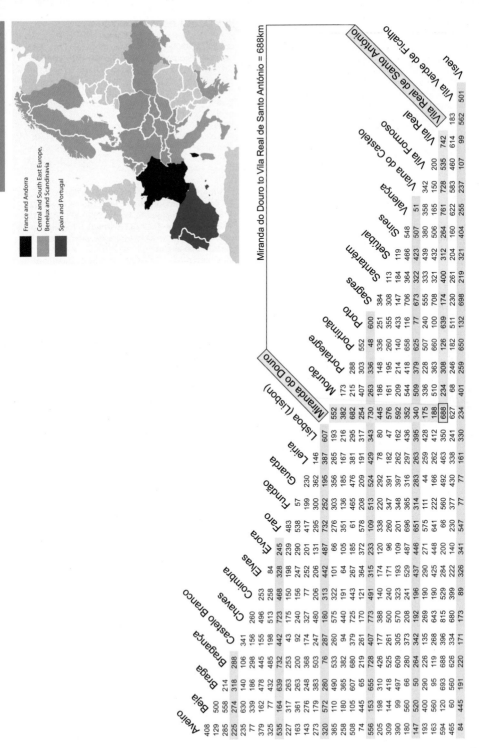

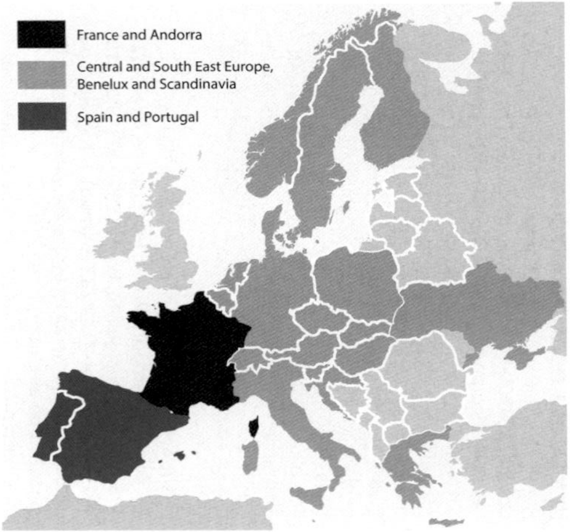

Miranda do Douro to Vila Real de Santo António = 688km

	Aveiro	Beja	Braga	Bragança	Castelo Branco	Chaves	Coimbra	Elvas	Évora	Faro	Fundão	Guarda	Leiria	Lisboa (Lisbon)	Miranda do Douro	Mourão	Portalegre	Portimão	Porto	Sagres	Santarém	Setúbal	Sines	Valença	Viana do Castelo	Vila Formoso	Vila Real	Vila Real de Santo António	Vila Verde de Ficalho
Beja	408																												
Braga	129	500																											
Bragança	285	558	214																										
Castelo Branco	225	274	318	288																									
Chaves	235	630	140	106	341																								
Coimbra	77	339	186	298	156	260																							
Elvas	379	162	478	445	155	496	253																						
Évora	325	77	432	485	198	513	258	84																					
Faro	535	164	639	732	442	723	468	328	245																				
Fundão	227	317	263	253	43	175	150	198	239	483																			
Guarda	163	361	263	200	92	240	156	247	290	538	57																		
Leiria	143	276	248	368	174	327	77	252	201	417	199	230																	
Lisboa (Lisbon)	273	179	383	503	247	480	206	206	131	295	300	362	146																
Miranda do Douro	320	572	280	76	287	180	313	442	487	732	252	195	387	607															
Mourão	365	110	490	533	260	575	322	101	66	276	303	356	265	193	552														
Portalegre	258	180	365	382	94	440	191	64	105	351	136	185	167	216	382	173													
Portimão	508	105	607	680	379	725	443	267	185	61	465	476	381	295	682	215	288												
Porto	74	445	65	219	261	170	121	364	372	578	208	209	191	317	254	407	303	552											
Sagres	556	153	655	728	407	773	491	315	233	109	513	524	429	343	730	263	336	48	600										
Santarém	205	198	310	426	177	388	140	174	120	338	220	292	78	80	445	186	148	336	251	384									
Setúbal	309	144	418	525	261	500	240	171	96	260	347	391	182	47	576	161	195	260	355	308	113								
Sines	390	99	497	609	305	570	323	193	109	201	348	397	262	162	592	209	214	140	433	147	184	119							
Valença	180	560	66	280	373	208	241	529	487	696	365	316	297	436	352	544	418	658	116	706	364	466	548						
Viana do Castelo	147	520	50	264	342	192	196	437	446	651	314	283	263	395	340	509	379	625	77	673	322	423	507	51					
Vila Formoso	193	400	290	226	135	269	190	290	271	575	111	44	259	428	175	336	228	507	240	555	333	439	380	358	342				
Vila Real	163	560	95	119	268	643	190	425	448	641	222	166	262	412	188	510	363	660	100	708	321	432	506	165	150	200			
Vila Real de Santo António	594	120	693	688	396	815	529	284	200	66	560	492	463	350	688	234	308	126	639	174	400	312	264	761	728	535	742		
Vila Verde de Ficalho	465	60	560	626	334	680	399	222	140	230	377	430	338	241	627	68	246	182	511	230	261	204	160	622	583	460	614	183	
Viseu	84	445	191	220	171	173	89	326	341	547	77	77	161	330	234	401	259	650	132	698	219	321	404	255	237	107	99	562	501

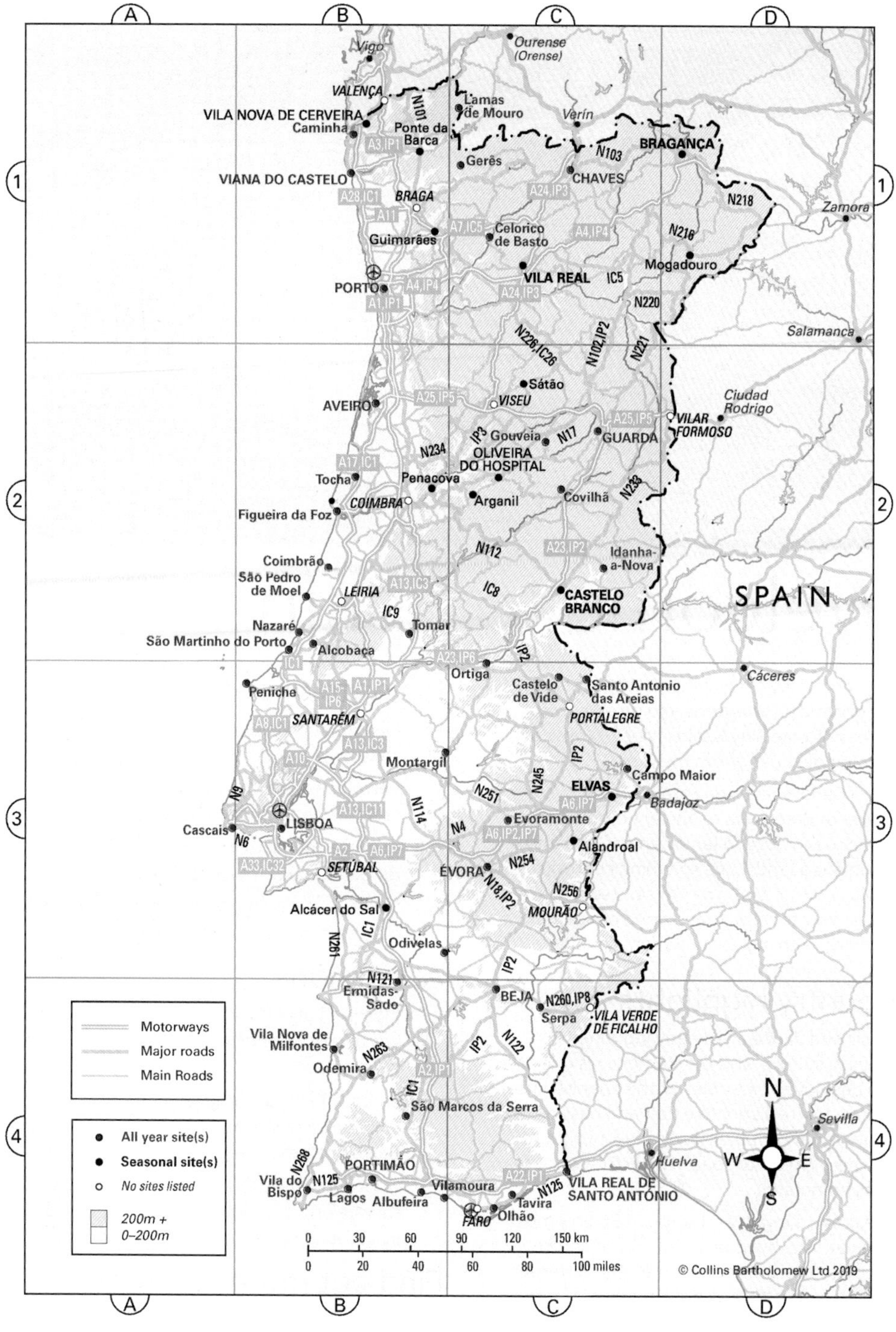
Ourense (Orense)
Vigo
VALENÇA
VILA NOVA DE CERVEIRA
Caminha
Lamas de Mouro
Ponte da Barca
Verín
BRAGANÇA
Gerês
VIANA DO CASTELO
CHAVES
BRAGA
N218
Zamora
Guimarães
Celorico de Basto
N216
Mogadouro
VILA REAL
PORTO
N220
Salamanca
N226,IC26
N102,IP2
N221
Sátão
AVEIRO
VISEU
Ciudad Rodrigo
VILAR FORMOSO
Gouveia
N17
GUARDA
N234
OLIVEIRA DO HOSPITAL
Tocha
Penacova
Arganil
Covilhã
N233
COIMBRA
Figueira da Foz
N112
Idanha-a-Nova
Coimbrão
São Pedro de Moel
LEIRIA
IC8
CASTELO BRANCO
SPAIN
IC9
Nazaré
Tomar
São Martinho do Porto
Alcobaça
Ortiga
Castelo de Vide
Santo Antonio das Areias
Cáceres
Peniche
SANTARÉM
PORTALEGRE
Montargil
Campo Maior
N251
N245
ELVAS
Badajoz
LISBOA
N114
Evoramonte
Cascais
N6
N9
N4
Alandroal
SETÚBAL
ÉVORA
N254
N18,IP2
N256
Alcácer do Sal
MOURÃO
N261
IC1
Odivelas
IP2
N121
Ermidas-Sado
BEJA
N260,IP8
Serpa
VILA VERDE DE FICALHO
Vila Nova de Milfontes
N263
N122
Odemira
São Marcos da Serra
N
W
E
S
Sevilla
N268
PORTIMÃO
Huelva
Vila do Bispo
N125
Lagos
Albufeira
Vilamoura
Tavira
FARO
Olhão
VILA REAL DE SANTO ANTÓNIO
Motorways
Major roads
Main Roads
All year site(s)
Seasonal site(s)
No sites listed
200m +
0–200m
0 30 60 90 120 150 km
0 20 40 60 80 100 miles
© Collins Bartholomew Ltd 2019

Salou

Welcome to Spain

Boasting lively cities, beautiful beaches, a rich history and an energetic and diverse culture, it's easy to see why Spain is one of the most popular destinations in the world.

From Gaudi's Sagrada Familia in the bustling centre of Barcelona to ancient monuments in the rocky, rugged landscape of Andalusia, Spain has a landscape spanning centuries to explore. After a long day of sightseeing, what better way to relax that on one of Spain's many beaches, with a glass of Sangria in hand.

Country highlights

Music and dance are deeply ingrained in Spanish culture, and the flamenco is one of the best loved examples of the Spanish arts. Known for its distinctive flair and passion, this dance is now popular worldwide, but there is nowhere better to soak in a performance than in its homeland.

Tapas and sangria and some of Spain's most popular fare, but there is plenty of choice for those looking to try something different. Orxata is a refreshing drink made of tigernuts, water and sugar, and is served ice-cold in the summer.

Major towns and cities

- Madrid – this vibrant capital is filled with culture.
- Barcelona – a city on the coast, filled with breathtaking architecture.
- Alicante – home to the Castle of Santa Barbara, a huge medieval fortress.
- Valencia – set on the Mediterranean sea, this city has numerous attractions on offer.

Attractions

- Mosque-Cathedral of Córdoba – this site has been a place of worship since the 8th century
- Sagrada Familia – designed by Catalan architect, Antoni Gaudí, this basilica is one of Barcelona's most famous sights.
- Guggenheim Museum, Bilbao – a museum for modern and contemporary art, housed in a ground-breaking 20th century building.

Find out more

www.spain.info
E: infosmile@tourspain.es T: (0)9 13 43 35 00

Country Information

Population (approx): 46.5 million

Capital: Madrid

Area: 506,000 sq km (inc Balearic & Canary Islands)

Bordered by: Andorra, France, Portugal

Terrain: High, rugged central plateau, mountains to north and south

Climate: Temperate climate; hot summers, cold winters in the interior; more moderate summers and cool winters along the northern and eastern coasts; very hot summers and mild/warm winters along the southern coast

Coastline: 4,964km

Highest Point (mainland Spain): Mulhacén (Granada) 3,478m

Languages: Castilian Spanish, Catalan, Galician, Basque

Local Time: GMT or BST + 1, i.e. 1 hour ahead of the UK all year

Currency: Euros divided into 100 cents; £1 = €1.13, €1 = £0.89 (November 2018)

Emergency numbers: Police 092; Fire brigade 080; Ambulance (SAMUR) 061. Operators speak English. Civil Guard 062. All services can be reached on 112.

Public Holidays 2019: Jan 1, 6; Apr 19; May 1; Aug 15; Oct 12; Nov 1; Dec 6 (Constitution Day), 8, 25.

Several other dates are celebrated for fiestas according to region. School summer holidays stretch from mid June to mid September.

Camping and Caravanning

There are more than 1,200 campsites in Spain with something to suit all tastes – from some of the best and biggest holiday parks in Europe, to a wealth of attractive small sites offering a personal, friendly welcome. Most campsites are located near the Mediterranean, especially on the Costa Brava and Costa del Sol, as well as in the Pyrenees and other areas of tourist interest. Campsites are indicated by blue road signs. In general pitch sizes are small at about 80 square metres.

Many popular coastal sites favoured for long winter stays may contain tightly packed pitches with long-term residents putting up large awnings, umbrellas and other structures. Many sites allow pitches to be reserved from year to year, which can result in a tight knit community of visitors who return every year.

If you're planning to stay on sites in the popular coastal areas between late spring and October, or in January and February, it is advisable to arrive early in the afternoon or to book in advance.

Although many sites claim to be open all year, if you're planning a visit out of season, always check first. It is common for many 'all year' sites to open only at weekends during the winter and facilities may be very limited.

Motorhomes

A number of local authorities now provide dedicated or short stay areas for motorhomes called 'Áreas de Servicio'.

For details see the websites www.lapaca.org or www.viajarenautocaravana.com for a list of regions and towns in Spain and Andorra which have at least one of these areas.

It is rare that yours will be the only motorhome staying on such areas, but take sensible precautions and avoid any that are isolated.

Some motorhome service points are situated in motorway service areas. Use these only as a last resort and do not be tempted to park overnight. The risk of a break-in is high.

Recent visitors to tourist areas on Spain's Mediterranean coast report that the parking of motorhomes on public roads and, in some instances, in public parking areas, may be prohibited in an effort to discourage 'wild camping'. Specific areas where visitors have encountered this problem include Alicante, Dénia, Palamós and the Murcian coast. Police are frequently in evidence moving parked motorhomes on and it is understood that a number of owners of motorhomes have been fined for parking on sections of the beach belonging to the local authority.

Cycling

There are around 2,200km of dedicated cycle paths in Spain, many of which follow disused railway tracks. Known as 'Vias Verdes' (Green Ways), they can be found mainly in northern Spain, in Andalucia, around Madrid and inland from the Costa Blanca. For more information see the website www.viasverdes.com or contact the Spanish Tourist Office.

There are cycle lanes in major cities and towns such as Barcelona, Bilbao, Córdoba, Madrid, Seville and Valencia. Madrid alone has over 100km of cycle lanes.

It is compulsory for all cyclists, regardless of age, to wear a safety helmet on all roads outside built-up areas. At night, in tunnels or in bad weather, bicycles must have front and rear lights and reflectors. Cyclists must also wear a reflective waistcoat or jacket while riding at night on roads outside built-up areas (to be visible from a distance of 150 metres) or when visibility is bad.

Strictly speaking, cyclists have right of way when motor vehicles wish to cross their path to turn left or right, but great care should always be taken. Do not proceed unless you are sure that a motorist is giving way.

Transportation of Bicycles

Spanish regulations stipulate that motor cycles or bicycles may be carried on the rear of a vehicle providing the rack to which the motorcycle or bicycle is fastened has been designed for the purpose. Lights, indicators, number plate and any signals made by the driver must not be obscured and the rack should not compromise the carrying vehicle's stability.

An overhanging load, such as bicycles, should not extend beyond the width of the vehicle but may exceed the length of the vehicle by up to 10% (up to 15% in the case of indivisible items). The load must be indicated by a 50cm x 50cm square panel with reflective red and white diagonal stripes. These panels may be purchased in the UK from motorhome or caravan dealers/accessory shops. There is currently no requirement for bicycle racks to be certified or pass a technical inspection.

If you are planning to travel from Spain to Portugal please note that slightly different official regulations apply. These are set out in the Portugal Country Introduction.

Electricity and Gas

The current on campsites should be a minimum of 4 amps but is usually more. Plugs have two round pins. Some campsites do not yet have CEE connections.

Campingaz is widely available in 901 and 907 cylinders. The Cepsa Company sells butane gas cylinders and regulators, which are available in large stores and petrol stations, and the Repsol Company sells butane cylinders at their petrol stations throughout the country.

French and Spanish butane and propane gas cylinders are understood to be widely available in Andorra.

Entry Formalities

Holders of valid British and Irish passports are permitted to stay up to three months without a visa. EU residents planning to stay longer are required to register in person at the Oficina de Extranjeros (Foreigners Office) in their province of residence or at a designated police station. You will be issued with a certificate confirming that the registration obligation has been fulfilled.

Dogs must be kept on a lead in public places and in a car they should be isolated from the driver by means of bars, netting or kept in a transport carrier.

Medical Services

Basic emergency health care is available free from practitioners in the Spanish National Health Service on production of a European Health Insurance Card (EHIC). Some health centres offer both private and state provided health care and you should ensure that staff are aware which service you require. In some parts of the country you may have to travel some distance to attend a surgery or health clinic operating within the state health service. It is probably quicker and more convenient to use a private clinic, but the Spanish health service will not refund any private health care charges.

In an emergency go to the casualty department (urgencias) of any major public hospital. Urgent treatment is free in a public ward on production of an EHIC; for other treatment you will have to pay a proportion of the cost.

Medicines prescribed by health service practitioners can be obtained from a pharmacy (farmacia) and there will be a charge unless you are an EU pensioner. In all major towns there is a 24 hour pharmacy.

Dental treatment is not generally provided under the state system and you will have to pay for treatment.

The Department of Health has two offices in Spain to deal with health care enquiries from British nationals visiting or residing in Spain. These are at the British Consultate offices in Alicante and Madrid, Tel: 965-21 60 22 or 917-14 63 00.

Opening Hours

Banks : Mon-Fri 8.30am/9am-2pm/2.30pm, Sat 9am-1pm (many banks are close Sat during summer).

Museums: Tue-Sat 9am/10am-1pm/2pm & 3pm/4pm-6pm/8pm. Sun 9am/10am-2pm; most close Mon.

Post Offices: Mon-Fri 8.30am-2.30pm & 5pm-8pm/8.30pm, Sat 9am/9.30am-1pm/1.30pm.

Shops: Mon-Sat 9am/10am-1.30pm/2pm & 4pm/4.30pm-8pm/8.30pm; department stores and shopping centres don't close for lunch.

Safety and Security

Street crime exists in many Spanish towns and holiday resorts. Keep all valuable personal items such as cameras or jewellery out of sight. The authorities have stepped up the police presence in tourist areas but nevertheless, you should remain alert at all times (including at airports, train and bus stations, and even in supermarkets and their car parks).

In Madrid particular care should be taken in the Puerto de Sol and surrounding streets, including the Plaza Mayor, Retiro Park and Lavapies, and on the metro. In Barcelona this advice also applies to the Ramblas, Monjuic, Plaza Catalunya, Port Vell and Olympic Port areas. Be wary of approaches by strangers either asking directions or offering help, especially around cash machines or at tills, as they may be trying to distract attention.

A few incidents have been reported of visitors being approached by a bogus police officer asking to inspect wallets for fake euro notes, or to check their identity by keying their credit card PIN into an official looking piece of equipment carried by the officer. If in doubt ask to see a police officer's official identification, refuse to comply with the request and offer instead to go to the nearest police station.

Spanish police have set up an emergency number for holidaymakers with English speaking staff and offering round the clock assistance - call 902 10 2 112. An English speaking operator will take a statement about the incident, translate it into Spanish and fax or email it to the nearest police station. You still have to report in person to a police station if you have an accident, or have been robbed or swindled, and the helpline operator will advise you where to find the nearest one.

Motorists travelling on motorways – particularly those north and south of Barcelona, in the Alicante region, on the M30, M40 and M50 Madrid ring roads and on the A4 and A5 – should be wary of approaches by bogus policemen in plain clothes travelling in unmarked cars. In all traffic related matters police officers will be in uniform. Unmarked vehicles will have a flashing electronic sign in the rear window reading 'Policía' or 'Guardia Civil' and will normally have blue flashing lights incorporated into their headlights, which are activated when the police stop you.

In non-traffic related matters police officers may be in plain clothes but you have the right to ask to see identification. Genuine officers may ask you to show them your documents but would not request that you hand over your bag or wallet. If in any doubt, converse through the car window and telephone the police on 112 or the Guardia Civil on 062 and ask them for confirmation that the registration number of the vehicle corresponds to an official police vehicle.

On the A7 motorway between the La Junquera and Tarragona toll stations be alert for 'highway pirates' who flag down foreign registered and hire cars (the latter have a distinctive number plate), especially those towing caravans. Motorists are sometimes targeted in service areas, followed and subsequently tricked into stopping on the hard shoulder of the motorway. The usual ploy is for the driver or passenger in a passing vehicle, which may be 'official-looking', to suggest by gesture that there is something seriously wrong with a rear wheel or exhaust pipe. If flagged down by other motorists or a motorcyclist in this way, be extremely wary. Within the Barcelona urban area thieves may also employ the 'punctured tyre' tactic at traffic lights.

In instances such as this, the Spanish Tourist Office advises you not to pull over but to wait until you reach a service area or toll station. If you do get out of your car when flagged down take care it is locked while you check outside, even if someone is left inside. Car keys should never be left in the ignition.

Spain shares with the rest of Europe an underlying threat from terrorism. Attacks could be indiscriminate and against civilian targets in public places including tourist areas.

The Basque terrorist organisation, ETA, has been less active in recent years and on 20 October 2011 announced a "definitive cessation of armed activity." However you should always be vigilant and follow the instructions of local police and other authorities.

Coast guards operate a beach flag system to indicate the general safety of beaches for swimming: red – danger / do not enter the water; yellow – take precautions; green – all clear. Coast guards operate on most of the popular beaches, so if in doubt, always ask. During the summer months stinging jellyfish frequent Mediterranean coastal waters.

There is a risk of forest fires during the hottest months and you should avoid camping in areas with limited escape routes. Take care to avoid actions that could cause a fire, e.g. disposal of cigarette ends.

Respect Spanish laws and customs. Parents should be aware that Spanish law defines anyone under the age of 18 as a minor, subject to parental control or adult supervision. Any unaccompanied minor coming to the attention of the local authorities for whatever reason is deemed to be vulnerable under the law and faces being taken into a minors centre for protection until a parent or suitable guardian can be found.

British Embassy & Consulate-General

TORRE ESPACIO, PASEO DE LA CASTELLANA 259D
28046 MADRID
Tel: 917 14 63 00
www.ukinspain.fco.gov.uk/en/

British Consulate-General

AVDA DIAGONAL 477-13, 08036 BARCELONA
Tel: 933 66 02 00

There are also British Consulates in Alicante and Málaga.

Irish Embassy

IRELAND HOUSE, PASEO DE LA CASTELLANA 46-4
28046 MADRID
Tel: 914 36 40 93
www.embassyofireland.es

There are also Irish Honorary Consulates in Alicante, Barcelona, Bilbao, El Ferrol, Málaga and Seville.

Customs Regulations

Under Spanish law the number of cigarettes which may be exported is set at eight hundred. Anything above this amount is regarded as a trade transaction which must be accompanied by the required documentation. Travellers caught with more than 800 cigarettes face seizure of the cigarettes and a large fine.

Documents

Driving Licence

The British EU format pink driving licence is recognised in Spain. Holders of the old style all green driving licence are advised to replace it with a photocard version. Alternatively, the old style licence may be accompanied by an International Driving Permit available from the AA, the RAC or selected Post Offices.

Passport

Visitors must be able to show some form of identity document if requested to do so by the police and you should carry your passport or photocard licence at all times.

Vehicle(s)

When driving in Spain it is compulsory at all times to carry your driving licence, vehicle registration certificate (V5C), insurance certificate and MOT certificate (if applicable). Vehicles imported by a person other than the owner must have a letter of authority from the owner.

Money

All bank branches offer foreign currency exchange, as do many hotels and travel agents.

The major credit cards are widely accepted as a means of payment in shops, restaurants and petrol stations. Smaller retail outlets in non commercial areas may not accept payments by credit card – check before buying. When shopping carry your passport or photocard driving licence if paying with a credit card as you will almost certainly be asked for photographic proof of identity.

Keep a supply of loose change as you could be asked for it frequently in shops and at kiosks.

Motoring in Spain

Drivers should take particular care as driving standards can be erratic, e.g. excessive speed and dangerous overtaking, and the accident rate is higher than in the UK. Pedestrians should take particular care when crossing roads (even at zebra crossings) or walking along unlit roads at night.

Accidents

The Central Traffic Department runs an assistance service for victims of traffic accidents linked to an emergency telephone network along motorways and some roads. Motorists in need of help should ask for 'auxilio en carretera' (road assistance). The special ambulances used are connected by radio to hospitals participating in the scheme.

It is not necessary to call the emergency services in case of light injuries. A European Accident Statement should be completed and signed by both parties and, if conditions allow, photos of the vehicles and the location should be taken. If one of the drivers involved does not want to give his/her details, the other should call the police or Guardia Civil.

Alcohol

The maximum permitted level of alcohol is 50 milligrams in 100 millilitres of blood, i.e. less than in the UK (80 milligrams) and it reduces to 30 milligrams for drivers with less than two years experience, drivers of vehicles with more than 8 passenger seats and for drivers of vehicles over 3,500kg. After a traffic accident all road users involved have to undergo a breath test. Penalties for refusing a test or exceeding the legal limit are severe and may include immobilisation of vehicles, a large fine and suspension of your driving licence. This limit applies to cyclists as well as drivers of private vehicles.

Breakdown Service

The motoring organisation, Real Automóvil Club de España (RACE), operates a breakdown service and assistance may be obtained 24 hours a day by telephoning the national centre in Madrid on 915 93 33 33. After hearing a message in Spanish press the number 1 to access the control room where English is spoken.

RACE's breakdown vehicles are blue and yellow and display the words 'RACE Asistencia' on the sides. This service provides on the spot minor repairs and towing to the nearest garage. Charges vary according to type of vehicle and time of day, but payment for road assistance must be made in cash.

Essential Equipment

Lights

Dipped headlights is now compulsory for all vehicles on all roads at night and in tunnels. Bulbs are more likely to fail with constant use and you are recommended to carry spares.

Dipped headlights must be used at all times on 'special' roads, e.g. temporary routes created at the time of road works such as the hard shoulder, or in a contra-flow lane.

Headlight flashing is only allowed to warn other road users about an accident or a road hazard, or to let the vehicle in front know that you intend to overtake.

Reflective Jacket/Waistcoat

If your vehicle is immobilised on the carriageway outside a built-up area at night, or in poor visibility, you must wear a reflective jacket or waistcoat when getting out of your vehicle. This rule also applies to passengers who may leave the vehicle, for example, to assist with a repair.

Reflectors/Marker Boards for Caravans

Any vehicle or vehicle combination, i.e. car plus caravan over 12 metres in length, must display at the rear of the towed vehicle two aluminium boards. These must have a yellow centre with a red outline, must be reflective and comply with ECE70 standards. These must be positioned between 50cm and 150cm off the ground and must be 500mm x 250mm or 565mm x 200mm in size. Alternatively a single horizontal reflector may be used measuring 1300mm x 250mm or 1130mm x 200mm.

To buy these aluminium marker boards (under Spanish regulations stickers are not acceptable) contact www.hgvdirect.co.uk, tel: 0845 6860008. Contact your local dealer or caravan manufacturer for advice on fitting them to your caravan.

Warning Triangles

All vehicles must carry warning triangles. They should be placed 50 metres behind and in front of broken down vehicles.

Child Restraint System

Children under the age of 12 years old and under the height of 1.35m must use a suitable child restraint system adapted for their size and weight (this does not apply in taxis in urban areas). Children measuring more than 1.35m in height may use an adult seatbelt.

Fuel

Credit cards are accepted at most petrol stations, but you should be prepared to pay cash if necessary in remote areas.

LPG (Autogas) can be purchased from some Repsol filling stations. Details of approximately 33 sales outlets throughout mainland Spain can be found on www.mylpg.eu.

Mountain Passes and Tunnels

Some passes are occasionally blocked in winter following heavy falls of snow. Check locally for information on road conditions.

Parking

Parking regulations vary depending on the area of a city or town, the time of day, the day of the week, and whether the date is odd or even. In many towns parking is permitted on one side of the street for the first half of the month and on the other side for the second half of the month. Signs marked '1-15' or '16-31' indicate these restrictions.

Yellow road markings indicate parking restrictions. Parking should be in the same direction as the traffic flow in one way streets or on the right hand side when there is two way traffic. Illegally parked vehicles may be towed away or clamped but, despite this, you will frequently encounter double and triple parking.

In large cities parking meters have been largely replaced by ticket machines and these are often located in areas known as 'zona azul', i.e. blue zones. The maximum period of parking is usually one and a half hours between 8am and 9pm. In the centre of some towns there is a 'zona O.R.A.' where parking is permitted for up to 90 minutes against tickets bought in tobacconists and other retail outlets.

In many small towns and villages it is advisable to park on the edge of town and walk to the centre, as many towns can be difficult to navigate due to narrow, congested streets.

Madrid

In Madrid, there is a regulated parking zone where parking spaces are shown by blue or green lines (called SER). Parking is limited to 1 or 2 hours in these areas for visitors and can be paid by means of ticket machines of by mobile phone.

Pedestrians

Jaywalking is not permitted. Pedestrians may not cross a road unless a traffic light is at red against the traffic, or a policeman gives permission. Offenders may be fined.

Priority and Overtaking

As a general rule traffic coming from the right has priority at intersections. When entering a main road from a secondary road drivers must give way to traffic from both directions. Traffic already on a roundabout (i.e. from the left) has priority over traffic joining it. Trams and emergency vehicles have priority at all times over other road users and you must not pass trams that are stationary while letting passengers on or off.

Motorists must give way to cyclists on a cycle lane, cycle crossing or other specially designated cycle track. They must also give way to cyclists when turning left or right.

You must use your indicators when overtaking. If a vehicle comes up behind you signalling that it wants to overtake and if the road ahead is clear, you must use your right indicator to acknowledge the situation.

Roads

There are approximately 16,200km of highways and dual carriageways. Roads marked AP (autopista) are generally toll roads and roads marked A (autovía) or N (nacional) are dual carriageways with motorway characteristics – but not necessarily with a central reservation – and are toll-free. In recent years some major national roads have been upgraded to Autovías and, therefore, have two identifying codes or have changed codes, e.g. the N-I from Madrid to Irún near the French border is now known as the A1 or Autovía del Norte. Autovías are often as fast as autopistas and are generally more scenic.

Roads managed by regional or local authorities are prefixed with the various identification letters such as C, CV, GR, L or T.

All national roads and roads of interest to tourists are generally in good condition, are well signposted, and driving is normally straightforward. Hills often tend to be longer and steeper than in parts of the UK and some of the coastal roads are very winding, so traffic flows at the speed of the slowest lorry.

As far as accidents are concerned the N340 coast road, especially between Málaga and Fuengirola, is notorious, as are the Madrid ring roads, and special vigilance is necessary.

Road humps are making an appearance on Spanish roads and recent visitors report that they may be high, putting low stabilisers at risk.

Andorra

The main road to Barcelona from Andorra is the C14/C1412/N141b via Ponts and Calaf. It has a good surface and avoids any high passes. The N260 along the south side of Andorra via Puigcerda and La Seo de Urgel also has a good surface.

Road Signs and Markings

Road signs conform to international standards. Lines and markings are white. Place names may appear both in standard (Castilian) Spanish and in a local form, e.g. Gerona/Girona, San Sebastián/Donostia, Jávea/Xàbio, and road atlases and maps usually show both.

You may encounter the following signs:

Spanish	English Translation
Carretera de peaje	Toll road
Ceda el paso	Give way
Cuidado	Caution
Curva peligrosa	Dangerous bend
Despacio	Slow
Desviación	Detour
Dirección única	One-way street
Embotellamiento	Traffic jam
Estacionamiento prohibido	No parking
Estrechamiento	Narrow lane
Gravillas	Loose chippings/gravel
Inicio	Start
Obras	Roadworks
Paso prohibido	No entry
Peligro	Danger
Prioridad	Right of way
Salida	Exit
Todas direcciones	All directions

Many non motorway roads have a continuous white line on the near (verge) side of the carriageway. Any narrow lane between this line and the side of the carriageway is intended primarily for pedestrians and cyclists and not for use as a hard shoulder.

A continuous line also indicates 'no stopping' even if it is possible to park entirely off the road and it should be treated as a double white line and not crossed except in a serious emergency. If your vehicle breaks down on a road where there is a continuous white line along the verge, it should not be left unattended as this is illegal and an on the spot fine may be levied.

Many road junctions have a continuous white centre line along the main road. This line must not be crossed to execute a left turn, despite the lack of any other 'no left turn' signs. If necessary, drive on to a 'cambio de sentido' (change of direction) sign to turn.

Traffic police are keen to enforce both the above regulations.

Watch out for traffic lights which may be mounted high above the road and hard to spot. The international three colour traffic light system is used in Spain. Green, amber and red arrows are used on traffic lights at some intersections.

Speed Limits

	Open Road (km/h)	Motorway (km/h)
Car Solo	90-100	120
Car towing caravan/trailer	70-80	80
Motorhome under 3500kg	80-90	100
Motorhome 3500-7500kg	80	90

In built-up areas speed is limited to 50km/h (31mph) except where signs indicate a lower limit. Reduce your speed to 20km/h (13mph) in residential areas. On motorways and dual carriageways in built-up areas, speed is limited to 80km/h (50mph) except where indicated by signs.

Outside built-up areas motorhomes under 3500kg are limited to 100km/h (62mph) and those over 3500kg are limited to 90km/h (56 mph) on motorways and dual carriageways. On other main roads motorhomes under 3500kg are limited to 80-90km/h (50-56mph) and those over 3500kg are limited to 80km/h (50mph)

It is prohibited to own, transport or use radar detectors.

Foreign Registered Vehicles

When a radar camera detects a foreign registered vehicle exceeding the speed limit, a picture of the vehicle and its number plate will be sent not only to the relevant traffic department, but also to the nearest Guardia Civil mobile patrol. The patrol will then stop the speeding vehicle and impose an on the spot fine which non-residents must pay immediately, otherwise the vehicle will be confiscated until the fine is paid.

This is to prevent offenders flouting the law and avoiding paying their fines, as pursuing them is proving costly and complicated for the Spanish authorities.

Towing

Motorhomes are prohibited from towing a car unless the car is on a special towing trailer with all four wheels of the car off the ground.

Any towing combination in excess of 10 metres in length must keep at least 50 metres from the vehicle in front except in built-up areas, on roads where overtaking is prohibited, or where there are several lanes in the same direction.

Traffic Jams

Roads around the large cities such as Madrid, Barcelona, Zaragoza, Valencia and Seville are extremely busy on Friday afternoons when residents leave for the mountains or coast, and again on Sunday evenings when they return. The coastal roads along the Costa Brava and the Costa Dorada may also be congested. The coast road south of Torrevieja is frequently heavily congested as a result of extensive holiday home construction.

Summer holidays extend from mid June to mid September and the busiest periods are the last weekend in July, the first weekend in August and the period around the Assumption holiday in mid August.

Traffic jams occur on the busy AP7 from the French border to Barcelona during the peak summer holiday period. An alternative route now exists from Malgrat de Mar along the coast to Barcelona using the C32 where tolls are lower than on the AP7.

The Autovía de la Cataluña Central (C25) provides a rapid east-west link between Gerona and Lleida via Vic, Manresa and Tàrrega. There is fast access from Madrid to La Coruña in the far north-west via the A6/AP6.

Information on road conditions, traffic delays, etc can be found on http://infocar.dgt.es/etraffic.

Violation of Traffic Regulations

The police are empowered to impose on the spot fines. Visiting motorists must pay immediately otherwise a vehicle will be confiscated until the fine is paid. An official receipt should be obtained. An appeal may be made within 15 days and there are instructions on the back of the receipt in English. RACE can provide legal advice - tel: 900 100 901.

Motorways

The Spanish motorway system has been subject to considerable expansion in recent years with more motorways under construction or planned. The main sections are along the Mediterranean coast, across the north of the country and around Madrid. Tolls are charged on most autopistas but many sections are toll-free, as are autovias. Exits on autopistas are numbered consecutively from Madrid. Exits on autovias are numbered according to the kilometre point from Madrid.

Many different companies operate within the motorway network, each setting their own tolls which may vary according to the time of day and classification of vehicles.

Avoid signposted 'Via T' lanes showing a circular sign with a white capital T on a blue background where toll collection is by electronic device only. Square 'Via T' signs are displayed above mixed lanes where other forms of payment are also accepted.

Rest areas with parking facilities, petrol stations and restaurants or cafés are strategically placed and are well signposted. Emergency telephones are located at 2km intervals.

Motorway signs near Barcelona are confusing. To avoid the city traffic when heading south, follow signs for Barcelona, but once signs for Tarragona appear follow these and ignore Barcelona signs.

Touring

A fixed price menu or 'menú del dia' invariably offers good value. Service is generally included in restaurant bills but a tip of approximately €1 per person up to 10% of the bill is appropriate if you have received good service. Smoking is not allowed in indoor public places, including bars, restaurants and cafés.

Spain is one of the world's top wine producers, enjoying a great variety of high quality wines of which cava, rioja and sherry are probably the best known. Local beer is low in alcohol content and is generally drunk as an aperitif to accompany tapas.

Perhaps due to the benign climate and long hours of sunshine, Spaniards tend to get up later and stay out later at night than their European neighbours. Out of the main tourist season and in 'non-touristy' areas it may be difficult to find a restaurant open in the evening before 9pm.

Taking a siesta is still common practice, although it is now usual for businesses to stay open during the traditional siesta hours.

Spain's many different cultural and regional influences are responsible for the variety and originality of fiestas held each year. Over 200 have been classified as 'of interest to tourists' while others have gained international fame, such as La Tomatina mass tomato throwing battle held each year in August in Buñol near Valencia. A full list of fiestas can be obtained from the Spanish Tourist Office, www.spain.info/uk or from provincial tourist offices. In addition, every year each town celebrates its local Saint's Day which is always a very happy and colourful occasion.

The Madrid Card, valid for one, two or three days, gives free use of public transport, free entry to various attractions and museums, including the Prado, Reina Sofia and Thyssen-Bornemisza collection, as well as free tours and discounts at restaurants and shows. You can buy the card from www.madridcard.com or by visiting the City Tourist Office in Plaza Mayor, or on Madrid Visión tour buses. Similar generous discounts can be obtained with the Barcelona Card, valid from two to five days, which can be purchased from tourist offices or online at www.barcelonacard.org. Other tourist cards are available in Burgos, Córdoba, Seville and Zaragoza.

The region of Valencia and the Balearic Islands are prone to severe storms and torrential rainfall between September and November and are probably best avoided at that time. Monitor national and regional weather on www.wmo.int.

Gibraltar

For information on Gibraltar contact:

GIBRALTAR GOVERNMENT TOURIST OFFICE
150 STRAND, LONDON WC2R 1JA
Tel: 020 7836 0777
www.gibraltar.gov.gi
info@gibraltar.gov.uk

There are no campsites on the Rock, the nearest being at San Roque and La Línea de la Concepción in Spain. The only direct access to Gibraltar from Spain is via the border at La Línea which is open 24 hours a day. You may cross on foot and it is also possible to take cars or motorhomes to Gibraltar.

A valid British passport is required for all British nationals visiting Gibraltar. Nationals of other countries should check entry requirements with the Gibraltar Government Tourist Office.

There is currently no charge for visitors to enter Gibraltar but Spanish border checks can cause delays and you should be prepared for long queues. As roads in the town are extremely narrow and bridges low, it is advisable to park on the outskirts. Visitors advise against leaving vehicles on the Spanish side of the border owing to the high risk of break-ins.

An attraction to taking the car into Gibraltar includes English style supermarkets and a wide variety of competitively priced goods free of VAT. The currency is sterling and British notes and coins circulate alongside Gibraltar pounds and pence, but note that Gibraltar notes and coins are not accepted in the UK. Scottish and Northern Irish notes are not generally accepted in Gibraltar. Euros are accepted but the exchange rate may not be favourable.

Disabled visitors to Gibraltar may obtain a temporary parking permit from the police station on production of evidence confirming their disability. This permit allows parking for up to two hours (between 8am and 10pm) in parking places reserved for disabled people.

Violence or street crime is rare but there have been reports of people walking from La Línea to Gibraltar at night being attacked and robbed.

If you need emergency medical attention while on a visit to Gibraltar, treatment at primary healthcare centres is free to UK passport holders under the local medical scheme.
Non UK nationals need a European Health Insurance Card (EHIC). You are not eligible for free treatment if you go to Gibraltar specifically to be treated for a condition which arose elsewhere, e.g in Spain.

Public Transport

Madrid boasts an extensive and efficient public transport network including a metro system, suburban railways and bus routes. You can purchase a pack of ten tickets which offer better value than single tickets. In addition, tourist travel passes for use on all public transport are available from metro stations, tourist offices and travel agencies and are valid for one to seven days – you will need to present your passport when buying them. Single tickets must be validated before travel. For more information see www.ctm-madrid.es.

Metro systems also operate in Barcelona, Bilbao, Seville and Valencia and a few cities operate tram services including La Coruña, Valencia, Barcelona and Bilbao.
The Valencia service links Alicante, Benidorm and Dénia.

Various operators run year round ferry services from Spain to North Africa, the Balearic Islands and the Canary Islands. All enquiries should be made through their UK agent:

SOUTHERN FERRIES
22 SUSSEX STREET, LONDON SW1V 4RW
www.southernferries.co.uk
mail@southernferries.co.uk

⊞ **AGUILAR DE CAMPOO** *1B4* (3km W Rural) *42.78694, -4.30222* **Monte Royal Camping, Carretera Virgen del Llano 34800 Aguilar de Campóo (Palencia) [979-18 10 07; info@campingmonteroyal.com; www.camping monteroyal.com]** App site fr S on N611 fr Palencia. At Aguilar de Campóo turn W at S end of rv bdge at S end of town. Site on L in 3km; sp at edge of reservoir. Fr N take 3rd exit fr rndabt on N611. Do not tow thro town. Lge, mkd, shd, pt sl, wc; chem disp; fam bthrm; shwrs; EHU (6A) inc; gas; lndry; shop nr; rest; bar; playgrnd; TV; 20% statics; dogs; phone; ccard acc; horseriding; watersports; fishing; CKE/CCI. "Useful, peaceful NH 2 hrs fr Santander; ltd/basic facs LS & poss stretched high ssn - in need of maintenance (2010); barking dogs poss problem; friendly staff; gd walking, cycling & birdwatching in National Park; unrel opening dates LS." ♦ € 22.00 2014*

⊞ **AGUILAS** *4G1* (2km SW Coastal) *37.3925, -1.61111* **Camping Bellavista, Ctra de Vera, Km 3, 30880 Águilas (Murcia) [968-44 91 51; info@campingbellavista.com; www.campingbellavista.com]** Site on N332 Águilas to Vera rd on R at top of sh, steep hill, 100m after R turn to El Cocon. Well mkd by flags. Fr S by N332 on L 400m after fuel stn, after v sharp corner. Sm, hdg, hdstg, pt shd, pt sl, wc; chem disp; mv service pnt; shwrs inc; EHU (10A) €5.20 or metered; gas; lndry; shop; rest nr; snacks; bbq; playgrnd; pool; beach sand 300m; red long stay; wifi; 10% statics; dogs €2.50; Eng spkn; adv bkg acc; quiet; ccard acc; bike hire; CKE/CCI. "Gd autumn/winter stay; clean, tidy site with excel facs; ltd pitches for lge o'fits; helpful owner; fine views; rd noise at 1 end; excel town & vg beaches; v secure site." € 33.60 2014*

⊞ **AGUILAS** *4G1* (11km NW Rural) *37.45387, -1.64488* **Camping La Quinta Bella, Finca El Charcon 31, 30889 Aguilas [968 43 85 35; info@quintabella.com; www.quintabella.com]** Fr AP7 exit at junc 878 onto RM11. Foll sp to Los Arejos. In 2km turn R foll sp to site. Med, mkd, hdstg, wc (htd); chem disp; mv service pnt; shwrs inc; EHU; gas; lndry; rest nr; bar; bbq; pool; twin axles; wifi; 30% statics; dogs; Eng spkn; adv bkg acc; quiet; ccard acc; CKE/CCI. "V lge pitches; ideal for carnival & Easter parades; v friendly English owners; boules; m'homes should have other transport; excel." ♦ € 20.00 2016*

⊞ **AINSA** *3B2* (3km N Rural) *42.43555, 0.13583* **Camping Peña Montañesa, Ctra Ainsa-Bielsa, Km 2.3, 22360 Labuerda (Huesca) [974-50 00 32; fax 974-50 09 91; info@penamontanesa.com; www.penamontanesa.com]** E fr Huesca on N240 for approx 50km, turn N onto N123 just after Barbastro twd Ainsa. In 8km turn onto A138 N for Ainsa & Bielsa. Or fr Bielsa Tunnel to A138 S to Ainsa & Bielsa, site sp. NB: Bielsa Tunnel sometimes clsd bet Oct & Easter due to weather. Lge, mkd, shd, wc (htd); chem disp; mv service pnt; fam bthrm; shwrs inc; EHU (6A) inc; gas; lndry; shop; rest; snacks; bar; bbq (elec, gas); playgrnd; pool (covrd, htd); sw nr; entmnt; wifi; TV; 30% statics; dogs €4.25; phone; adv bkg acc; quiet; ccard acc; tennis; games rm; bike hire; sauna; fishing; canoeing; games area; horseriding; CKE/CCI. "Situated by fast-flowing rv; v friendly staff; Eng spkn; gd, clean san facs; pitching poss diff due trees; no o'fits over 10m; nr beautiful medieval town of Ainsa & Ordesa National Park; excel." ♦ € 39.00 2014*

See advertisement

AINSA *3B2* (1km E Rural) *42.41944, 0.15111* **Camping Ainsa, Ctra Ainsa-Campo, 22330 Ainsa (Huesca) [974-50 02 60; fax 974-50 03 61; info@campingainsa.com; www.campingainsa.com]** Fr Ainsa take N260 E dir Pueyo de Araguás, cross rv bdge, site sp L in 200m. Foll lane to site. Sm, pt shd, terr, wc; fam bthrm; shwrs inc; EHU €4.75; gas; lndry; shop nr; rest; snacks; bar; playgrnd; pool; wifi; TV; 50% statics; dogs €2.20; phone; ccard acc; games rm; CKE/CCI. "Pleasant, welcoming, well-maintained site; fine view of old city & some pitches mountain views; vg san facs; not suitable lge o'fits; gd pool." Holy Week-30 Oct. € 24.00 2014*

SPAIN

ALBARRACIN *3D1* (2km E Rural) *40.41228, -1.42788* **Camp Municipal Ciudad de Albarracín, Camino de Gea s/n, 44100 Albarracín (Teruel) [978-71 01 97 or 657-49 84 33 (mob); campingalbarracin5@hotmail.com; www.campingalbarracin.com]** Fr Teruel take A1512 to Albarracín. Go thro vill, foll camping sps. Med, pt shd, pt sl, wc; chem disp; fam bthrm; shwrs inc; EHU (16A) €3.85; gas; lndry; shop; snacks; bar; bbq; playgrnd; wifi; 10% statics; dogs; phone; adv bkg acc; quiet; ccard acc; CKE/CCI. "Gd site; immac san facs; pool adj in ssn; narr pitches poss diff for lge o'fits; sports cent adj; gd touring base & gd walking fr site; rec; friendly staff; gd rest & shop; views striking, gd Sierras." 4 Mar-27 Nov. € 20.00 2016*

ALBERCA, LA *1D3* (6km N Rural) *40.52181, -6.13762* **Camping Sierra de Francia, Ctra Salamanca-La Alberca, Km 73, 37623 El Caserito (Salamanca) [923-45 40 81; fax 923-45 40 01; info@campingsierradefrancia.com; www.campingsierradefrancia.com]** Fr Cuidad Rodrigo take C515. Turn R at El Cabaco, site on L in approx 2km. Med, mkd, hdg, shd, wc; mv service pnt; shwrs; EHU (3-6A) €3.75; gas; lndry; shop; rest; bar; bbq; playgrnd; pool; paddling pool; wifi; 10% statics; dogs free; quiet; ccard acc; horseriding; bike hire. "Conv 'living history' vill of La Alberca & Monasterio San Juan de la Peña; excel views." ♦ 1 Apr-15 Sep. € 23.50 2017*

⊞ **ALCALA DE LOS GAZULES** *2H3* (4km E Rural) *36.46403, -5.66482* **Camping Los Gazules, Ctra de Patrite, Km 4, 11180 Alcalá de los Gazules (Cádiz) [956-42 04 86; fax 956-42 03 88; camping@losgazules.e.telefonica.net; www.campinglosgazules.com]** Fr N exit A381 at 1st junc to Alcalá, proceed thro town to 1st rndabt & turn L onto A375/A2304 dir Ubriqu, site sp strt ahead in 1km onto CA2115 dir Patrite on v sharp L. Fr S exit A381 at 1st sp for Acalá. At rndabt turn R onto A375/A2304 dir Ubrique. Then as above. Med, mkd, pt shd, pt sl, wc; chem disp; mv service pnt; shwrs inc; EHU (10A) €5.25 (poss rev pol); lndry; shop; rest; bar; playgrnd; pool; red long stay; TV; 90% statics; dogs €2; phone; adv bkg acc; bike hire; CKE/CCI. "Well-maintained, upgraded site; take care canopy frames; sm pitches & tight turns & kerbs on site; friendly, helpful staff; ltd facs LS; ltd touring pitches; attractive town with v narr rds, leave car in park at bottom & walk; gd walking, birdwatching." € 35.00 2013*

⊞ **ALCANAR** *3D2* (4km NE Coastal) *40.53986, 0.52071* **Camping Estanyet, Paseo del Marjal s/n 43870, Les Cases d'Alcanar (Catalonia) [977-73 72 68; http://fr.campings.com/camping-estanyet-les-cases-dalcanar]** Leave Alcanar to N340 at lLes Cases D'Alcanar foll camping signs. Fr AP7 Junc 41 fr N or Junc 43 fr S. Med, hdg, hdstg, pt shd, wc (htd); chem disp; fam bthrm; shwrs inc; EHU (10A) €6.50; lndry; shop; rest; snacks; bar; bbq (charcoal); playgrnd; pool; beach shgl adj; 10% statics; dogs €3.50; phone; adv bkg acc; ccard acc; CKE/CCI. "Gd; friendly owners; but ltd facs esp LS." € 41.00 2013*

⊞ **ALCARAZ** *4F1* (6km E Rural) *38.67301, -2.40462* **Camping Sierra de Peñascosa, Ctra Peñascosa-Bogarra, Km 1, 02313 Peñascosa (Albacete) [967-38 25 21; informacion@campingpenascosa.com; www.campingsierrapenascosa.com]** Fr N322 turn E bet km posts 279 & 280 sp Peñascosa. In vill foll site sp for 1km beyond vill. Gravel access track & narr ent. Sm, mkd, hdstg, shd, terr, wc; chem disp; shwrs; EHU (6A) €4; gas; lndry; shop; rest; snacks; bar; playgrnd; pool; dogs €2; quiet; ccard acc; bike hire; CKE/CCI. "Not suitable lge o'fits or faint-hearted; pitches sm, uneven & amongst trees - care needed when manoeuvring; open w/end in winter; historical sites nr." ♦ € 21.00 2016*

⊞ **ALCOSSEBRE** *3D2* (3km NE Rural/Coastal) *40.27016, 0.30646* **Camping Ribamar, Partida Ribamar s/n, 12579 Alcossebre (Castellón) [964 76 1601; fax 964 99 4082; info@campingribamar.com; www.campingribamar.com]** Exit AP7 at junc 44 into N340 & foll sp to Alcossebre, then dir Sierra de Irta & Las Fuentes. Turn in dir of sea & foll sp to site in 2km - pt rough rd. Med, mkd, hdstg, hdg, pt shd, pt sl, terr, wc; chem disp; mv service pnt; fam bthrm; shwrs inc; EHU (6/10A) €4.50-6.50 (metered for long stay); gas; lndry; shop; rest; bar; playgrnd; pool; paddling pool; beach sand 100m; red long stay; entmnt; wifi; TV; 25% statics; dogs €1.90; Eng spkn; adv bkg acc; quiet; tennis; games area; games rm; CKE/CCI. "Excel, refurbished tidy site in 'natural park'; warm welcome; realistic pitch sizes; variable prices; excel san facs; beware caterpillars in spring - poss dangerous for dogs." ♦ € 47.00 2017*

See advertisement opposite

⊞ **ALCOSSEBRE** *3D2* (3km S Coastal) *40.22138, 0.26888* **Camping Playa Tropicana, Camino de l'atall s/n, 12579 Alcossebre (Castellón) [964-41 24 63; fax 964 41 28 05; info@playatropicana.com; www.playatropicana.com]** Fr AP7 exit junc 44 onto N340 dir Barcelona. After 3km at km 1018 turn on CV142 twd Alcossebre. Just bef ent town turn R sp 'Platjes Capicorb', turn R at beach in 2.5km, site on R. Lge, mkd, pt shd, terr, serviced pitches; wc (htd); chem disp; fam bthrm; shwrs inc; EHU (10A) €4.50; gas; lndry; shop; rest; snacks; bar; playgrnd; pool; beach sand adj; red long stay; entmnt; wifi; TV; 10% statics; Eng spkn; adv bkg rec; quiet; ccard acc; bike hire; car wash; watersports; games area; kayak hire. "Excel facs & security; beauty salon; superb well-run site; vg LS; ACSI acc; poss rallies Jan-Apr; management v helpful; poss flooding after heavy rain; pitch access poss diff lge o'fits due narr access rds & high kerbs; cinema rm; take fly swat!" ♦ € 56.00 2013*

⊞ **ALGAMITAS** *2G3* (3km SW Rural) *37.01934, -5.17440* **Camping El Peñon, Ctra Algámitas-Pruna, Km 3, 41661 Algámitas (Sevilla) [955-85 53 00; info@algamitasaventura.es; www.algamitas aventura.es]** Fr A92 turn S at junc 41 (Arahal) to Morón de la Frontera on A8125. Fr Morón take A406 & A363 dir Pruna. At 1st rndabt at ent Pruna turn L onto SE9225 to Algámitas. Site on L in approx 10km - steep app rd. Sm, hdg, hdstg, mkd, pt shd, wc; chem disp; mv service pnt; shwrs inc; EHU (16A) €3.32; gas; lndry; shop nr; rest; bar; bbq; playgrnd; pool; 50% statics; dogs; adv bkg acc; quiet; ccard acc; site clsd 13-24 Nov; games area; CKE/CCI. "Conv Seville, Ronda & white vills; walking, hiking & horseriding fr site; excel rest; excel, clean san facs; vg site - worth effort to find." ♦ € 15.00 2016*

⊞ **ALHAURIN DE LA TORRE** *2H4* (4km W Rural) *36.65174, -4.61064* **Camping Malaga Monte Parc, 29130 Alhaurín de la Torre (Málaga) [951-29 60 28; info@malagamonteparc.com; www.malagamonte parc.com]** W fr Málaga on AP7 or N340 take exit for Churriana/Alhaurín de la Torre. Thro Alhuarín de la Torre take A404 W sp Alhaurín el Grande, site on R, sp. Sm, hdstg, hdg, mkd, shd, pt sl, wc (htd); chem disp; shwrs inc; EHU (6A) inc; lndry; shop nr; rest; snacks; bar; bbq; pool; wifi; TV; 10% statics; dogs €1.70; bus 200m; Eng spkn; adv bkg acc; quiet; ccard acc; golf nr; CKE/CCI. "Vg site; well-appointed, clean san facs; friendly Welsh owner; all facs open all year; sm pitches; gd position to tour Costa Del Sol." ♦ € 25.00 2014*

⊞ **ALICANTE** *4F2* (13km NE Coastal) *38.41333, -0.40556* **Camping Bon Sol, Camino Real de Villajoyosa 35, Playa Muchavista, 03560 El Campello (Alicante) [965-94 13 83; info@campingbonsol.es; www.campingbonsol.es]** Exit AP7 N of Alicante at junc 67 onto N332 sp Playa San Juan; on reaching coast rd turn N twds El Campello; site sp. Sm, hdstg, mkd, pt shd, serviced pitches; wc; chem disp; shwrs; EHU (10A); lndry; shop; rest; bar; beach sand; red long stay; wifi; adv bkg acc; ccard acc; CKE/CCI. "Diff ent for long o'fits; helpful friendly staff; poss cold shwrs; vg." ♦ € 20.00 2016*

⊞ **ALLARIZ** *1B2* (2km W Rural) *42.18443, -7.81811* **Camping Os Invernadeiros, Ctra Allariz-Celanova, Km 3, 32660 Allariz (Ourense) [988-44 20 06; reatur@allariz.com]** Well sp off N525 Orense-Xinzo rd & fr A52. Steep descent to site off rd OU300. Height limit 2.85m adj recep - use gate to R. Sm, pt shd, wc; shwrs inc; EHU €4.50; gas; lndry; shop; snacks; bar; playgrnd; red long stay; 10% statics; dogs €2; Eng spkn; quiet; ccard acc; horseriding; bike hire; CKE/CCI. "Vg; steep slope into site, level exit is avail; pool 1.5km; site combined with horseriding stable; rv walk adj." € 23.00 2017*

⊞ **ALMERIA** *4G1* (23km SE Rural/Coastal) *36.80187, -2.24471* **Camping Cabo de Gata, Ctra Cabo de Gata s/n, Cortijo Ferrón, 04150 Cabo de Gata (Almería) [950-16 04 43; fax 950-91 68 21; info@campingcabo degata.com; www.campingcabodegata.com]** Exit m'way N340/344/E15 junc 460 or 467 sp Cabo de Gata, foll sp to site. Lge, mkd, hdg, shd, wc; chem disp; fam bthrm; shwrs inc; EHU (6-16A) €4.60; gas; lndry; shop; rest; snacks; bar; bbq; playgrnd; pool; beach sand 900m; red long stay; wifi; TV; 10% statics; dogs €2.80; bus 1km; Eng spkn; adv bkg acc; quiet; ccard acc; tennis; bike hire; games area; excursions; games rm; CKE/CCI. "M'vans with solar panels/TV aerials take care sun shades; gd cycling, birdwatching esp flamingoes; diving cent; popular at w/end; isolated, dry area of Spain with many interesting features; warm winters; excel site." ♦ € 41.00 2014*

⊞ **ALMERIA** *4G1* (18km W Coastal) *36.79738, -2.59128* **Camping Roquetas, Ctra Los Parrales s/n, 04740 Roquetas de Mar (Almería) [950-34 90 85 or 950-34 38 09; fax 950-34 25 25; info@camping roquetas.com; www.campingroquetas.com]** Fr A7 take exit 429; ahead at rndabt A391 sp Roquetas. Turn L at rndabt sp camping & foll sp to site. V lge, pt shd, wc; chem disp; mv service pnt; shwrs inc; EHU (10-16A) €6.35-7.45; gas; lndry; shop; snacks; bar; pool; paddling pool; beach shgl 400m; red long stay; wifi; TV; 10% statics; dogs €2.25; phone; bus 1km; Eng spkn; adv bkg rec; quiet; ccard acc; tennis; CKE/CCI. "Double-size pitches in winter; helpful staff; gd clean facs; tidy site but poss dusty; artificial shd; many long term visitors in winter; gd dedicated cycle path along sea front." ♦ € 24.00 2017*

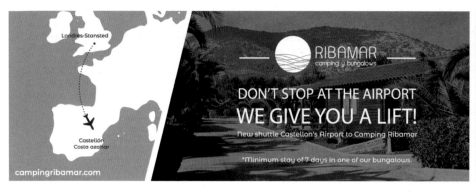

SPAIN

⊞ **ALMERIA** *4G1* (4km W Coastal) *36.82560, -2.51685* **Camping La Garrofa, Ctra N340a, Km 435.4, 04002 Almería [950-23 57 70; info@lagarrofa.com; www.lagarrofa.com]** Site sp on coast rd bet Almería & Aguadulce. Med, mkd, shd, pt sl, wc; chem disp; mv service pnt; shwrs inc; EHU (6-10A) €4.30-4.90; gas; lndry; shop; rest; snacks; bar; playgrnd; beach shgl adj; red long stay; wifi; 10% statics; dogs €2.40; phone; bus adj; quiet; sep car park; games area; CKE/CCI. "V pleasant site adj eucalyptus grove; helpful staff; modern, clean facs; sm pitches, not rec lge o'fits; vg." ♦ € 20.50 2014*

⊞ **ALQUEZAR** *3B2* (2km SW Rural) *42.16454, 0.01527* **Camping Alquézar, Ctra. Barbastro s/n. 22145 Alquézar, Huesca [34 974 318 300; fax 34 974 318 434; camping@alquezar.com; campingalquezar.com]** A22 Huesca - Barbastro. Take N240 W of Barbastro, dir Barbastro. Then A1232 & A1233 to Alquezar, foll camping sp. Med, hdstg, mkd, pt shd, terr, wc; chem disp; mv service pnt; fam bthrm; shwrs; EHU (10A) inc; lndry; shop; rest; snacks; bar; bbq; playgrnd; twin axles; wifi; dogs; Eng spkn; adv bkg acc; quiet; ccard acc. "Tricky access; hard 2km walk up hill to old town; interesting town & selection of walking rtes; gd." € 28.00 2016*

⊞ **AMETLLA DE MAR, L'** *3C2* (3km S Coastal) *40.86493, 0.77860* **Camping L'Ametlla Village Platja, Paratge de Santes Creus s/n, 43860 L'Ametlla de Mar (Tarragona) [977-26 77 84; fax 977-26 78 68; info@campingametlla.com; www.campingametlla.com]** Exit AP7 junc 39, fork R as soon as cross m'way. Foll site sp for 3km - 1 v sharp, steep bend. Lge, hdstg, mkd, hdg, pt shd, terr, wc (htd); chem disp; mv service pnt; fam bthrm; shwrs inc; EHU (5-10A) inc; gas; lndry; shop; rest; snacks; bar; bbq; playgrnd; pool; paddling pool; beach shgl 400m; red long stay; entmnt; wifi; TV; 10% statics; dogs free; phone; Eng spkn; adv bkg acc; ccard acc; games rm; bike hire; fitness rm; games area; CKE/CCI. "Conv Port Aventura & Ebro Delta National Park; excel site & facs; can cycle into vill with mkt; diving cent; delightful site." ♦ € 41.50 2016*

AMPOLLA, L' *3C2* (2km SW Coastal) *40.79940, 0.69974* **Camping L'Ampolla Playa, Playa Arenal s/n, 43895 L'Ampolla [977-46 05 35; reservas@campingampolla.es; reservas@campingampolla.es]** Exit AP7 at junc 39A twds S on N340 to km 1098. Turn L onto TV3401 sp L'Ampolla. At rndabt after1 km go L twds L'Ampolla and at next rndabt take 1st exit alongside campsite to ent. Med, hdstg, hdg, mkd, shd, wc; chem disp; mv service pnt; shwrs; EHU (5-10A); lndry; shop; rest; bar; bbq; playgrnd; beach 50m; twin axles; wifi; dogs; train/bus 1km; Eng spkn; adv bkg acc; games area; CCI. "Kite surfing; natural park of Ebro Delta; cycling; historic ctrs; vg." ♦ 4 Mar-1 Nov. € 34.00 2016*

⊞ **ARANDA DE DUERO** *1C4* (3km N Rural) *41.70138, -3.68666* **Camping Costajan, Ctra A1/E5, Km 164-165, 09400 Aranda de Duero (Burgos) [947-50 20 70; fax 947-51 13 54; campingcostajan@camping-costajan.com]** Sp on A1/E5 Madrid-Burgos rd, N'bound exit km 164 Aranda Norte, S'bound exit km 165 & foll sp to Aranda & site 500m on R. Med, shd, pt sl, wc (htd); chem disp; mv service pnt; shwrs inc; EHU (10A) €5 (poss rev pol &/or no earth); gas; lndry; shop; rest; snacks; bar; bbq; playgrnd; pool; wifi; 10% statics; dogs €2; phone; Eng spkn; adv bkg acc; quiet; tennis; games area; CKE/CCI. "Lovely site under pine trees; poultry farm adj; diff pitch access due trees & sandy soil; friendly, helpful owner; site poss clsd LS - phone ahead to check; many facs clsd LS & gate clsd o'night until 0800; recep poss open evening only LS; poss cold/tepid shwrs LS; gd winter NH; vg site for dogs." € 27.00 2014*

⊞ **ARANJUEZ** *1D4* (3km NE Rural) *40.04222, -3.59944* **Camping International Aranjuez, Calle Soto del Rebollo s/n, 28300 Aranjuez (Madrid) [918-91 13 95; fax 918-92 04 06; info@campingaranjuez.com; www.campingaranjuez.com]** Fr N (Madrid) turn off A4 exit 37 onto M305. After ent town turn L bef rv, after petrol stn on R. Take L lane & watch for site sp on L, also mkd M305 Madrid. Site in 500m on R. (If missed cont around cobbled rndabt & back twd Madrid.) Fr S turn off A4 for Aranjuez & foll Palacio Real sp. Join M305 & foll sp for Madrid & camping site. Site on Rv Tajo. Warning: rd surface rolls, take it slowly on app to site & ent gate tight. Lge, hdg, mkd, unshd, pt sl, serviced pitches; wc (htd); chem disp; mv service pnt; fam bthrm; shwrs inc; EHU (16A) €4 (poss no earth, rev pol); gas; lndry; shop; rest; snacks; bar; playgrnd; pool; paddling pool; red long stay; entmnt; wifi; 10% statics; dogs free; phone; quiet; ccard acc; rv fishing; bike hire; games area; canoe hire; CKE/CCI. "Well-maintained site; gd san facs; rest vg value; some lge pitches - access poss diff due trees; hypmkt 3km; some uneven pitches - care req when pitching; pleasant town - World Heritage site; conv Madrid by train; excel site; free train to Royal Palace each morning; shop & rest clsd Tuesdays." ♦ € 35.00 2016*

See advertisement opposite

ARBIZU *3B1* (2km S Rural) *42.89860, -2.03444* **Camping Arbizu eko, NA 7100 km 5, 31839 Arbizu [848-47 09 22; info@campingarbizu.com; www.campingarbizu.com]** Fr A10 Irurtzun to Altsasu exit 17 onto NA-7100. Site on R in 1km. Med, wc; chem disp; fam bthrm; shwrs; EHU (16A); lndry; shop; rest; snacks; bar; bbq; playgrnd; paddling pool; beach sand; twin axles; entmnt; wifi; TV; dogs; Eng spkn; adv bkg acc; quiet; ccard acc; fishing; games rm. "Stunning views of mountains fr site; excel, clean shwr block; v helpful, friendly staff; lots to do in area; gd size pitches; best site." 7 Jan-23 Dec. € 28.00 2015*

SPAIN

ARENAS, LAS *1A4* (1km E Rural) *43.29973, -4.80321* **Camping Naranjo de Bulnes, Ctra Cangas de Onís-Panes, Km 32.5, 33554 Arenas de Cabrales (Asturias) [985-84 65 78; info@campingnaranjodebulnes.com; www.campingnaranjodebulnes.com]** Fr Unquera on N634, take N621 S to Panes, AS114 23km to Las Arenas. Site E of vill of Las Arenas de Cabrales, both sides of rd. V lge, mkd, pt shd, pt sl, terr, wc; chem disp; fam bthrm; shwrs inc; EHU (10A) €3.50 (poss rev pol); gas; lndry; shop; rest; snacks; bar; playgrnd; wifi; TV; bus 100m; ccard acc. "Beautifully-situated site by rv; delightful vill; attractive, rustic-style, clean san facs; wcs up steps; poss poor security; conv Picos de Europa; mountain-climbing school; excursions; walking; excel cheese festival last Sun in Aug; excel rest, bars in vill; lovely." 2 Apr-9 Oct. € 32.00 2017*

"There aren't many sites open at this time of year"

If you're travelling outside peak season remember to call ahead to check site opening dates – even if the entry says 'open all year'.

ARIJA *1A4* (1km N Rural) *43.00064, -3.94492* **Camping Playa de Arija, Avda Gran Via, 09570 Arija (Burgos) [942-77 33 00; fax 942-77 32 72; info@campingplayadearija.com; www.campingplayadearija.com]** Fr W on A67 at Reinosa along S side of Embalse del Ebro. Go thro Arija & take 1st L after x-ing bdge. Go under rlwy bdge, site well sp on peninsula N of vill on lakeside. Or fr E on N623 turn W onto BU642 to Arija & turn R to peninsula & site. NB Rd fr W under repair 2009 & in poor condition. Lge, shd, wc; chem disp; mv service pnt; fam bthrm; shwrs inc; EHU (5A) €3; lndry; shop; rest; bar; bbq; playgrnd; sw; 10% statics; dogs; phone; bus 1km; quiet; watersports; games area; CKE/CCI. "Gd new site; gd birdwatching; LS phone ahead for site opening times." Easter-15 Sep. € 14.00 2013*

⊞ **ARNES** *3C2* (1km NE Rural) *40.91860, 0.26780* **Camping Els Ports, Ctra Tortosa T330, Km 2, 43597 Arnes (Tarragona) [977-43 55 60; info@camping-elsports.com; www.camping-elsports.com]** Exit AP7 at junc Tortosa onto C12 sp Gandesa. Turn W onto T333 at El Pinell de Brai, then T330 to site. Med, pt shd, wc (htd); shwrs inc; EHU (6A) €5.20; lndry; rest; bar; pool; paddling pool; entmnt; TV; 10% statics; dogs; phone; bus 1km; quiet; ccard acc; bike hire; games area; horseriding 3km. "Nr nature reserve & many sports activities; excel walking/mountain cycling; basic san fac; rock pegs req; nice site; great views; elc kept tripping fuses so low amps." ♦ € 27.00 2016*

AURITZ *3A1* (3km SW Rural) *42.97302, -1.35248* **Camping Urrobi, Ctra Pamplona-Valcarlos, Km 42, 31694 Espinal-Aurizberri (Navarra) [948-76 02 00; info@campingurrobi.com; www.campingurrobi.com]** NE fr Pamplona on N135 twd Valcarlos thro Erro; 1.5km after Auritzberri (Espinal) turn R on N172. Site on N172 at junc with N135 opp picnic area. Med, pt shd, wc; chem disp; mv service pnt; shwrs inc; EHU (5A) €4.90; gas; lndry; shop; rest; snacks; bar; bbq; playgrnd; pool; wifi; 20% statics; phone; Eng spkn; quiet; ccard acc; horseriding; tennis; bike hire; CKE/CCI. "Excel, busy site & facs; solar htd water - hot water to shwrs only; walks in surrounding hills; ltd facs LS; poss youth groups; overprices; poor wifi; clean; gd local walks." ♦ 1 Apr-1 Nov. € 29.00 2017*

⊞ **AYERBE** *3B2* (1km NE Rural) *42.28211, -0.67536* **Camping La Banera, Ctra Loarre Km.1, 22800 Ayerbe (Huesca) [974-38 02 42 or 659-16 15 90 (mob); labanera@gmail.com; www.campinglabanera.com]** Take A132 NW fr Huesca dir Pamplona. Turn R at 1st x-rds at ent to Ayerbe sp Loarre & Camping. Site 1km on R on A1206. Med, mkd, pt shd, terr, wc; chem disp; fam bthrm; shwrs inc; EHU (6A) €2.60; gas; lndry; shop nr; rest; snacks; bar; cooking facs; red long stay; TV; dogs €2; Eng spkn; adv bkg acc; quiet; ccard acc; CKE/CCI. "Friendly, pleasant, well-maintained, peaceful, family-run site; facs clean; pitches poss muddy after rain; helpful owners; wonderful views; close to Loarre Castle; care req by high o'fits as many low trees; area famous for Griffon Vultures which inhabit tall cliffs nr Loarre." ♦ € 17.50 2014*

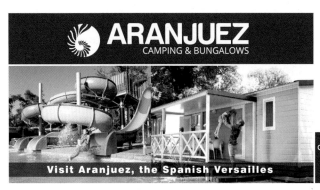

⊞ **AYERBE** *3B2* (10km NE Rural) *42.31989, -0.61848* **Camping Castillo de Loarre, Ctra del Castillo s/n, 22809 Loarre (Huesca) [974-38 27 22; info@campingloarre.com; www.campingloarre.com]** NW on A132 fr Huesca, turn R at ent to Ayerbe to Loare sp Castillo de Loarre. Pass 1st site on R (La Banera) & foll sp to castle past Loarre vill on L; site on L. App rd steep & twisting. Med, pt shd, pt sl, wc; chem disp; shwrs inc; EHU (6A) €4.50; gas; lndry; shop; rest; snacks; bar; playgrnd; pool; 10% statics; dogs; phone; Eng spkn; quiet; ccard acc; bike hire; CKE/CCI. "Elevated site in almond grove; superb scenery & views, esp fr pitches on far L of site; excel birdwatching - many vultures/eagles; site open w/end in winter; busy high ssn & w/ends; pitching poss diff lge o'fits due low trees; worth the journey; site clsd Feb; v pleasant, well maint site; san facs clean but dated." ♦ € 16.00 2015*

BAIONA *1B2* (5km NE Urban/Coastal) *42.13861, -8.80916* **Camping Playa América, Ctra Vigo-Baiona, Km 9.250,Aptdo. Correos 3105 - 36350 Nigrán (Pontevedra) [986-36 54 03 or 986-36 71 61; fax 986-36 54 04; oficina@campingplayaamerica.com; www.campingplayaamerica.com]** Sp on rd PO552 fr all dirs (Vigo/Baiona) nr beach. Med, mkd, pt shd, wc; chem disp; mv service pnt; fam bthrm; shwrs inc; EHU (6A) €5; gas; lndry; shop; rest; snacks; bar; bbq; playgrnd; pool; paddling pool; beach sand 300m; 60% statics; dogs; bus 500m; Eng spkn; adv bkg acc; bike hire; CKE/CCI. "Friendly staff; pleasant, wooded site; gd." ♦ 16 Mar-15 Oct. € 25.50 2013*

⊞ **BAIONA** *1B2* (1km E Coastal) *42.11416, -8.82611* **Camping Bayona Playa, Ctra Vigo-Baiona, Km 19, Sabarís, 36393 Baiona (Pontevedra) [986-35 00 35; fax 986-35 29 52; campingbayona@campingbayona.com; www.campingbayona.com]** Fr Vigo on PO552 sp Baiona. Or fr A57 exit Baiona & foll sp Vigo & site sp. Lge, mkd, pt shd, wc; chem disp; mv service pnt; shwrs inc; EHU (3A) €4.80; gas; lndry; shop; rest; snacks; bar; playgrnd; pool; beach sand adj; red long stay; 50% statics; dogs; phone; adv bkg req; quiet; waterslide; CKE/CCI. "Area of outstanding natural beauty with sea on 3 sides; well-organised site; excel, clean san facs; avoid access w/end as v busy; ltd facs LS; tight access to sm pitches high ssn; gd cycle track to town; replica of ship 'La Pinta' in harbour." ♦ € 33.00 2017*

BAIONA *1B2* (8km SW Coastal) *42.08642, -8.89129* **Camping Mougás (Naturist), As Mariñas 20B, Ctra Baiona-A Guarda, Km 156, 36309 Mougás (Pontevedra) [986-38 50 11; fax 986-36 16 20; info@campingmuino.com]** Fr Baiona take coastal rd PO552 S; site sp. Med, mkd, pt shd, wc; chem disp; mv service pnt; fam bthrm; shwrs; EHU €4.65; lndry; shop; rest; snacks; bar; bbq; playgrnd; pool; entmnt; wifi; 80% statics; phone; Eng spkn; quiet; ccard acc; fishing; games area; tennis; CKE/CCI. "Excel staff; lovely site on rocky coast; gd for watching sunsets; gd NH; o'looks beach; vg." ♦ 18 Mar-27 Mar & 15 May-15 Sep. € 29.00 2016*

⊞ **BALAGUER** *3C2* (8km N Rural) *41.86030, 0.83250* **Camping La Noguera, Partida de la Solana s/n, 25615 Sant Llorenç de Montgai (Lleida) [973-42 03 34; fax 973-42 02 12; info@campinglanoguera.com; www.campinglanoguera.com]** Fr Lleida, take N11 ring rd & exit at km 467 onto C13 NE dir Andorra & Balaguer. Head for Balaguer town cent, cross rv & turn R onto LV9047 dir Gerb. Site on L in 8km thro Gerb. App fr Camarasa not rec. Lge, mkd, hdstg, pt shd, terr, wc; chem disp; mv service pnt; fam bthrm; shwrs inc; EHU (6A) €5.15; gas; lndry; shop; rest; snacks; bar; bbq; playgrnd; pool; red long stay; TV; 80% statics; dogs €3.50; phone; Eng spkn; adv bkg acc; quiet; ccard acc; games area; CKE/CCI. "Next to lake & nature reserve; gd cycling; poss diff lge o'fits; friendly warden; gd facs." ♦ € 35.50 2017*

⊞ **BALERMA** *2H4* (1km S Coastal) *36.72202, -2.87838* **Camping Mar Azul, Ctra de Guardias Viejas, S/N 04712 Balerma [950-93 76 37; info@campingbalerma.com; www.campingbalerma.com]** Exit junc 403 off A7/E15 for Balerma. Site in 6 km. Lge, mkd, hdg, hdstg, shd, wc; chem disp; mv service pnt; fam bthrm; shwrs; EHU (16A) €0.35 per kw; lndry; shop; snacks; bar; bbq; playgrnd; pool; paddling pool; beach 100m; red long stay; twin axles; wifi; TV; 4% statics; dogs (€2.60); bus 1km; Eng spkn; adv bkg acc; games area. "Lge car park; excel." ♦ € 28.60 2016*

⊞ **BANYOLES** *3B3* (2km W Rural) *42.12071, 2.74690* **Camping Caravaning El Llac, Ctra Circumvallació de l'Estany s/n, 17834 Porqueres (Gerona) [972-57 03 05; info@campingllac.com; www.campingllac.com]** Exit AP7 junc 6 to Banyoles. Go strt thro town (do not use by-pass) & exit town at end of lake in 1.6km. Use R-hand layby to turn L sp Porqueres. Site on R in 2.5km. Lge, mkd, pt shd, wc (htd); chem disp; shwrs; EHU €4.60; lndry; shop; snacks; bar; pool; sw; red long stay; wifi; 80% statics; dogs €2.30; bus 1km; site clsd mid-Dec to mid-Jan. "Immac, ltd facs LS & stretched high ssn; sm pitches bet trees; pleasant walk around lake to town; site muddy when wet." ♦ € 30.00 2016*

⊞ **BEAS DE GRANADA** *2G4* (1km N Rural) *37.22416, -3.48805* **Camping Alto de Viñuelas, Ctra de Beas de Granada s/n, 18184 Beas de Granada (Granada) [958-54 60 23; fax 958-54 53 57; info@campingaltodevinuelas.com; www.campingaltodevinuelas.com]** E fr Granada on A92, exit junc 256 & foll sp to Beas de Granada. Site well sp on L in 1.5km. Sm, mkd, pt shd, terr, wc (htd); chem disp; mv service pnt; shwrs inc; EHU (5A) €3.50; lndry; shop; rest; snacks; bar; bbq; playgrnd; pool; red long stay; wifi; 10% statics; dogs; bus to Granada at gate; Eng spkn; CKE/CCI. "In beautiful area; views fr all pitches; 4X4 trip to adj natural park; gd; conv for night halt." € 26.00 2014*

⊞ **BECERREA** *1B2* (16km E Rural) *42.83315, -7.06112* **Camping Os Ancares, Ctra NV1, Liber, 27664 Mosteiro-Cervantes (Lugo) [982-36 45 56]** Fr A6 exit Becerreá S onto LU722 sp Navia de Suarna. After 10km in Liber turn R onto LU723 sp Doiras, site in 7km just beyond Mosteiro hamlet; site sp. Site ent steep & narr - diff lge o'fits & lge m'vans. Med, shd, terr, wc; shwrs inc; EHU (6A) €3; gas; lndry; rest; snacks; bar; playgrnd; pool; 10% statics; dogs €1; quiet; fishing; horseriding; CKE/CCI. "Isolated, scenic site; gd rest & san facs; ltd facs LS; low trees some pitches; gd walking; friendly owner; 17km fr nearest town." € 21.00 2015*

⊞ **BEJAR** *1D3* (15km SW Rural) *40.28560, -5.88182* **Camping Las Cañadas, Ctra N630, Km 432, 10750 Baños de Montemayor (Cáceres) [927-48 11 26; fax 927-48 13 14; info@campinglascanadas.com; www.campinglascanadas.com]** Fr S turn off A630 m'way at 437km stone to Heruns then take old N630 twd Béjar. Site at 432km stone, behind 'Hervas Peil' (leather goods shop). Fr N exit A66 junc 427 thro Baños for 3km to site at km432 on R. Lge, mkd, shd, pt sl, wc (htd); chem disp; mv service pnt; fam bthrm; shwrs inc; EHU (5A) €4; gas; lndry; shop; rest; snacks; bar; playgrnd; pool; paddling pool; red long stay; TV; 60% statics; dogs; Eng spkn; quiet; ccard acc; fishing; bike hire; games area; tennis; CKE/CCI. "Gd san facs but poss cold shwrs; high vehicles take care o'hanging trees; gd walking country; NH/sh stay." ♦ € 25.00 2016*

BENABARRE *3B2* (1km N Urban) *42.1103, 0.4811* **Camping Benabarre, 22580 Benabarre (Huesca) [974-54 35 72; fax 974-54 34 32; aytobenabarre@aragon.es]** Fr N230 S, turn L after 2nd camping sp over bdge & into vill. Ignore brown camping sp (pt of riding cent). Med, hdstg, pt shd, wc; shwrs inc; EHU (10A) inc; shop nr; bar; pool; phone; bus 600m; quiet; tennis. "Excel, friendly, simple site; gd facs; gd value for money; v quiet LS; warden calls 1700; mkt on Fri; lovely vill with excel chocolate shop; conv Graus & mountains - a real find." 1 Apr-30 Sep. € 14.50 2012*

⊞ **BENAJARAFE** *2H4* (3km E Coastal) *36.71962, -4.16467* **Camping Valle Niza Playa, Ctra N340, km 264,1 ES-29792 Benajarafe [952-51 31 81; info@campingvalleniza.es; www.campingvalleniza.es]** On N340 (old coast rd), bet Torre Del Mar and Benajarafe. Lge, mkd, hdg, hdstg, pt shd, wc; chem disp; mv service pnt; shwrs; EHU (10-16A); lndry; shop; rest; snacks; bar; bbq; playgrnd; pool (htd); beach 50 mtrs; red long stay; twin axles; entmnt; wifi; TV; 20% statics; phone; bus 100m; Eng spkn; adv bkg acc; bike hire; games rm. "Gd site; gymnasium; free yoga twice a week." ♦ € 33.00 2017*

⊞ **BENICARLO** *3D2* (3km NE Urban/Coastal) *40.42611, 0.43777* **Camping La Alegría del Mar, Ctra N340, Km 1046, Calle Playa Norte, 12580 Benicarló (Castellón) [964-47 08 71; info@campingalegria.com; www.campingalegria.com]** Sp off main N340 app Benicarló. Take slip rd mkd Service, go under underpass, turn R on exit & cont twd town, then turn at camp sp by Peugeot dealers. Sm, hdg, hdstg, mkd, pt shd, wc (htd); chem disp; mv service pnt; fam bthrm; shwrs; EHU (16A) €4.90; gas; lndry; shop; rest; snacks; bar; bbq; playgrnd; pool; paddling pool; beach adj; red long stay; twin axles; entmnt; wifi; 40% statics; dogs; phone; bus 800m; Eng spkn; quiet; ccard acc; games rm; CKE/CCI. "Friendly British owners; access to pitches variable, poss diff in ssn; vg, clean san facs; Xmas & New Year packages; phone ahead to reserve pitch; excel; well run site." € 30.00 2016*

⊞ **BENICASSIM** *3D2* (1km E Coastal) *40.05709, 0.07429* **Camping Bonterra Park, Avda de Barcelona 47, 12560 Benicàssim (Castellón) [964 30 00 07; fax 964 10 06 69; info@bonterrapark.com; www.bonterrapark.com]** Fr N exit AP7 junc 45 onto N340 dir Benicàssim. In approx 7km turn R to Benicàssim/Centro Urba; strt ahead to traff lts, then turn L, site on L 500m after going under rlwy bdge. Lge, mkd, hdstg, shd, pt sl, serviced pitches; wc (htd); chem disp; mv service pnt; fam bthrm; shwrs inc; EHU (6-10) inc; gas; lndry; shop; rest; snacks; bar; bbq; playgrnd; pool (covrd, htd); paddling pool; beach sand 300m; red long stay; entmnt; wifi; TV; 15% statics; dogs €2.24 (not acc Jul/Aug); phone; train; Eng spkn; adv bkg acc; ccard acc; gym; games rm; tennis; bike hire; games area; CKE/CCI. "Fabulous, excel site in gd location; excel cycle tracks & public trans; lovely beach; reasonable sized pitches; no o'fits over 10m; well-kept & well-run; clean modern san facs; access to some pitches poss diff due to trees; sun shades some pitches; winter festival 3rd wk Jan; Harley Davidson rallies Jan & Sep, check in adv; highly rec; flat rd to town; excel facs; sep car park; ACSI card acc; site organises trips out; gd rest; excel." ♦ € 60.00 2018*

⊞ **BENICASSIM** *3D2* (5km NW Coastal) *40.05908, 0.08515* **Camping Azahar, Ptda Villaroig s/n, 12560 Benicàssim (Castellón) [964-30 35 51 or 964-30 31 96; campingazahar.benicasim@gmail.com; www.campingazahar.es]** Fr AP7 junc 45 take N340 twd València; in 5km L at top of hill (do not turn R to go-karting); foll sp. Turn R under rlwy bdge opp Hotel Voramar. Lge, mkd, unshd, pt sl, terr, wc (htd); chem disp; mv service pnt; fam bthrm; shwrs inc; EHU (4-6A) €2.90 (long leads poss req); gas; lndry; rest; snacks; bar; playgrnd; pool; beach sand 300m; red long stay; 25% statics; dogs €4.07; phone; bus adj; Eng spkn; adv bkg acc; ccard acc; bike hire; CKE/CCI. "Popular site, esp in winter; access poss diff for m'vans & lge o'fits; poss uneven pitches; organised events; tennis at hotel; gd walking & cycling; gd touring base." ♦ € 38.00 2017*

SPAIN

⊞ **BENIDORM** *4F2* (6km N Coastal) *38.56926, -0.09328* **Camping Almafrá, Partida de Cabut 25, 03503 Benidorm (Alicante) [965-88 90 75; info@campingalmafra.es; www.campingalmafra.es]** Exit AP7/E15 junc 65 onto N332 N. Foll sp Alfaz del Pi, site sp. V lge, mkd, hdstg, hdg, pt shd, wc (htd); chem disp; mv service pnt; fam bthrm; shwrs inc; EHU (16A); lndry; shop; rest; snacks; bar; bbq; playgrnd; pool (covrd, htd); paddling pool; twin axles; red long stay; entmnt; wifi; TV (pitch); 20% statics; dogs; bus; Eng spkn; adv bkg acc; games rm; jacuzzi; sauna; gym; games area; tennis; CKE/CCI. "Tennis, Alfaz del Pi a sh walk away; private san facs avail; wellness/fitness cent; reg bus into Benidorm; excel." ♦ € 15.00 2016*

⊞ **BENIDORM** *4F2* (2km NE Urban) *38.54833, -0.09851* **Camping El Raco, Avda Dr Severo Ochoa, 19 Racó de Loix, 03503 Benidorm (Alicante) [965-86 85 52; fax 965-86 85 44; info@campingraco.com; www.campingraco.com]** Turn off A7 m'way at junc 65 then L onto A332; take turning sp Benidorm Levante Beach; ignore others; L at 1st traff lghts; strt on at next traff lts, El Raco 1km on R. Lge, mkd, hdstg, hdg, pt shd, pt sl, serviced pitches; wc; chem disp; fam bthrm; shwrs inc; EHU (10A) metered; gas; lndry; shop; rest; snacks; bar; bbq; playgrnd; pool (covrd, htd); beach 1.5km; wifi; TV; 30% statics; dogs €1.15; bus; Eng spkn; quiet; games area; CKE/CCI. "Excel site; popular winter long stay but strictly applied rules about leaving c'van unoccupied; EHU metered for long stay; friendly helpful staff; two pin adaptor needed for elec conn." ♦ € 34.00 2014*

⊞ **BENIDORM** *4F2* (4km NE Urban) *38.55564, -0.09754* **Camping Villamar, Ctra del Albir, Km 0.300, 03503 Benidorm (Alicante) [966-81 12 55; fax 966-81 35 40; camping@ampingvillamar.com; www.campingvillamar.com]** Exit AP7 junc 65. Down hill twd town, turn L at traff lts into Ctra Valenciana, turn R where 2 petrol stns either side of rd, site on L. V lge, mkd, pt shd, terr, serviced pitches; wc; chem disp; shwrs; EHU (16A) €3.50; gas; lndry; shop; rest; bar; playgrnd; pool (covrd, htd); beach sand 2km; red long stay; entmnt; TV (pitch); phone; adv bkg acc; quiet; games rm. "Excel site, esp winter; gd security; v welcoming; gd walking area; excel food; spotless san facs; great value; excel food." ♦ € 28.00 2016*

"That's changed – Should I let the Club know?"

If you find something on site that's different from the site entry, fill in a report and let us know. See camc.com/europereport.

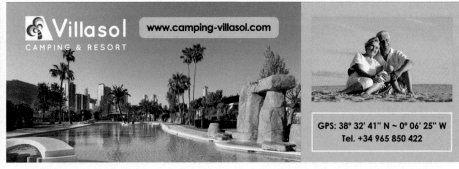

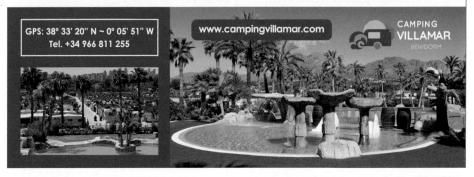

⊞ **BENIDORM** *4F2* (1km E Coastal) *38.5449, -0.10696* **Camping Villasol, Avda Bernat de Sarriá 13, 03500 Benidorm (Alicante) [965-85 04 22; fax 966-80 64 20; info@camping-villasol.com; www.camping-villasol.com]** Leave AP7 at junc 65 onto N332 dir Alicante; take exit into Benidorm sp Levante. Turn L at traff lts just past Camping Titus, then in 200m R at lts into Avda Albir. Site on R in 1km. Care - dip at ent, poss grounding. V lge, hdstg, mkd, shd, wc (htd); chem disp; fam bthrm; shwrs inc; EHU (5A) €4.28; lndry; shop; rest; snacks; bar; playgrnd; pool (covrd, htd); beach sand 300m; red long stay; wifi; TV (pitch); 5% statics; phone; Eng spkn; adv bkg acc; quiet; ccard acc; games area. "Excel, well-kept site espec in winter; medical service; currency exchange; some sm pitches; friendly staff." ♦ € 32.00 2013*

See advertisement opposite

BENQUERENCIA *2E3* (2km SW Rural) *39.29626, -06.10109* **Camping Las Grullas (Naturist), Camino Valdefuentes 4, 10185 Benquerencia [34 634 264 504; info@lasgrullas.es; www.campinglasgrullas.es]** Take EX-206 to Miajadas. After 30km take 2nd turning to Benquerencia (close to Valdefuentes vill). Site on R after 2.5 km. Sm, hdg, mkd, pt shd, wc; chem disp; shwrs; EHU (4-6A) inc; pool (htd); twin axles; wifi; Eng spkn; adv bkg acc; quiet. "Over 16's only; close historic town; no childrens facs; v helpful owners; gd walks fr site; vg." 1 Apr-16 Oct. € 25.00 2016*

BIELSA *3B2* (8km W Rural) *42.65176, 0.14076* **Camping Pineta, Ctra del Parador, Km 7, 22350 Bielsa (Huesca) [974-50 10 89; fax 974-50 11 84; info@campingpineta.com; www.campingpineta.com]** Fr A138 in Beilsa turn W & foll sp for Parador Monte Perdido & Valle de Pineta. Site on L after 8km (ignore previous campsite off rd). Lge, pt shd, pt sl, terr, wc; chem disp; mv service pnt; fam bthrm; shwrs inc; EHU (6A) €5 (poss rev pol); gas; lndry; shop; rest; snacks; bar; bbq; playgrnd; pool; 10% statics; dogs €2.50; phone; ccard acc; bike hire; games area; CKE/CCI. "Well-maintained site; clean facs; glorious location in National Park." 25 Mar-2 Oct. € 27.00 2016*

⊞ **BIESCAS** *3B2* (3km SE Rural) *42.61944, -0.30416* **Camping Gavín, Ctra N260, Km 502.5, 22639 Gavín (Huesca) [974-48 50 90 or 659-47 95 51; fax 974-48 50 17; info@campinggavin.com; www.campinggavin.com]** Take N330/A23/E7 N fr Huesca twd Sabiñánigo then N260 twd Biescas & Valle de Tena. Ignore all sp to Biescas on N260 until R turn at g'ge. Drive over blue bdge & foll sp Gavin & site. Site is at km 502.5 fr Huesca, bet Biescas & Gavín. Lge, mkd, pt shd, terr, wc (htd); chem disp; mv service pnt; fam bthrm; shwrs inc; EHU (10A) inc; gas; lndry; shop; rest; snacks; bar; playgrnd; pool; wifi; TV; dogs inc; phone; bus 1km; Eng spkn; adv bkg acc; quiet; tennis; CKE/CCI. "Wonderful, scenic site nr Ordesa National Park; poss diff access to pitches for lge o'fits & m'vans; superb htd san facs; bike hire in National Park; immac kept, excel, superb site; pitches with views, gd for walking." ♦ € 37.00 2016*

⊞ **BILBAO** *1A4* (18km N Coastal) *43.38916, -2.98444* **Camping Sopelana, Ctra Bilbao-Plentzia, Km 18, Playa Atxabiribil 30, 48600 Sopelana (Vizcaya) [946-76 19 81 or 649-11 57 51; fax 944-21 50 10; recepcion@campingsopelana.com; www.campingsopelana.com]** In Bilbao cross rv by m'way bdge sp to airport, foll 637/634 N twd & Plentzia. Cont thro Sopelana & foll sp on L. Med, hdg, sl, wc; own san req; chem disp; mv service pnt; fam bthrm; shwrs; EHU (10A) €4.50; gas; lndry; shop; rest; snacks; bar; playgrnd; pool; beach sand 200m; red long stay; 70% statics; Eng spkn; adv bkg req; CKE/CCI. "Poss strong sea winds; ltd space for tourers; pitches sm, poss flooded after heavy rain & poss diff due narr, steep site rds; ltd facs LS; helpful manager; site used by local workers; poss clsd LS - phone ahead to check; gd NH/sh stay only." ♦ € 36.00 2016*

"I like to fill in the reports as I travel from site to site"

You'll find report forms at the back of this guide, or you can fill them in online at camc.com/europereport.

BILBAO *1A4* (3km W Urban) *43.25960, -2.96351* **Motorhome Parking Bilbao, Monte Kobeta 31, 48001 Bilbao [688 809 399 or 944 655 789; kobetamendi@suspertu.net; Bilbao.net]** A8 m'way, Balmaseda exit, dir Altamira - Alto de Kastrexana. Med, unshd, terr, wc; mv service pnt; EHU inc; wifi; bus adj; ccard acc. "MH's only; on hill o'looking Bilbao; ehu & water to each pitch; v conv for visiting city by bus (every 1/2 hr); max stay 2 days; clsd 1st week of Jul; vg; full security." 17 Mar-9 Jan. € 15.00 2017*

⊞ **BLANES** *3C3* (1km S Coastal) *41.65933, 2.77000* **Camping Blanes, Avda Vila de Madrid 33, 17300 Blanes (Gerona) [972-33 15 91; fax 972-33 70 63; info@campingblanes.com; www.campingblanes.com]** Fr N on AP7/E15 exit junc 9 onto NII dir Barcelona & foll sp Blanes. Fr S to end of C32, then NII dir Blanes. On app Blanes, foll camping sps & Playa S'Abanell - all campsites are sp at rndabts; all sites along same rd. Site adj Hotel Blau-Mar. Lge, mkd, shd, wc; chem disp; mv service pnt; shwrs inc; EHU (5A) inc; gas; lndry; shop; snacks; bar; playgrnd; pool; beach sand; entmnt; wifi; dogs; phone; bus; Eng spkn; quiet; ccard acc; solarium; bike hire; watersports; games rm. "Excel site, espec LS; helpful owner; narr site rds; easy walk to town cent; dir access to beach; trains to Barcelona & Gerona." ♦ € 37.00 2017*

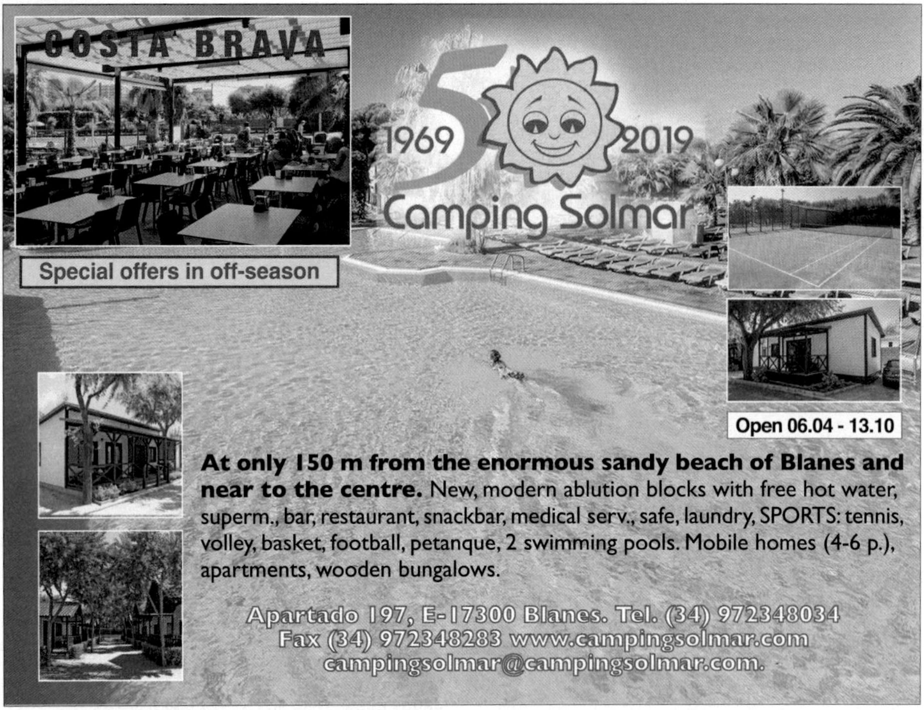

BLANES *3C3* (2km SW Coastal) *41.66206, 2.78046* **Camping Solmar, Calle Cristòfor Colom 48, 17300 Blanes (Gerona) [972-34 80 34; fax 972-34 82 83; campingsolmar@campingsolmar.com; www.campingsolmar.com]** Fr N on AP7/E15 exit junc 9 onto NII dir Barcelona & foll sp Blanes. Fr S to end of C32, then NII dir Blanes. On app Blanes, foll camping sps. Lge, hdg, mkd, shd, wc; chem disp; mv service pnt; fam bthrm; shwrs inc; EHU (6A) inc; lndry; shop; rest; snacks; bar; bbq; playgrnd; pool; paddling pool; beach sand 150m; red long stay; entmnt; wifi; 10% statics; dogs free; bus 100m; adv bkg acc; quiet; ccard acc; games area; games rm; tennis; CKE/CCI. "Excel site & facs." ♦ 2 Apr-12 Oct. € 39.40 2017*

See advertisement

⊞ **BOCAIRENT** *4F2* (9km E Rural) *38.75332, -0.54957* **Camping Mariola, Ctra Bocairent-Alcoy, Km 9, 46880 Bocairent (València) [962-13 51 60; info@campingmariola.com; www.campingmariola.com]** Fr N330 turn E at Villena onto CV81. N of Banyeres & bef Bocairent turn E sp Alcoi up narr, steep hill with some diff turns & sheer drops; site sp. Lge, hdstg, pt shd, wc (htd); chem disp; mv service pnt; fam bthrm; shwrs inc; EHU (6A) inc; lndry; shop; rest; snacks; bar; bbq; cooking facs; playgrnd; pool; paddling pool; twin axles; entmnt; wifi; TV; 50% statics; phone; Eng spkn; adv bkg acc; ccard acc; games area; games rm; CKE/CCI. "In Mariola mountains; gd walking; bicycles; superb tranquil location in Sierra Mariola National Park, excel walking & cycling, great for dogs, friendly family atmosphere." ♦ € 22.00 2013*

⊞ **BOSSOST** *3B2* (3km SE Rural) *42.74921, 0.70071* **Camping Prado Verde, Ctra de Lleida a Francia, N230, Km 173, 25551 Era Bordeta/La Bordeta de Vilamòs (Lleida) [973-64 71 72; info@campingpradoverde.es; www.campingpradoverde.es]** On N230 at km 173 on banks of Rv Garona. Med, shd, wc (htd); mv service pnt; fam bthrm; shwrs; EHU (6A) €5.50; lndry; shop; rest; snacks; bar; playgrnd; pool; paddling pool; wifi; TV; 10% statics; dogs; bus; quiet; ccard acc; bike hire; fishing; CKE/CCI. "V pleasant NH." € 22.00 2016*

BROTO *3B2* (1km W Rural) *42.59779, -0.13072* **Camping Oto, Afueras s/n, 22370 Oto-Valle De Broto (Huesca) [974-48 60 75; fax 974-48 63 47; info@campingoto.com; www.campingoto.com]** On N260 foll camp sp on N o'skts of Broto. Diff app thro vill but poss. Lge, pt shd, pt sl, wc; chem disp; fam bthrm; shwrs inc; EHU (10A) €3.80 (poss no earth); gas; lndry; shop; snacks; bar; bbq; playgrnd; pool; paddling pool; entmnt; adv bkg acc; quiet; ccard acc. "Excel, clean san facs; excel bar & café; friendly owner; pitches below pool rec; some noise fr adj youth site; conv Ordesa National Park; gd site, pleasant." 5 Mar-15 Oct. € 30.00 2013*

⊞ **BROTO** *3B2* (7km W Rural) *42.61576, -0.15432* **Camping Viu, Ctra N260, Biescas-Ordesa, Km 484.2, 22378 Viu de Linás (Huesca) [974-48 63 01; fax 974-48 63 73; info@campingviu.com; www.campingviu.com]** Lies on N260, 4km W of Broto. Fr Broto, N for 2km on rd 135; turn W twd Biesca at junc with Torla rd; site approx 4km on R. Med, pt shd, sl, wc (htd); chem disp; mv service pnt; shwrs inc; EHU (5-8A) €4.20; gas; lndry; shop; rest; bbq; playgrnd; phone; adv bkg acc; quiet; ccard acc; games rm; bike hire; horseriding; CKE/CCI. "Friendly owners; gd home cooking; walking adj; skiing adj; car wash; fine views; highly rec; climbing adj; clean, modern san facs; poss not suitable for lge o'fits." € 17.40 2016*

⊞ **BURGOS** *1B4* (4km E Rural) *42.34111, -3.65777* **Camp Fuentes Blancas, Ctra Cartuja Miraflores, Km 3.5, 09193 Burgos [947-48 60 16; info@camping burgos.com; www.campingburgos.com]** E or W on A1 exit junc 238 & cont twd Burgos. Strt over 1st rndabt, turn R sp Cortes & then L. Look for yellow sps to site. Fr N (N627 or N623) on entering Burgos keep in R hand lane. Foll signs Cartuja miraflores & yellow camp signs. Lge, mkd, shd, wc (htd); chem disp; mv service pnt; fam bthrm; shwrs inc; EHU (6A) inc; gas; lndry; shop; rest; snacks; bar; playgrnd; pool; wifi; 10% statics; dogs €2.17; phone; bus at gate; Eng spkn; quiet; ccard acc; games area. "Neat, roomy, adj woodland; some sm pitches; ltd facs LS; poss v muddy in wet; easy access town car parks or cycle/rv walk; Burgos lovely town; gd NH; San Rafael dated; gd bus to town; excel, well maintained busy site; gd shd; gd facs." € 28.00 2018*

⊞ **CABRERA, LA** *1D4* (1km SW Rural) *40.85797, -3.61580* **Camping Pico de la Miel, Ctra A-1 Salida 57, 28751 La Cabrera (Madrid) [918-68 80 82 or 918-68 95 07; fax 918-68 85 41; info@picodelamiel.com; www.picodelamiel.com]** Fr Madrid on A1/E5, exit junc 57 sp La Cabrera. Turn L at rndabt, site sp. Lge, mkd, pt shd, pt sl, wc (htd); chem disp; shwrs inc; EHU (10A) €4.45; gas; lndry; shop; rest; snacks; bar; playgrnd; pool; paddling pool; red long stay; 75% statics; dogs; phone; Eng spkn; adv bkg acc; quiet; ccard acc; tennis; games area; sailing; windsurfing; squash; fishing; CKE/CCI. "Attractive walking country; conv Madrid; mountain-climbing; car wash; ltd touring area not v attractive; some pitches have low sun shades; excel san facs; ltd facs LS." ♦ € 33.50 2017*

⊞ **CACERES** *2E3* (6km NW Urban) *39.48861, -6.41277* **Camp Municipal Ciudad de Cáceres, Ctra N630, Km 549.5, 10005 Cáceres [927-23 31 00; fax 927- 23 58 96; reservas@campingcaceres.com; www.campingcaceres.com]** Fr Cáceres ring rd take N630 dir Salamanca. At 1st rndbt turn R sp Via de Servicio with camping symbol. Foll sp 500m to site. Or fr N exit A66 junc 545 onto N630 twd Cáceres. At 2nd rndabt turn L sp Via de Servicio, site on L adj football stadium. Med, hdstg, mkd, unshd, terr, wc; chem disp; mv service pnt; shwrs inc; EHU (10A) €4.50; gas; lndry; shop; rest; snacks; bar; bbq; playgrnd; pool; paddling pool; wifi; TV; 15% statics; dogs; bus 500m over footbdge; Eng spkn; adv bkg acc; ccard acc; games area; CKE/CCI. "Vg, well-run site; excel facs; ACSI acc; vg value rest; gd bus service to and fr interesting old town with many historical bldgs; excel site with ensuite facs at each pitch; individual san facs each pitch; location not pretty adj to football stadium & indus est; town too far to walk; v lush, free use of spa; Lydl 1.6km." € 26.00 2017*

CADAQUES *3B3* (1km N Coastal) *42.29172, 3.28260* **Camping Cadaqués, Ctra Port Lligat 17, 17488 Cadaqués (Girona) [972-25 81 26; fax 972-15 93 83; info@campingcadaques.com; www.spain.info]** At ent to town, turn L at rdbt (3rd exit) sp thro narr rds, site in about 1.5km on L. NB App to Cadaqués on busy, narr mountain rds, not suitable lge o'fits. If raining, rds only towable with 4x4. Lge, mkd, hdstg, pt shd, sl, wc; chem disp; shwrs; EHU (5A) €5.95; gas; lndry; shop; rest; snacks; bar; playgrnd; pool; paddling pool; beach shgl 600m; Eng spkn; quiet; ccard acc; sep car park. "Cadaqués home of Salvador Dali; sm pitches; medical facs high ssn; san facs poss poor LS; fair, red facs in LS next to m'way so poss noisy but gd for en-route stop." Easter-17 Sep. € 35.00 2013*

⊞ **CALATAYUD** *3C1* (15km N Rural) *41.44666, -1.55805* **Camping Saviñan Parc, Ctra El Frasno-Mores, Km 7, 50299 Saviñan (Zaragoza) [976-82 54 23; info@campingsavinan.com; www.campingsavinan.com]** Exit A2/E90 (Zaragoza-Madrid) at km 255 to T-junc. Turn R to Saviñan for 6km, foll sps to site 1km S. Lge, hdstg, pt shd, terr, wc; chem disp; mv service pnt; shwrs inc; EHU (6-10A) €4.60; gas; lndry; shop; playgrnd; pool; 15% statics; dogs €2.70; phone; quiet; ccard acc; site clsd Jan; tennis; horseriding; CKE/CCI. "Beautiful scenery & views; some sm narr pitches; rec identify pitch location to avoid stop/start on hill; terr pitches have steep, unfenced edges; many pitches with sunscreen frames & diff to manoeuvre long o'fits; modern facs block but cold in winter & poss stretched high ssn; hot water to some shwrs only; gates poss clsd LS - use intercom; site poss clsd Feb." € 20.00 2017*

SPAIN

CALELLA *3C3* (2km NE Coastal) *41.61774, 2.67680* **Camping Caballo de Mar, Passeig Maritim s/n, 08397 Pineda de Mar (Barcelona) [937-67 17 06; fax 937-67 16 15; info@caballodemar.com; www.caballodemar.com]** Fr N exit AP7 junc 9 & immed turn R onto NII dir Barcelona. Foll sp Pineda de Mar & turn L twd Paseo Maritimo. Fr S on C32 exit 122 dir Pineda de Mar & foll dir Paseo Maritimo. Lge, mkd, shd, wc; chem disp; fam bthrm; shwrs inc; EHU (3-6A) €3.40-4.40; gas; lndry; shop; rest; snacks; bar; bbq; playgrnd; pool; beach sand adj; red long stay; entmnt; wifi; 10% statics; dogs €2.20; Eng spkn; adv bkg acc; quiet; ccard acc; games area; games rm; CKE/CCI. "Excursions arranged; gd touring base & conv Barcelona; gd, modern facs; rlwy stn 2km (Barcelona 30 mins); excel; pitches sm for twin axles; noise fr locals on site." ♦ 31 Mar-30 Sep. € 34.00 2012*

CALIG *3D2* (1km NW Rural) *40.45183, 0.35211* **Camping L'Orangeraie, Camino Peniscola-Calig, 12589 Càlig [34 964 765 059; fax 34 964 765 460; info@camping-lorangeraie.es]** On AP7 exit 43 Benicarlo-Peniscola. 1st R at rndabt to Calig then foll sp to campsite. Fr N340 exit N232 to Morella, then after 1.5km turn L to Calig CV135, foll sp to campsite. Med, mkd, hdg, pt shd, terr, wc; chem disp; mv service pnt; fam bthrm; EHU (10A); lndry; shop; snacks; bar; bbq; playgrnd; pool; paddling pool; twin axles; entmnt; wifi; 15% statics; dogs €2.50; bus 1km; Eng spkn; adv bkg acc; quiet; waterslide; games area. "Excel site." ♦ 1 Apr-12 Oct. € 41.00 2014*

⊞ **CALPE** *4F2* (0km NE Urban/Coastal) *38.64488, 0.05604* **Camping CalpeMar, Calle Eslovenia 3, 03710 Calpe (Alicante) [965-87 55 76; info@campingcalpemar.com; www.campingcalpemar.com]** Exit AP7/E15 junc 63 onto N332 & foll sp, take slip rd sp Calpe Norte & foll dual c'way CV746 round Calpe twd Peñón d'Ifach. At rndabt nr police stn with metal statues turn L, then L at next rndabt, over next rndabt, site 200m on R. Med, hdstg, mkd, hdg, unshd, serviced pitches; wc (htd); chem disp; fam bthrm; shwrs inc; EHU (10A) inc (metered for long stay); lndry; shop nr; rest; snacks; bar; bbq; playgrnd; pool; beach sand 300m; red long stay; entmnt; wifi; TV; 3% statics; dogs free (dog wash); phone; bus adj; Eng spkn; adv bkg acc; quiet; ccard acc; ice; games area; games rm; CKE/CCI. "High standard site; well-kept & laid out; Spanish lessons; gd security; excel; extra lge pitches avail at additional charge; car wash; gd for long stay, friendly staff; sep car park; close to beach and Lidl." ♦ € 34.00 2018*

See advertisement above

CAMBRILS *3C2* (2km N Urban/Coastal) *41.06500, 1.08361* **Camping Playa Cambrils Don Camilo, Carrer Oleastrum 2, Ctra Cambrils-Salou, Km 1.5, 43850 Cambrils (Tarragona) [977-36 14 90; fax 977-36 49 88; camping@playacambrils.com; www.playacambrils.com]** Exit A7 junc 37 dir Cambrils & N340. Turn L onto N340 then R dir port then L onto coast rd. Site sp on L at rndabt after rv bdge 100m bef watch tower on R, approx 2km fr port. V lge, hdg, mkd, shd, wc; chem disp; fam bthrm; shwrs inc; EHU (6A) inc; gas; lndry; shop; rest; snacks; bar; playgrnd; pool (htd); paddling pool; beach sand adj; red long stay; entmnt; wifi; TV; 25% statics; dogs €4.35; bus 200m; Eng spkn; adv bkg req; ccard acc; tennis; watersports; games rm; bike hire; boat hire; CKE/CCI. "Helpful, friendly staff; children's club; cash machine; doctor; cinema; sports activities avail; Port Aventura 5km; 24-hr security; vg site." ♦ 15 Mar-16 Oct. € 46.00 2018*

"I need an on-site restaurant"

We do our best to make sure site information is correct, but it is always best to check any must-have facilities are still available or will be open during your visit.

CAMBRILS *3C2* (11km SW Coastal) *41.02512, 0.95906* **Camping Miramar, Ctra N340, km 1134 43892 Mont-roig del Camp [977-81 12 03; recepcio@camping-miramar.com; www.camping-miramar.es]** Fr the AP7 take exit 37 to the N340. Turn L at KM134 to campsite. Sm, pt shd, wc; chem disp; fam bthrm; shwrs; EHU (6A) €5.20; lndry; shop; snacks; bar; bbq; playgrnd; beach; red long stay; twin axles; wifi; 75% statics; dogs; bus; Eng spkn; adv bkg acc. "Site on beach, walking, sw, snorkelling; vg." ♦ 1 Jan-30 Nov. € 35.00 2017*

CAMBRILS *3C2* (8km SW Coastal) *41.03333, 0.96777* **Playa Montroig Camping Resort, N340, Km1.136, 43300 Montroig (Tarragona) [977 810 637; fax 977 811 411; info@playamontroig.com; www.playamontroig.com]** Exit AP7 junc 37, W onto N340. Site has own dir access onto N340 bet Cambrils & L'Hospitalet de L'Infant, well sp fr Cambrils. V lge, mkd, shd, pt sl, serviced pitches; wc (htd); chem disp; mv service pnt; fam bthrm; shwrs inc; EHU (10A) inc; gas; lndry; shop; rest; snacks; bar; playgrnd; pool (htd); beach sand adj; entmnt; wifi; 30% statics; phone; Eng spkn; adv bkg acc; ccard acc; games rm; games area; golf 3km; tennis; bike hire; CKE/CCI. "Magnificent, clean, secure site; private, swept beach; skateboard track; some sm pitches & low branches; cash machine; doctor; 4 grades pitch/price; highly rec." ♦ 1 Apr-30 Oct. € 53.00 2017*

See advertisement above

⊞ **CAMBRILS** *3C2* (2km W Urban/Coastal) *41.06550, 1.04460* **Camping La Llosa, Ctra N340 Barcelona a Valencia, Km 1143, 43850 Cambrils (Tarragona) [977-36 26 15; fax 977-79 11 80; info@camping-lallosa.com; www.camping-lallosa.com]** Exit A7/E15 at junc 37 & join N340 S. Head S into Cambrils (ignore L turn to cent) & at island turn R. Site sp on L within 100m. Fr N exit junc 35 onto N340. Strt over at x-rds, then L over rlwy bdge at end of rd, strt to site. V lge, hdstg, shd, wc; shwrs inc; EHU (5A) €5; gas; lndry; shop; rest; snacks; bar; playgrnd; pool; beach sand; red long stay; entmnt; 50% statics; dogs €3.50; phone; bus 500m; Eng spkn; ccard acc; car wash. "Interesting fishing port; gd facs; excel pool; gd supmkt nrby; poss diff siting for m'vans due low trees; excel winter NH." ♦ € 51.00 2014*

SPAIN

CAMBRILS *3C2* (8km W Coastal) *41.03717, 0.97622* **Camping La Torre del Sol, Ctra N340, Km 1.136, Miami-Playa, 43300 Montroig Del Camp (Tarragona) [977 810 486; fax 977-81 13 06; info@latorredelsol.com; www.latorredelsol.com]** Leave A7 València/Barcelona m'way at junc 37 & foll sp Cambrils. After 1.5km join N340 coast rd S for 6km. Watch for site sp 4km bef Miami Playa. Fr S exit AP7 junc 38, foll sp Cambrils on N340. Site on R 4km after Miami Playa. Site ent narr, alt ent avail for lge o'fits. V lge, mkd, hdg, shd, wc; chem disp; mv service pnt; fam bthrm; shwrs inc; EHU (6A) inc (10A avail); gas; lndry; shop; rest; snacks; bar; bbq; playgrnd; pool (htd); paddling pool; beach sand; entmnt; wifi; TV; 40% statics; Eng spkn; adv bkg acc; ccard acc; squash; games rm; gym; tennis; sauna; bike hire; golf 4km. "Attractive well-guarded site for all ages; sandy pitches; gd, clean san facs; steps to facs for disabled; whirlpool; jacuzzi; access to pitches poss diff lge o'fits due trees & narr site rds; skateboard zone; radios/TVs to be used inside vans only; conv Port Aventura, Aquaparc, Aquopolis; cinema; disco; highly rec, can't praise site enough; excel." ♦ 15 Mar-30 Oct. € 52.00 2017*

See advertisement opposite

CANGAS DE ONIS *1A3* (3km SE Rural) *43.34715, -5.08362* **Camping Covadonga, 33589 Soto de Cangas (Asturias) [985-94 00 97; campingcovadonga@hotmail.es; www.camping-covadonga.com]** N625 fr Arriondas to Cangas de Onis, then AS114 twds Covadonga & Panes, cont thro town sp Covadonga. At rndabt take 2nd exit sp Cabrales, site on R in 100m. Access tight. Med, mkd, pt shd, wc; chem disp; shwrs; EHU (10A) €3.50 (no earth); lndry; shop; rest; snacks; bar; red long stay; wifi; bus adj; adv bkg acc; quiet; CKE/CCI. "Sm pitches; take care with access; site rds narr; 17 uneven steps to san facs; conv for Picos de Europa." Holy Week & 15 Jun-30 Sep. € 21.00 2017*

⊞ **CARBALLO** *1A2* (15km N Coastal) *43.29556, -8.65528* **Camping Baldayo, San Salvador de Rebordelos, 15105 Carballo [981-73 95 29; campingbaldayo@yahoo.es; www.campingbaldayo.com]** Access via DP-1916 fr Carballo. Turn L in vill and foll sp. App rds are unmade. On site access rds are narr with sharp corners, diff for long o'fits. Med, hdg, mkd, hdstg, pt shd, pt sl, terr, wc; chem disp; shwrs; EHU (3A) €1.50; lndry; shop; snacks; bar; bbq; playgrnd; beach sand 200m; twin axles; wifi; 75% statics; dogs; phone; adv bkg acc; quiet. "Sm pitches & narr camp rds poss diff lge o'fits; poss unkempt LS; nrby lagoon with boardwalk and bird hides; beach popular for surfing & watersports." ♦ € 15.00 2017*

CARIDAD, LA (EL FRANCO) *1A3* (1km SE Coastal) *43.54795, -6.80701* **Camping Playa de Castelló, Ctra N634, Santander-La Coruña, Km 532, 33758 La Caridad (El Franca) (Asturias) [985-47 82 77; contacto@campingcastello.com; www.campingcastello.com]** On N634/E70 Santander dir La Coruña, turn N at km 532. Site in 200m fr N634, sp fr each dir. Sm, mkd, pt shd, wc; chem disp; fam bthrm; shwrs inc; EHU (2-5A) €3; gas; lndry; shop; bar; bbq; playgrnd; beach shgl 800m; red long stay; wifi; 10% statics; dogs €1; bus 200m; Eng spkn; adv bkg acc; quiet; CKE/CCI. "A green oasis with character; gd." Holy Week & 1 Jun-30 Sep. € 23.00 2012*

⊞ **CARIDAD, LA (EL FRANCO)** *1A3* (4km W Rural) *43.55635, -6.86218* **Camping A Grandella, Ctra N634, Km 536.9 (Desvío San Juan de Prendonés), 33746 Valdepares (Asturias) [607-85 49 00 or 661-35 28 70 (mob); camping@campingagrandella.com]** Sp fr N634/E70. Med, pt shd, wc; chem disp; shwrs inc; EHU €3.50; lndry; snacks; bar; playgrnd; 10% statics; dogs €1; bus 200m; quiet; site clsd mid-Dec to mid-Jan. "Attractive little site; well-situated." € 18.00 2012*

"Satellite navigation makes touring much easier"

Remember most sat navs don't know if you're towing or in a larger vehicle – always use yours alongside maps and site directions.

CARRION DE LOS CONDES *1B4* (0km W Rural) *42.33506, -04.60447* **Camping El Edén, Ctra Vigo-Logroño, Km 200, 34120 Carrión de los Condes (Palencia) [979-88 07 14]** Fr N231 take junc 85, go S then L at 3rd rndabt to ctr. Site sp on L. Med, mkd, pt shd, wc; mv service pnt; shwrs; EHU (6A) €3.50; gas; lndry; rest; bar; playgrnd; wifi; dogs; bus 500m; Eng spkn; quiet. "Pleasant walk to town; basic rvside site; recep in bar/rest; site open w/ends only LS; fair NH; quite lively in high ssn; nice site in interesting town on pilgrim rte; no lts in san block." ♦ 1 Apr-30 Oct. € 19.80 2017*

CASPE *3C2* (12km NE Rural) *41.28883, 0.05733* **Lake Caspe Camping, Ctra N211, Km 286.7, 50700 Caspe (Zaragoza) [976-63 41 74 or 689-99 64 30 (mob); fax 976-63 41 87; lakecaspe@lakecaspe.com; www.campinglakecaspe.com]** Fr E leave AP2 or N11 at Fraga & foll N211 dir Caspe to site. Fr W take N232 fr Zaragoza then A1404 & A221 E thro Caspe to site in 16km on L at km 286.7, sp. Med, hdg, mkd, hdstg, pt shd, wc; chem disp; fam bthrm; shwrs inc; EHU (5-10A) €5.60; gas; lndry; shop; rest; snacks; bar; playgrnd; pool; 10% statics; dogs €3.75; phone; Eng spkn; adv bkg acc; quiet; fishing; sailing; CKE/CCI. "Gd, well-run, scenic site but isolated (come prepared); avoid on public hols; site rds gravelled but muddy after rain; sm pitches nr lake; gd watersports; mosquitoes; beware low branches." 1Mar-10 Nov. € 37.00 2013*

SPAIN

"There aren't many sites open at this time of year"

If you're travelling outside peak season remember to call ahead to check site opening dates – even if the entry says 'open all year'.

CASTELLO D'EMPURIES *3B3* (4km NE Rural) *42.26460, 3.10160* **Camping Mas Nou, Ctra Mas Nou 7, Km 38, 17486 Castelló d'Empúries (Gerona) [972-45 41 75; fax 972-45 43 58; info@campingmasnou.com; www.campingmasnou.com]** On m'way A7 exit 3 if coming fr France & exit 4 fr Barcelona dir Roses (E) C260. Site on L at ent to Empuriabrava - use rndabt to turn. Lge, mkd, shd, wc (htd); chem disp; mv service pnt; fam bthrm; shwrs inc; EHU (10A) €4.90; lndry; shop nr; rest; snacks; bar; bbq; playgrnd; pool; beach 2.5km; red long stay; entmnt; wifi; TV; 5% statics; dogs €2.35; phone; Eng spkn; ccard acc; tennis; games area; CKE/CCI. "Aqua Park 4km, Dali Museum 10km; gd touring base; helpful staff; well-run site; excel, clean san facs; sports activities & children's club; gd cycling; excel; vg rest." ♦ 8 Apr-24 Sep. € 47.50 2017*

CASTELLO D'EMPURIES *3B3* (4km SE Coastal) *42.20725, 3.10026* **Camping Nautic Almatá, Aiguamolls de l'Empordà, 17486 Castelló d'Empúries (Gerona) [972-45 44 77; fax 972-45 46 86; info@almata.com; www.almata.com]** Fr A7 m'way exit 3; foll sp to Roses. After 12km turn S for Sant Pere Pescador & site on L in 5km. Site clearly sp on rd Castelló d'Empúries-Sant Pere Pescador. Lge, pt shd, wc; chem disp; shwrs inc; EHU (10A) inc; gas; lndry; shop; rest; playgrnd; pool; beach sand adj; entmnt; TV; dogs €6.40; adv bkg acc; quiet; sailing school; tennis; bike hire; horseriding; games area. "Excel, clean facs; ample pitches; sports facs inc in price; disco bar on beach; helpful staff; direct access to nature reserve; waterside pitches rec." ♦ 16 May-20 Sep. € 59.00 2016*

See advertisement

"That's changed – Should I let the Club know?"

If you find something on site that's different from the site entry, fill in a report and let us know. See camc.com/europereport.

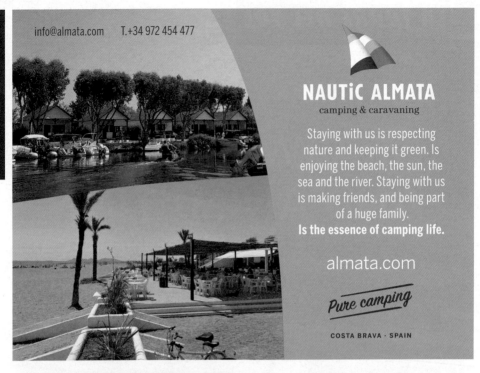

CASTELLO D'EMPURIES *3B3* (1km S Coastal) *42.25563, 3.13791* **Camping Castell Mar, Ctra Roses-Figueres, Km 40.5, Playa de la Rubina, 17486 Castelló d'Empúries (Gerona) [972-45 08 22; fax 972-45 23 30; cmar@campingparks.com; www.campingparks.com]** Exit A7 at junc 3 sp Figueres; turn L onto C260 sp Roses, after traff lts cont twd Roses, turn R down side of rest La Llar for 1.5km, foll sp Playa de la Rubina. Lge, mkd, hdg, pt shd, serviced pitches; wc; chem disp; fam bthrm; shwrs inc; EHU (6-10A) inc; gas; lndry; shop; rest; snacks; bar; bbq; playgrnd; pool; paddling pool; beach sand 100m; entmnt; TV (pitch); 30% statics; dogs; phone; Eng spkn; adv bkg acc; quiet; games rm; CKE/CCI. "Pitches poss unsuitable lge o'fits; gd location; excel for families." ♦ 22 May-19 Sep. € 52.00 2013*

CASTRO URDIALES *1A4* (1km N Coastal) *43.39000, -3.24194* **Camping de Castro, Barrio Campijo, 39700 Castro Urdiales (Cantabria) [942-86 74 23; fax 942-63 07 25; info@campingdecastro.com]** Fr Bilbao turn off A8 at 2nd Castro Urdiales sp, km 151. Camp sp on R by bullring. V narr, steep lanes to site - no passing places, great care req. Lge, unshd, pt sl, terr, wc; shwrs inc; EHU (6A) €3; lndry; shop; rest; bar; playgrnd; pool; beach sand 1km; 90% statics; dogs; phone; bus; Eng spkn; adv bkg acc; quiet; CKE/CCI. "Gd, clean facs; conv NH for ferries; ltd touring pitches; narr, long, steep single track ent; great views over Bilbao bay." ♦ 13 Feb-10 Dec. € 39.60 2014*

CASTROJERIZ *1B4* (1km NE Rural) *42.29102, -4.13165* **Camping Camino de Santiago, Calle Virgen del Manzano s/n, 09110 Castrojeriz (Burgos) [947-37 72 55 or 658-96 67 43 (mob); fax 947-37 72 36; info@campingcamino.com; www.campingcamino.com]** Fr N A62/E80 junc 40 dir Los Balbases, Vallunquera & Castrojeriz - narr, uneven rd. In 16 km ent Castrojeriz, sp fr BU400 where you turn onto BU404. Once on BU404 proceed for approx 500yds to next rndabt, take 2nd exit. Site 1m on the L. Fr S A62, exit 68 twds Torquemada, then take P412, then BU4085 to Castrojeriz. Do not go thro town, as rd are narr. Med, mkd, hdg, shd, pt sl, wc; chem disp; shwrs inc; EHU (5-10A) €4 (poss no earth); lndry; shop nr; rest; snacks; bar; wifi; TV; dogs €2; bus 200m; Eng spkn; quiet; games area; games rm; CKE/CCI. "Lovely site; helpful owner; pilgrims' refuge on site; some diff sm pitches; vg; site ent narr; excel bird watching tours on req; san facs dated." ♦ 15 Mar-15 Nov. € 32.00 2014*

CERVERA DE PISUERGA *1B4* (1km W Rural) *42.87135, -4.50332* **Camping Fuentes Carrionas, La Bárcena s/n, 34840 Cervera de Pisuerga (Palencia) [979-87 04 24; fax 979-12 30 76; campingfuentescarrionas@hotmail.com; campingfuentescarrionas.com]** Fr Aguilar de Campóo on CL626 pass thro Cervera foll sp CL627 Potes. Site sp on L bef rv bdge. Med, mkd, pt shd, wc; chem disp; shwrs inc; EHU €3.50; lndry; shop nr; rest nr; bar; 80% statics; bus 100m; quiet; games area; tennis; CKE/CCI. "Gd walking in nature reserve; conv Casa del Osos bear info cent; rest/bar area busy at wkend; new gd clean san facs; site busy with wkenders in statics or cabins; owners v helpful and pleasant; well maintained." ♦ Holy Week-30 Sep. € 21.00 2017*

⊞ **CIUDAD RODRIGO** *1D3* (2km S Rural) *40.59206, -6.53445* **Camping La Pesquera, Ctra Cáceres-Arrabal, Km 424, Huerta La Toma, 37500 Ciudad Rodrigo (Salamanca) [923-48 13 48; campinglapesquera@hotmail.com; www.campinglapesquera.es]** Fr Salamanca on A62/E80 exit junc 332. Look for tent sp on R & turn R, then 1st L & foll round until site on rvside. Med, mkd, pt shd, wc; shwrs inc; EHU (6A) inc; lndry; shop; snacks; sw nr; wifi; TV; dogs free; phone; quiet; ccard acc; fishing adj; CKE/CCI. "Medieval walled city worth visit - easy walk over Roman bdge; gd san facs; vg, improved site; friendly nice sm site next to rv, gd for NH; lovely town; v helpful staff." ♦ € 17.00 2017*

⊞ **CLARIANA** *3B3* (6km NE Rural) *41.95900, 1.60384* **Camping La Ribera, Pantà de Sant Ponç, 25290 Clariana de Cardener (Lleida) [973-48 25 52 or 973-48 15 57; info@campinglaribera.com; www.campinglaribera.com]** Fr Solsona S on C55, turn L onto C26 at km 71. Go 2.7km, site sp immed bef Sant Ponç Dam. Lge, mkd, hdstg, pt shd, wc; chem disp; fam bthrm; shwrs; EHU (4-10A) €4.60-8.45; lndry; shop; snacks; bar; playgrnd; pool; paddling pool; sw nr; TV; 95% statics; dogs; phone; quiet; games area; tennis. "Excel facs; gd site; narr pitches, can be diff to site o'fit." ♦ € 38.00 2016*

⊞ **COLOMBRES** *1A4* (3km SW Rural) *43.37074, -4.56799* **Camping Colombres, Ctra El Peral A Noriega Kml - 33590 Colombres (Ribadedeva) [985 412 244; fax 985 413 056; campingcolombres@hotmail.com; www.campingcolombres.com]** Turn off A8/E70 at J277. 3rd exit then 2nd exit fr rndabts. East on N634 then R opp petrol stn. Site to L 1km. Med, mkd, pt shd, terr, wc (htd); chem disp; mv service pnt; fam bthrm; shwrs inc; EHU (6A) €4.20; lndry; shop; rest; snacks; bar; bbq (sep area); playgrnd; pool; beach sand 3km; twin axles; wifi; 10% statics; dogs free; Eng spkn; quiet; ccard acc; games area; CKE/CCI. "Quiet, peaceful site in rural setting with fine mountain views; v helpful owners; nice pool; excel san facs; well kept & clean; immac, modern san facs; gd walking." ♦ € 31.50 2017*

COLUNGA *1A3* (12km E Coastal) *43.47160, -5.18434* **Camping Arenal de Moris, Ctra de la Playa s/n, 33344 Caravia Alta (Asturias) [985-85 30 97; fax 985-85 31 37; camoris@desdeasturias.com; www.arenaldemoris.com]** Fr E70/A8 exit junc 337 onto N632 to Caravia Alta, site clearly sp. Lge, mkd, pt shd, terr, wc; chem disp; shwrs inc; EHU (5A) €4.50; lndry; shop; rest; snacks; bar; playgrnd; pool; beach sand 500m; 10% statics; adv bkg acc; quiet; ccard acc; tennis; CKE/CCI. "Lovely views to mountains & sea; well-kept, well-run site; excel, clean san facs." 23 Mar-15 Sep. € 35.00 2015*

COMILLAS *1A4* (1km E Coastal) *43.38583, -4.28444* **Camping de Comillas, 39520 Comillas (Cantabria) [942-72 00 74; info@campingcomillas.com; www.campingcomillas.com]** Site on coast rd CA131 at E end of Comillas by-pass. App fr Santillana or San Vicente avoids town cent & narr rds. Lge, hdg, mkd, pt shd, pt sl, wc; chem disp; shwrs inc; EHU (5A) €3.85; lndry; shop; rest nr; snacks; bar; playgrnd; beach sand 800m; TV; dogs; phone; adv bkg acc; quiet; CKE/CCI. "Clean, ltd facs LS (hot water to shwrs only); vg site in gd position on coast with views; easy walk to interesting town; gd but rocky beach across rd; helpful owner; pitches inbetween 2 rds." Holy Week & 1 Jun-30 Sep. € 29.50 2017*

COMILLAS *1A4* (5km E Rural) *43.38328, -4.24689* **Camping El Helguero, 39527 Ruiloba (Cantabria) [942-72 21 24; fax 942-72 10 20; reservas@campingelhelguero.com; www.campingelhelguero.com]** Exit A8 junc 249 dir Comillas onto CA135 to km 7. Turn dir Ruiloba onto CA359 & thro Ruiloba & La Iglesia, fork R uphill. Site sp. Lge, mkd, pt shd, pt sl, wc (htd); chem disp; mv service pnt; fam bthrm; shwrs inc; EHU (6A) €4.35; lndry; shop; rest; snacks; bar; playgrnd; pool; paddling pool; beach sand 3km; wifi; 80% statics; dogs; Eng spkn; ccard acc; tennis 300m; bike hire; CKE/CCI. "Attractive site, gd touring cent; clean facs but some in need of refurb; helpful staff; night security; sm pitches poss muddy in wet; v gd." ♦ 1 Apr-30 Sep. € 30.00 2016*

COMILLAS *1A4* (3km W Rural/Coastal) *43.3858, -4.3361* **Camping Rodero, Ctra Comillas-St Vicente, Km 5, 39528 Oyambre (Cantabria) [942-72 20 40; fax 942-72 26 29; rodero@campingrodero-oyambre.es; www.campingrodero-oyambre.es]** Exit A8 dir San Vicente de la Barquera, cross bdge over estuary & take R fork nr km27.5. Site just off C131 bet San Vicente & Comillas, sp. Lge, mkd, pt shd, pt sl, terr, wc; chem disp; mv service pnt; shwrs inc; EHU (6A) €3; gas; lndry; shop; rest; snacks; bar; playgrnd; pool; beach sand 200m; wifi; 10% statics; phone; bus 200m; adv bkg acc; ccard acc; games area; CKE/CCI. "Lovely views; on top of hill; friendly owners; site noisy but happy - owner puts Dutch/British in quieter pt; sm pitches; poss run down LS." ♦ 15 Mar-30 Sep. € 29.00 2014*

⊞ **CONIL DE LA FRONTERA** *2H3* (3km NE Rural) *36.31061, -6.11276* **Camping Roche, Carril de Pilahito s/n, N340 Km 19.2, 11149 Conil de la Frontera (Cádiz) [956-44 22 16; fax 956-44 26 24; info@campingroche.com; www.campingroche.com]** Exit A48 junc 15 Conil Norte. Site sp on N340 dir Algeciras. Lge, mkd, hdstg, pt shd, wc; chem disp; mv service pnt; EHU (10A) €5; lndry; shop; rest; snacks; bar; playgrnd; pool; paddling pool; beach sand 2.5km; red long stay; TV; 20% statics; dogs €3.75; Eng spkn; adv bkg acc; quiet; ccard acc; games area; games rm; tennis. "V pleasant, peaceful site in pine woods; all-weather pitches; friendly, helpful staff; special monthly rates; clean san facs; superb beaches nr; excel facs; lack of adequate management." ♦ € 36.60 2014*

⊞ **CONIL DE LA FRONTERA** *2H3* (1km NW Rural/Coastal) *36.29340, -6.09626* **Camping La Rosaleda, Ctra del Pradillo, Km 1.3, 11140 Conil de la Frontera (Cádiz) [956-44 33 27; fax 956-44 33 85; info@campinglarosaleda.com; www.campinglarosaleda.com]** Exit A48 junc 26 dir Conil. In 2km at rndabt foll sp Puerto Pesquero along CA3208. Site sp on R. Lge, hdstg, mkd, pt shd, pt sl, terr, wc; chem disp; mv service pnt; shwrs inc; EHU (5-10A) inc; gas; lndry; shop; rest; snacks; bar; playgrnd; pool; beach sand 1.3km; red long stay; entmnt; wifi; 10% statics; dogs (except 15 Jun-15 Sep, otherwise in sep area €5); phone; Eng spkn; adv bkg acc; quiet; car wash; CKE/CCI. "Well-run site; friendly, helpful staff; gd social atmosphere; sm pitches not suitable lge o'fits but double-length pitches avail; poss travellers; pitches soft/muddy when wet; lge rally on site in winter; gd walking & cycling; sea views; historical, interesting area; conv Seville, Cádiz, Jerez, day trips Morocco; If low occupancy, facs maybe clsd and excursions cancelled." ♦ € 38.50 2013*

⊞ **CORDOBA** *2F3* (8km N Rural) *37.96138, -4.81361* **Camping Los Villares, Ctra Los Villares, Km 7.5, 14071 Córdoba (Córdoba) [957-33 01 45; fax 957-33 14 55; campingvillares@latinmail.com]** Best app fr N on N432: turn W onto CP45 1km N of Cerro Muriano at km 254. Site on R after approx 7km shortly after golf club. Last 5-6km of app rd v narr & steep, but well-engineered. Badly sp, easy to miss. Or fr city cent foll sp for Parador until past municipal site on R. Shortly after, turn L onto CP45 & foll sp Parque Forestal Los Villares, then as above. Sm, hdstg, shd, sl, wc; chem disp; shwrs inc; EHU (15A) €4.30 (poss rev pol); gas; lndry; shop; rest; bar; red long stay; 10% statics; bus 1km; quiet; CKE/CCI. "In nature reserve; peaceful; cooler than Córdoba city with beautiful walks, views & wildlife; sm, close pitches; basic facs (v ltd & poss unclean LS); mainly sl site in trees; strictly run; suitable as NH; take care electrics; poss no drinking water/hot water." ♦ € 17.70 2014*

⊞ **CORDOBA** *2F3* (2km NW Urban) *37.89977, -04.78725* **Camp Municipal El Brillante, Avda del Brillante 50, 14012 Córdoba [957-40 38 36; fax 957-28 21 65; elbrillante@campings.net; www.campingelbrillante.com]** Fr N1V take Badejoz turning N432. Take rd Córdoba N & foll sp to Parador. Turn R into Paseo del Brilliante which leads into Avda del Brilliante; white grilleblock wall surrounds site. Alt, foll sp for 'Macdonalds Brilliante.' Site on R 400m beyond Macdonalds on main rd going uphill away fr town cent. Site poorly sp. Med, mkd, hdstg, hdg, pt shd, serviced pitches; wc; chem disp; mv service pnt; fam bthrm; shwrs inc; EHU (6-10A) €5.50 (poss no earth); gas; lndry; shop; snacks; bar; bbq; playgrnd; wifi; dogs free; phone; bus adj; Eng spkn; CKE/CCI. "Well-run, busy, clean site; rec arr bef 1500; friendly staff; easy walk/gd bus to town; quiet but traff noise & barking dogs off site; poss cramped pitches - diff lge o'fits; poss travellers LS (noisy); pool adj in ssn; gd for wheelchair users; highly rec; easy walk to beautiful city; pitches easier to access with motor mover; Aldi nrby." ♦ € 32.50 2018*

CORUNA, LA *1A2* (11km E Rural) *43.348194, -8.335745* **Camping Los Manzanos, Olieros, 15179 Santa Cruz (La Coruña) [981-61 48 25; informacion@campinglosmanzanos.com; www.campinglosmanzanos.com]** App La Coruña fr E on NVI, bef bdge take AC173 sp Santa Cruz. Turn R at 2nd traff lts in Santa Cruz cent (by petrol stn), foll sp, site on L. Fr AP9/E1 exit junc 3, turn R onto NVI dir Lugo. Take L fork dir Santa Cruz/La Coruña, then foll sp Meiras. Site sp. Lge, pt shd, wc; chem disp; shwrs; EHU (6A) €4.80; gas; lndry; shop; rest; snacks; bar; playgrnd; pool; wifi; TV; 10% statics; dogs free; phone; Eng spkn; adv bkg req; ccard acc; CKE/CCI. "Lovely site; steep slope into site, level exit is avail; helpful owners; hilly 1km walk to Santa Cruz for bus to La Coruña; gd rest; conv for Santiago de Compostela; excel." 22 Mar-30 Sep. € 32.00 2018*

"I like to fill in the reports as I travel from site to site"

You'll find report forms at the back of this guide, or you can fill them in online at camc.com/europereport.

COTORIOS *4F1* (2km E Rural) *38.05255, -2.83996* **Camping Llanos de Arance, Ctra Sierra de Cazorla/Beas de Segura, Km 22, 23478 Cotoríos (Jaén) [953-71 31 39; fax 953-71 30 36; arancell@inicia.es; www.llanosdearance.com]** Fr Jaén-Albecete rd N322 turn E onto A1305 N of Villanueva del Arzobispo sp El Tranco. In 26km to El Tranco lake, turn R & cross over embankment. Cotoríos at km stone 53, approx 25km on shore of lake & Río Guadalaquivir. App fr Cazorla or Beas definitely not rec if towing. Lge, shd, wc; shwrs; EHU (5A) €3.21; gas; shop nr; rest; snacks; bar; bbq; playgrnd; pool; 2% statics; phone; quiet; ccard acc; CKE/CCI. "Lovely site; excel walks & bird life, boar & wild life in Cazorla National Park." € 21.40 2014*

COVARRUBIAS *1B4* (1km E Rural) *42.05944, -3.51527* **Camping Covarrubias, Ctra Hortigüela, 09346 Covarrubias (Burgos) [947-40 64 17; fax 983-29 58 41; proatur@proatur.com; www.proatur.com]** Take N1/E5 or N234 S fr Burgos, turn onto BU905 after approx 35km. Site sp on BU905. Lge, mkd, pt shd, pt sl, wc; shwrs; EHU (12A) €3.90; gas; lndry; shop nr; rest; bar; playgrnd; pool; paddling pool; 90% statics; phone. "Ltd facs LS; pitches poss muddy after rain; charming vill; poss vultures; phone to confirm if open." € 18.50 2013*

CREIXELL *3C3* (3km E Coastal) *41.16512, 1.45800* **Camping La Plana, Ctra N340, Km 1182, 43839 Creixell (Tarragona) [977-80 03 04; fax 977-66 36 63; info@campinglaplana.com; www.campinglaplana.com]** Site sp at Creixell off N340. Med, hdstg, shd, wc; chem disp; shwrs inc; EHU €6; gas; lndry; shop; rest; snacks; bar; beach sand adj; wifi; dogs €1.10; Eng spkn; adv bkg acc. "Vg, v clean site; v helpful & pleasant owners." 3 Mar-30 Sep. € 23.25 2017*

CREIXELL *3C3* (3km SW Coastal) *41.14851, 1.41821* **Camping La Noria, Passeig Miramar 278, 43830 Torredembarra [977-64 04 53; fax 977-64 52 72; info@camping-lanoria.com; www.camping-lanoria.com]** Just outside Torredembara, going N on the old coastal N340 rd. Lge, hdstg, hdg, mkd, pt shd, wc; chem disp; mv service pnt; fam bthrm; shwrs; EHU (6A) €5; lndry; shop; rest; snacks; bar; bbq; playgrnd; beach adj; twin axles; red long stay; entmnt; wifi; TV; 60% statics; dogs; bus adj; Eng spkn; adv bkg rec; games area; games rm. "Adj to Els Muntanyans Nature reserve, with walks & birdwatching; rail 1.6km; beach has naturist area; some noise fr train line bet beach & site; vg; conv for town; sep beach area for naturist." ♦ 1 Apr-1 Oct. € 41.00 2018*

See advertisement

⊞ **CREVILLENT** *4F2* (8km S Rural) *38.17770, -0.80876* **Marjal Costa Blanca Eco Camping Resort, AP-7 Salida 730, 03330 Crevillent (Comunidad Valenciana) [965-48 49 45; camping@marjalcostablanca.com; www.marjalcostablanca.com]** Fr A7/E15 merge onto AP7 (sp Murcia), take exit 730; site sp fr exit. V lge, hdg, mkd, hdstg, pt shd, serviced pitches; wc; chem disp; fam bthrm; shwrs; EHU (16A) inc; gas; lndry; shop; rest; snacks; bar; bbq; playgrnd; pool (htd); sw; twin axles; entmnt; wifi; TV; 30% statics; dogs €2.20; phone; Eng spkn; adv bkg acc; ccard acc; tennis; bike hire; games area; games rm; CCI. "Superb site; car wash; hairdresser; doctor's surgery; gd security; excel facs; wellness cent with fitness studio, htd pools, saunas, physiotherapy & spa; tour ops; new site, trees and hedges need time to grow; lge pitches extra charge; excel, immac san facs; nr to Elfondo birdwatching; v helpful staff." ♦ € 47.50 2016*

⊞ **CUBILLAS DE SANTA MARTA** *1C4* (4km S Rural) *41.80511, -4.58776* **Camping Cubillas, Ctra N620, Km 102, 47290 Cubillas de Santa Marta (Valladolid) [983-58 50 02; fax 983-58 50 16; info@campingcubillas.com; www.campingcubillas.com]** A-62 Exit 102 Cubillas de Santa Marta. Fr N foll slip rd and cross rd to Cubillas de Santa Marta the site is on the R in 200m. Fr S take exit 102 take 5th exit off rndabt, cross over m'way and then 1st L. Site on R in 200m. Lge, hdg, mkd, unshd, pt sl, wc; chem disp; mv service pnt; shwrs inc; EHU (6-10A) inc; gas; lndry; shop; rest; bar nr; bbq; playgrnd; pool; red long stay; entmnt; 50% statics; dogs €2; phone; ccard acc; site clsd 18 Dec-10 Jan; CKE/CCI. "Ltd space for tourers; conv visit Palencia & Valladolid; rd & m'way, rlwy & disco noise at w/end until v late; v ltd facs LS; NH only; 2.5h fr Santander; adequate o'night stop." ♦ € 26.00 2014*

CUDILLERO *1A3* (3km SE Rural) *43.55416, -6.12944* **Camping Cudillero, Ctra Playa de Aguilar, Aronces, 33150 El Pito (Asturias) [985-59 06 63; info@campingcudillero.com; www.campingcudillero.com]** Exit N632 (E70) sp El Pito. Turn L at rndabt sp Cudillero & in 300m at end of wall turn R at site sp, cont for 1km, site on L. Do not app thro Cudillero; rds v narr & steep; much traff. Med, mkd, hdg, pt shd, wc; chem disp; fam bthrm; shwrs inc; EHU (6A) €4.50; gas; lndry; shop; snacks; bar; playgrnd; pool (htd); beach sand 1.2km; entmnt; wifi; TV; dogs €2.15; phone; bus 1km; adv bkg acc; quiet; games area; CKE/CCI. "Excel, well-maintained, well laid-out site; some generous pitches; gd san facs; steep walk to beach & vill; v helpful staff; excel facs; vill worth a visit, parking on quay but narr rds; diff to get to vill, very steep; sm crowded pitches." ♦ 18 Mar-27 Mar & 30 Apr-15 Sep. € 18.00 2016*

CUDILLERO *1A3* (2km S Rural) *43.55555, -6.13777* **Camping L'Amuravela, El Pito, 33150 Cudillero (Asturias) [985-59 09 95; camping@lamuravela.com; www.lamuravela.com]** Exit N632 (E70) sp El Pito. Turn L at rndabt sp Cudillero & in approx 1km turn R at site sp. Do not app thro Cudillero; rds v narr & steep; much traff. Med, mkd, unshd, pt sl, wc; chem disp; mv service pnt; shwrs inc; EHU €4.10; gas; shop; snacks; bar; pool; paddling pool; beach sand 2km; wifi; 50% statics; dogs €1; ccard acc. "Pleasant, well-maintained site; gd clean facs; hillside walks into Cudillero, attractive fishing vill with gd fish rests; red facs LS & poss only open w/ends, surroundings excel." Holy Week & 1 Jun-30 Sep. € 22.00 2017*

CUENCA *3D1* (8km N Rural) *40.12694, -2.14194* **Camping Cuenca, Ctra Tragacete, Km8, 16147 Cuenca [969-23 16 56; info@campingcuenca.com; www.campingcuenca.com]** Fr Madrid take N400/A40 dir Cuenca & exit sp 'Ciudad Encantada' & Valdecabras on CM2110. In 7.5km turn R onto CM2105, site on R in 1.5km. Foll sp 'Nalimiento des Rio Jucar'. Lge, pt shd, pt sl, terr, wc; chem disp; mv service pnt; shwrs inc; EHU (6-10A) €4; gas; lndry; shop; snacks; bar; playgrnd; pool; 15% statics; dogs €1; phone; Eng spkn; adv bkg acc; quiet; games area; tennis; jacuzzi; CKE/CCI. "Pleasant, well-kept, green site; gd touring cent; friendly, helpful staff; excel san facs but ltd LS; interesting rock formations at Ciudad Encantada." ♦ 19 Mar-11 Oct. € 21.00 2013*

DEBA *3A1* (6km E Coastal) *43.29436, -2.32853* **Camping Itxaspe, N634, Km 38, 20829 Itziar (Guipúzkoa) [943-19 93 77; itxaspe@hotmail.es; www.campingitxaspe.com]** Exit A8 junc 13 dir Deba; at main rd turn L up hill, in 400m at x-rds turn L, site in 2km - narr, winding rd. NB Do not go into Itziar vill. Sm, mkd, pt shd, pt sl, wc; chem disp; fam bthrm; shwrs; EHU (5A) €4; gas; shop; rest nr; bar nr; bbq; playgrnd; pool; beach shgl 4km; wifi; 10% statics; adv bkg acc; quiet; solarium; CKE/CCI. "Excel site; helpful owner; w/ends busy; sea views; Coastal geology is UNESCO site, walking fr site superb." ♦ 1 Apr-30 Sep. € 35.60 2012*

DEBA *3A1* (6km W Coastal) *43.30577, -2.37789* **Camping Aitzeta, Ctra Deba-Guernica, Km. 3.5, C6212, 20930 Mutriku (Guipúzkoa) [943-60 33 56; fax 943-60 31 06; aitzeta@hotmail.com; www.campingseuskadi.com/aitzeta]** On N634 San Sebastián-Bilbao rd thro Deba & on o'skts turn R over rv sp Mutriku. Site on L after 3km on narr & winding rd up sh steep climb. Med, mkd, pt shd, terr, wc; chem disp; shwrs inc; EHU (4A) €3; gas; lndry; shop; rest nr; snacks; bar; playgrnd; beach sand 1km; dogs; phone; bus 500m; quiet; CKE/CCI. "Easy reach of Bilbao ferry; sea views; gd, well-run, clean site; not suitable lge o'fits; ltd pitches for tourers; helpful staff; walk to town; basic facs but gd NH." ♦ 1 May-31 Oct. € 20.00 2016*

DELTEBRE *3D2* (8km E Coastal) *40.72041, 0.84849* **Camping L'Aube, Afores s/n, 43580 Deltebre (Tarragona) [977-26 70 66; fax 977-26 75 05; campinglaube@hotmail.com; www.campinglaube.com]** Exit AP7 junc 40 or 41 onto N340 dir Deltebre. Fr Deltebre foll T340 sp Riumar for 8km. At info kiosk branch R, site sp 1km on R. Lge, hdstg, mkd, pt shd, wc; chem disp; mv service pnt; shwrs inc; EHU (3-10A) €2.80-5; lndry; shop; rest; snacks; bar; playgrnd; pool; beach sand adj; red long stay; 40% statics; phone; CKE/CCI. "At edge of Ebro Delta National Park; excel birdwatching; ltd facs in winter; sm pitches; pricey; interesting local rest." ♦ 1 Mar-31 Oct. € 26.00 2018*

⊞ **DENIA** *4E2* (4km SE Coastal) *38.82968, 0.14767* **Camping Los Pinos, Ctra Dénia-Les Rotes, Km 3, Les Rotes, 03700 Dénia (Alicante) [965-78 26 98; lospinosdenia@gmail.com; www.lospinosdenia.com]** Fr N332 foll sp to Dénia in dir of coast. Turn R sp Les Rotes/Jávea, then L twrds Les Rotes. Foll site sp turn L into narr access rd poss diff lge o'fits. Med, mkd, pt shd, wc; chem disp; shwrs inc; EHU (6-10A) €3.20; gas; lndry; shop nr; bbq; cooking facs; playgrnd; beach shgl adj; red long stay; wifi; TV; 25% statics; dogs €3; phone; bus 100m; Eng spkn; adv bkg acc; quiet; ccard acc; CKE/CCI. "Friendly, well-run, clean, tidy site but san facs tired (Mar 09); excel value; access some pitches poss diff due trees - not suitable lge o'fits or m'vans; many long-stay winter residents; cycle path into Dénia; social rm with log fire; naturist beach 1km, private but rocky shore." ♦ € 26.40 2012*

⊞ **DENIA** *4E2* (7km W Coastal) *38.87264, -0.02031* **Camping Los Patos, Playa de Les Deveses, Vergel, 03700 Dénia (Alicante) [965-75 52 93; info@camping-lospatos.com; www.camping-lospatos.com]** Exit A7/E15 junc 61 onto N332. Foll site sp. Med, hdg, mkd, pt shd, wc (htd); chem disp; mv service pnt; shwrs inc; EHU (6A); gas; lndry; shop; rest; snacks; bar; bbq; playgrnd; beach sand adj; red long stay; twin axles; wifi; dogs; Eng spkn; adv bkg acc; quiet; golf 1km. "Gd site." € 14.50 2015*

DOS HERMANAS *2G3* (3km SW Urban) *37.27731, -5.93722* **Camping Villsom, Ctra Sevilla/Cádiz A4, Km 554.8, 41700 Dos Hermanas (Sevilla) [954-72 08 28; campingvillsom@hotmail.com; campingvillsom.blogspot.co.uk]** On main Seville-Cádiz NIV rd travelling fr Seville take exit at km. 555 sp Dos Hermanos-Isla Menor. At the rndabt turn R (SE-3205 Isla Menor) to site 80 m. on R. Lge, hdg, mkd, hdstg, pt shd, pt sl, wc; chem disp; shwrs inc; EHU (8A) inc (poss no earth); gas; lndry; shop; snacks; bar; playgrnd; pool; wifi; dogs; bus to Seville 300m (over bdge & rndabt); Eng spkn; adv bkg acc; ccard acc; CKE/CCI. "Adv bkg rec Holy Week; helpful staff; clean, tidy, well-run site; vg, san facs, ltd LS; height barrier at Carrefour hypmkt - ent via deliveries; no twin axles; wifi only in office & bar area; pitches long but narr, no rm for awnings; dusty, rough, tight turns; c'vans parked too cls." 10 Jan-23 Dec. € 29.00 2018*

EL BARRACO *1D4* (7km SW Rural) *40.428630, -4.616408* **Camping Pantano del Burguillo, AV-902 Km 16 400, El Barraco [678-48 20 69; info@pantanodelburguillo.com; www.pantanodelburguillo.com]** 31km S of Avila on N403, take AV902 W for 5km. Site on L beside reservoir. Sm, hdstg, mkd, pt shd, wc; chem disp; shwrs inc; EHU inc; lndry; snacks; bar; bbq (sep area); dogs €2; quiet; ccard acc. "Mainly residential c'vans, but staff very welcoming; a bit scruffy; open hg ssn & w'ends only; fair." 22 Jun-10 Sep. € 27.00 2018*

"I need an on-site restaurant"

We do our best to make sure site information is correct, but it is always best to check any must-have facilities are still available or will be open during your visit.

⊞ **ELCHE** *4F2* (10km SW Urban) *38.24055, -0.81194* **Camping Las Palmeras, Ctra Murcia-Alicante, Km 45.3, 03330 Crevillent (Alicante) [965-40 01 88 or 966-68 06 30; fax 966-68 06 64; laspalmeras@laspalmeras-sl.com; www.laspalmeras-sl.com]** Exit A7 junc 726/77 onto N340 to Crevillent. Immed bef traff lts take slip rd into rest parking/service area. Site on R, access rd down side of rest. Med, hdstg, mkd, pt shd, wc; chem disp; shwrs inc; EHU (6A) inc; lndry; shop nr; rest; snacks; bar; pool; paddling pool; 10% statics; dogs free; ccard acc; CKE/CCI. "Useful NH; report to recep in hotel; helpful staff; gd cent for touring Murcia; gd rest in hotel; gd, modern san facs; excel." € 28.00 2017*

SPAIN

ESCALA, L' *3B3* (1km S Urban/Coastal) *42.1211, 3.1346* **Camping L'Escala, Camí Ample 21, 17130 L'Escala (Gerona) [972-77 00 84; fax 972-77 00 08; info@campinglescala.com; www.campinglescala.com]** Exit AP7 junc 5 onto GI623 dir L'Escala; at o'skts of L'Escala, at 1st rndabt (with yellow sp GI623 on top of rd dir sp) turn L dir L'Escala & Ruïnes Empúries; at 2nd rndabt go str on dir L'Escala-Riells, then foll site sp. Do not app thro town. Med, mkd, hdg, pt shd, serviced pitches; wc; chem disp; fam bthrm; shwrs inc; EHU (6A) inc; gas; lndry; shop; rest nr; snacks; bar; bbq; playgrnd; beach 300m; TV; 20% statics; phone; Eng spkn; adv bkg acc; quiet; car wash; CKE/CCI. "Access to sm pitches poss diff lge o'fits; helpful, friendly staff; vg, modern san facs; Empúrias ruins 5km; vg." ♦ 12 Apr-21 Sep. € 44.50 2014*

"Satellite navigation makes touring much easier"

Remember most sat navs don't know if you're towing or in a larger vehicle – always use yours alongside maps and site directions.

ESCALA, L' *3B3* (2km S Coastal) *42.11027, 3.16555* **Camping Illa Mateua, Avda Montgó 260, 17130 L'Escala (Gerona) [972-77 02 00 or 77 17 95; fax 972-77 20 31; info@campingillamateua.com; www.campingillamateua.com]** On N11 thro Figueras, approx 3km on L sp C31 L'Escala; in town foll sp for Montgó & Paradis. Lge, pt shd, terr, wc; chem disp; mv service pnt; fam bthrm; shwrs inc; EHU (5A) inc; gas; lndry; shop; rest; bar; playgrnd; pool; beach sand adj; red long stay; entmnt; 5% statics; dogs €3.60; Eng spkn; adv bkg req; quiet; watersports; games area; tennis; CKE/CCI. "V well-run site; spacious pitches; excel san facs; gd beach; no depth marking in pool; excel rest." ♦ 11 Mar-20 Oct. € 58.40 2013*

ESCALA, L' *3B3* (3km S Coastal) *42.10512, 3.15843* **Camping Neus, Cala Montgó, 17130 L'Escala (Gerona) [638-65 27 12; fax 972-22 24 09; info@campingneus.com; www.campingneus.com]** Exit AP7 junc 5 twd L'Escala then turn R twd Cala Montgó & foll sp. Med, mkd, shd, pt sl, terr, wc; chem disp; mv service pnt; fam bthrm; shwrs inc; EHU (6A) €4; gas; lndry; shop; snacks; bar; playgrnd; pool; paddling pool; beach sand 850m; red long stay; entmnt; wifi; TV; 15% statics; dogs €2; phone; bus 500m; Eng spkn; adv bkg acc; quiet; ccard acc; fishing; tennis; car wash; CKE/CCI. "Pleasant, clean site in pine forest; gd san facs; lge pitches; vg." 14 May-18 Sep. € 48.00 2016*

⊞ **ESCORIAL, EL** *1D4* (6km NE Rural) *40.62630, -4.09970* **Camping-Caravaning El Escorial, Ctra Guadarrama a El Escorial, Km 3.5, 28280 El Escorial (Madrid) [918 90 24 12 or 02 01 49 00; fax 918 96 10 62; info@campingelescorial.com; www.campingelescorial.com]** Exit AP6 NW of Madrid junc 47 El Escorial/Guadarrama, onto M505 & foll sp to El Escorial, site on L at km stone 3,500 - long o'fits rec cont to rndabt (1km) to turn & app site on R. V lge, mkd, hdstg, pt shd, wc (htd); chem disp; fam bthrm; shwrs inc; EHU (5A) inc (long cable rec); gas; lndry; shop; rest; snacks; bar; bbq; playgrnd; pool; entmnt; wifi; TV; 80% statics; dogs free; Eng spkn; adv bkg acc; ccard acc; tennis; games rm; horseriding 7km. "Excel, busy site; mountain views; o'fits over 8m must reserve lge pitch with elec, water & drainage; helpful staff; clean facs; gd security; sm pitches poss diff due trees; o'head canopies poss diff for tall o'fits; facs ltd LS; trains & buses to Madrid nr; Valle de Los Caídos & Palace at El Escorial well worth visit; cash machine; easy parking in town for m'vans if go in early; mkt Wed; stunning scenery, nesting storks; well stocked shop; v well run; 20min walk to bus stop for El Escorial or Madrid; rec; gd for families; gd pool." ♦ € 39.00 2017*

ESPOT *3B2* (1km SE Rural) *42.57223, 1.09677* **Camping Sol I Neu, Ctra Sant Maurici s/n, 25597 Espot (Lleida) [973-62 40 01; fax 973-62 41 07; camping@solineu.com; www.solineu.com]** N fr Sort on C13 turn L to Espot on rd LV5004, site on L in approx 6.5km by rvside. Med, mkd, pt shd, wc; chem disp; fam bthrm; shwrs inc; EHU (6-10A) €5.80; gas; lndry; shop; bar; playgrnd; pool; paddling pool; TV; dogs; quiet; ccard acc; CKE/CCI. "Excel facs; beautiful site nr National Park (Landrover taxis avail - no private vehicles allowed in Park); suitable sm o'fits only; poss unrel opening dates; 10 mins walk to vill, many bars & rest." 1 Jul-31 Aug. € 25.60 2015*

"There aren't many sites open at this time of year"

If you're travelling outside peak season remember to call ahead to check site opening dates – even if the entry says 'open all year'.

ESTARTIT, L' *3B3* (1km S Coastal) *42.04972, 3.18416* **Camping El Molino, Camino del Ter, 17258 L'Estartit (Gerona) [972-75 06 29]** Fr N11 junc 5, take rd to L'Escala. Foll sp to Torroella de Montgri, then L'Estartit. Ent town & foll sp, site on rd GI 641. V lge, hdg, pt shd, pt sl, wc; mv service pnt; shwrs; EHU (6A) €3.60; gas; lndry; shop; rest; bar; playgrnd; beach sand 1km; wifi; bus 1km; adv bkg rec; games rm. "Site in 2 parts - 1 in shd, 1 at beach unshd; gd facs; quiet location outside busy town." 1 Apr-30 Sep. € 28.00 2014*

ESTARTIT, L' *3B3* (2km S Coastal) *42.04250, 3.18333* **Camping Les Medes, Paratge Camp de l'Arbre s/n, 17258 L'Estartit (Gerona) [972-75 18 05; fax 972-75 04 13; info@campinglesmedes.com; www.campinglesmedes.com]** Fr Torroella foll sp to L'Estartit. In vill turn R at town name sp (sp Urb Estartit Oeste), foll rd for 1.5km, turn R, site well sp. Lge, mkd, shd, serviced pitches; wc (htd); chem disp; mv service pnt; fam bthrm; shwrs inc; EHU (6A) €4.60; gas; lndry; shop; rest; snacks; bar; playgrnd; pool (htd, indoor); beach sand 800m; red long stay; entmnt; wifi; TV; 7% statics; dogs (except high ssn, otherwise €2.60); phone; Eng spkn; adv bkg acc; quiet; sauna; games area; tennis; horseriding 400m; car wash; games rm; watersports; solarium; bike hire; CKE/CCI. "Excel, popular, family-run & well organised site; helpful, friendly staff; gd clean facs & constant hot water; gd for children; no twin axle vans high ssn - by arrangement LS; conv National Park; well mkd foot & cycle paths; ACSI acc." ♦ € 47.00 2017*

"That's changed – Should I let the Club know?"

If you find something on site that's different from the site entry, fill in a report and let us know. See camc.com/europereport.

ESTEPAR *1B4* (2km NE Rural) *42.29233, -3.85097* **Quinta de cavia, A62, km 17, 09196 Cavia, Burgos [947-41 20 78; reservas@quintadecavia.es]** Site 15km SW of Burgos on N side of A62/E80, adj Hotel Rio Cabia. Ent via Campsa petrol stn, W'bound exit 17, E'bound exit 16, cross over & re-join m'way. Ignore camp sp at exit 18 (1-way). Site at 17km stone. Med, pt shd, wc; chem disp; shwrs inc; EHU (6A) inc; rest; bar; playgrnd; 10% statics; dogs; ccard acc; CKE/CCI. "Friendly, helpful owner; gd rest; conv for m'way for Portugal but poorly sp fr W; poss v muddy in winter; NH only; food in rest vg and cheap; can get to Bilbao ferry in morn if ferry is mid-aft; elec security gate; lit at night; gd NH; new san facs (2018); site 2723 feet abv sea level, cold at night." € 20.00 2018*

ESTEPONA *2H3* (7km E Coastal) *36.45436, -5.08105* **Camping Parque Tropical, Ctra N340, Km 162, 29680 Estepona (Málaga) [952-79 36 18; parquetropicalcamping@hotmail.com; www.campingparquetropical.com]** On N side of N340 at km 162, 200m off main rd. Med, mkd, hdg, pt shd, terr, serviced pitches; wc; chem disp; mv service pnt; shwrs inc; EHU (10A) €4; gas; lndry; shop; rest; snacks; bar; playgrnd; pool (covrd, htd); beach shgl 1km; red long stay; 10% statics; dogs €2; phone; bus 400m; Eng spkn; adv bkg acc; golf nr; horseriding nr; CKE/CCI. "Tropical plants thro out; clean facs; tropical paradise swimming pool; wildlife park 1km; helpful owners." ♦ € 27.00 2017*

ETXARRI ARANATZ *3B1* (2km N Rural) *42.91255, -2.07919* **Camping Etxarri, Parase Dambolintxulo, 31820 Etxarri-Aranatz (Navarra) [948-46 05 37; info@campingetxarri.com; www.campingetxarri.com]** Fr N exit A15 at junc 112 to join A10 W dir Vitoria/Gasteiz. Exit at junc 19 onto NA120; go thro Etxarri vill, turn L & cross bdge, then take rd over rlwy. Turn L, site sp. Med, hdg, pt shd, wc; chem disp; shwrs inc; EHU (6A) €5.50; gas; lndry; shop; rest; bar; bbq; playgrnd; pool; entmnt; wifi; 90% statics; dogs €2.15; phone; Eng spkn; ccard acc; games area; archery; cycling; horseriding; CKE/CCI. "Gd, wooded site; gd walks; interesting area; helpful owner; conv NH to/fr Pyrenees; youth hostel & resident workers on site; san facs gd; various pitch sizes & shapes, some diff lge o'fits; NH only." 1 Mar-5 Oct. € 38.00 2014*

SPAIN

FIGUERES *3B3* (12km N Rural) *42.37305, 2.91305* **Camping Les Pedres, Calle Vendador s/n, 17750 Capmany (Gerona) [972-54 91 92 or 686 01 12 23 (mob); info@campinglespedres.net; www.campinglespedres.net]** S fr French border on N11, turn L sp Capmany, L again in 2km at site sp & foll site sp. Med, mkd, pt shd, pt sl, wc (htd); chem disp; shwrs inc; EHU (6-10A) €4.50; lndry; shop nr; rest; snacks; bar; pool; 20% statics; dogs; phone; Eng spkn; adv bkg acc; quiet; ccard acc; CKE/CCI. "Helpful Dutch owner; lovely views; gd touring & walking cent; gd winter NH." ♦ € 39.40 2013*

FIGUERES *3B3* (8km NE Rural) *42.33902, 3.06758* **Camping Vell Empordà, Ctra Roses-La Jonquera s/n, 17780 Garriguella (Gerona) [972-53 02 00 or 972-57 06 31 (LS); fax 972-55 23 43; vellemporda@vellemporda.com; www.vellemporda.com]** On A7/E11 exit junc 3 onto N260 NE dir Llançà. Nr km 26 marker, turn R sp Garriguella, then L at T-junc N twd Garriguella. Site on R shortly bef vill. Lge, hdg, mkd, hdstg, shd, terr, wc (htd); chem disp; mv service pnt; fam bthrm; shwrs inc; EHU (6-10A) inc; gas; lndry; shop; rest; snacks; bar; bbq; playgrnd; pool; paddling pool; red long stay; entmnt; wifi; TV; 20% statics; dogs €4.50; phone; Eng spkn; adv bkg acc; quiet; ccard acc; games rm; games area; CKE/CCI. "Conv N Costa Brava away fr cr beaches & sites; 20 mins to sea at Llançà; o'hanging trees poss diff high vehicles; excel." ♦ 1 Feb-15 Dec. € 40.00 2018*

FIGUERUELA DE ARRIBA *1B3* (2km E Rural) *41.86563, -6.41937* **Camping Sierra de la Culebra, Carretera Riomanzanas, S/N 49520 Figueruela de Arriba [980-68 30 20 or 630-66 13 29 (mob); info@campingsierradelaculebra.com; www.campingsierradelaculebra.com]** Fr S: At Alcanices on the N122 (E82) Zamora-Braganca rd, turn N onto ZA9112. In 21km at Mahide turn L. Site sp on L in 3km. Fr N: Exit A52 at junc 49 onto N631 dir Zamora. In 4km turn R onto ZA912. In 28km at Mahide turn R. Site sp. Med, pt shd, wc; chem disp; fam bthrm; shwrs; EHU (6A) €4.10; rest; snacks; bar; bbq; playgrnd; pool; paddling pool; twin axles; wifi; TV; 10% statics; dogs; Eng spkn; adv bkg acc; quiet; games area. "Fam owned; many interesting old vill; gd bird-watching; tennis; gd walking; gd for Braganca in Portugal; excel." ♦ 5 Mar-2 Nov. € 26.60 2016*

SPAIN

⊞ **FORTUNA** *4F1* (3km N Rural) *38.20562, -1.10712* **Camping Fuente, Camino de la Bocamina s/n, 30709 Baños de Fortuna (Murcia) [968-68 50 17; fax 968 68 51 25; info@campingfuente.com; www.campingfuente.com]** Fr Murcia on A7/E15 turn L onto C3223 sp Fortuna. After 19km turn onto A21 & foll sp Baños de Fortuna, then sp 'Complejo Hotelero La Fuente'. Avoid towing thro vill, if poss. Med, mkd, hdstg, unshd, pt sl, wc (htd); chem disp; shwrs; EHU (10-16A) €2.20 or metered, poss rev pol; gas; lndry; shop; rest; snacks; bar; bbq; playgrnd; pool (htd); red long stay; wifi; 10% statics; dogs €1.10; phone; bus 200m; adv bkg acc; ccard acc; CKE/CCI. "Gd san facs; excel pool & rest; secure o'flow parking area; many long-stay winter visitors - adv bkg rec; private san facs some pitches; ltd recep hrs LS; spa; jacuzzi; poss sulphurous smell fr thermal baths." ♦ € 19.00 2015*

FOZ *1A2* (11km E Coastal) *43.56236, -7.20761* **Camping Poblado Gaivota, Playa de Barreiros, 27790 Barreiros Lugo [982-12 44 51; www.campingpobladogaivota.com]** Junc 516 on A8, on to N634 to Barreiros. Foll sp to the R in Barreiros. Rd winds down to coast for 2km. Site on L, parallel with sea. Sm, hdg, shd, wc; chem disp; mv service pnt; shwrs; EHU (6A); lndry; shop; rest; snacks; bar; bbq; playgrnd; beach opp; entmnt; wifi; Eng spkn. "Excel." 21 Mar-15 Oct. € 30.00 2016*

FRANCA, LA *1A4* (1km NW Coastal) *43.39250, -4.57722* **Camping Las Hortensias, Ctra N634, Km 286, 33590 Colombres/Ribadedeva (Asturias) [985-41 24 42; fax 985-41 21 53; lashortensias@campinglashortensias.com; www.campinglashortensias.com]** Fr N634 on leaving vill of La Franca, at km286 foll sp 'Playa de la Franca' & cont past 1st site & thro car park to end of rd. Med, mkd, pt shd, pt sl, terr, wc; chem disp; fam bthrm; shwrs inc; EHU (6-10A) €5; gas; lndry; shop; rest nr; bar nr; playgrnd; beach sand adj; dogs (but not on beach) €5; phone; bus 800m; Eng spkn; adv bkg acc; ccard acc; bike hire; tennis; CKE/CCI. "Beautiful location nr scenic beach; sea views fr top terr pitches; vg." 5 Jun-30 Sep. € 28.50 2017*

FRESNEDA, LA *3C2* (3km SW Rural) *40.90705, 0.06166* **Camping La Fresneda, Partida Vall del Pi, 44596 La Fresneda (Teruel) [978-85 40 85; info@campinglafresneda.com; www.campinglafresneda.com]** Fr Alcañiz S on N232 dir Morella; in 15km turn L onto A231 thro Valjunquera to La Fresneda; cont thro vill; in 2.5km turn R onto site rd. Site sp fr vill. Sm, mkd, hdg, pt shd, terr, wc; chem disp; fam bthrm; shwrs inc; EHU (6A) inc; gas; lndry; rest; snacks; bar; pool; red long stay; wifi; phone; Eng spkn; quiet; ccard acc; CKE/CCI. "Narr site rds, poss diff lge o'fits; various pitch sizes; gd; adv bkg adv; v helpful owners." ♦ 15 Mar-15 Oct. € 24.50 2012*

⊞ **FUENTE DE PIEDRA** *2G4* (1km S Rural) *37.12905, -4.73315* **Camping Fuente de Pedra, Calle Campillos 88-90, 29520 Fuente de Piedra (Málaga) [952-73 52 94; info@campingfuentedepiedra.com; www.camping-rural.com]** Turn off A92 at km 132 sp Fuente de Piedra. Sp fr vill cent. Or to avoid town turn N fr A384 just W of turn for Bobadilla Estación, sp Sierra de Yeguas. In 2km turn R into nature reserve, cont for approx 3km, site on L at end of town. Sm, hdstg, mkd, pt shd, pt sl, terr, wc; shwrs inc; EHU (10A) €5; gas; lndry; shop; rest; snacks; bar; bbq; playgrnd; pool; red long stay; wifi; 25% statics; dogs €3; phone; Eng spkn; ccard acc; CKE/CCI. "Mostly sm, narr pitches, but some avail for o'fits up to 7m; gd rest; san facs dated & poss stretched; poss noise fr adj public pool; adj lge lake with flamingoes (visible but access is 4.8km away); gd." ♦ € 26.00 2017*

FUENTE DE SAN ESTABAN, LA *1D3* (1km E Rural) *40.79128, -6.24384* **Camping El Cruce, 37200 La Fuente de San Estaban (Salamanca) [923-44 01 30; campingelcruce@yahoo.es; www.campingelcruce.com]** On A62/E80 (Salamanca-Portugal) take exit 293 into vill & foll signs immed behind hotel on S side of rd. Fr E watch for sp 'Cambio de Sentido' to cross main rd. Med, pt shd, wc; chem disp; shwrs; EHU (6A) €3.50 (poss no earth); rest nr; snacks; bar; playgrnd; wifi; Eng spkn; ccard acc; CKE/CCI. "Conv NH/sh stay en rte Portugal; friendly." ♦ 1 May-30 Sep. € 15.00 2013*

⊞ **FUENTEHERIDOS** *2F3* (1km SW Rural) *37.9050, -6.6742* **Camping El Madroñal, Ctra Fuenteheridos-Castaño del Robledo, Km 0.6, 21292 Fuenteheridos (Huelva) [959-50 12 01; castillo@campingelmadronal.com; www.campingelmadronal.com]** Fr Zafra S on N435n turn L onto N433 sp Aracena, ignore first R to Fuenteheridos vill, camp sp R at next x-rd 500m on R. At rndabt take 2nd exit. Avoid Fuenteheridos vill - narr rds. Med, mkd, pt shd, pt sl, wc; chem disp; shwrs; EHU €3.20; gas; lndry; shop; snacks; bar; bbq; pool; 80% statics; dogs; phone; bus 1km; quiet; horseriding; car wash; bike hire; CKE/CCI. "Tranquil site in National Park of Sierra de Aracena; pitches among chestnut trees - poss diff lge o'fits or m'vans & poss sl & uneven; o'hanging trees on site rds; scruffy, pitches not clearly mkd; beautiful vill 1km away, worth a visit." € 13.00 2014*

⊞ **GALLARDOS, LOS** *4G1* (4km N Rural) *37.18448, -1.92408* **Camping Los Gallardos, 04280 Los Gallardos (Almería) [950-52 83 24; fax 950-46 95 96; reception@campinglosgallardos.com; www.campinglosgallardos.com]** Fr N leave A7/E15 at junc 525; foll sp to Los Gallardos; take 1st R after approx 800m pass under a'route; turn L into site ent. Med, hdstg, mkd, pt shd, serviced pitches; wc; chem disp; mv service pnt; shwrs inc; EHU (10A) €3; gas; lndry; shop; rest; snacks; bar; pool; red long stay; 40% statics; dogs €2.25; adv bkg acc; ccard acc; tennis adj; golf; CKE/CCI. "British owned; 90% British clientele LS; gd social atmosphere; sep drinking water supply nr recep; prone to flooding wet weather; 2 grass bowling greens; facs tired; poss cr in winter; friendly staff." ♦ € 23.50 2017*

⊞ **GANDIA** *4E2* (4km NE Coastal) *38.98613, -0.16352* **Camping L'Alqueria, Avda del Grau s/n; 46730 Grao de Gandía (València) [962-84 04 70; fax 962-84 10 63; lalqueria@lalqueria.com; www.lalqueria.com]** Fr N on A7/AP7 exit 60 onto N332 dir Grao de Gandía. Site sp on rd bet Gandía & seafront. Fr S exit junc 61 & foll sp to beaches. Lge, mkd, hdstg, pt shd, wc (htd); chem disp; mv service pnt; fam bthrm; shwrs inc; EHU (10A) €5.94; gas; lndry; shop; rest nr; snacks; bar; playgrnd; pool (covrd, htd); beach sand 1km; red long stay; entmnt; wifi; 30% statics; dogs (only sm dogs, under 10kg) €1.90; phone; bus; adv bkg acc; quiet; ccard acc; games area; jacuzzi; bike hire; CKE/CCI. "Pleasant site; helpful, friendly family owners; lovely pool; easy walk to town & stn; excel beach nrby; bus & train to Valencia; shop & snacks not avail in Jul; gd biking; htd pool avail; m'van friendly; site scruffy but san facs gd; gd location; rec; gd location bet old town & beach; gd cycling & walking." ♦ € 42.00 2018*

⊞ **GARGANTILLA DEL LOZOYA** *1C4* (2km SW Rural) *40.9503, -3.7294* **Camping Monte Holiday, Ctra C604, Km 8.8, 28739 Gargantilla del Lozoya (Madrid) [918-69 52 78; monteholiday@monteholiday.com; www.monteholiday.com]** Fr N on A1/E5 Burgos-Madrid rd turn R on M604 at km stone 69 sp Rascafría; in 8km turn R immed after rlwy bdge & then L up track in 300m, foll site sp. Do not ent vill. Lge, pt shd, pt sl, terr, wc; chem disp; mv service pnt; fam bthrm; shwrs inc; EHU (7A) €4.30 (poss rev pol); lndry; rest; bar; pool; wifi; 80% statics; phone; bus 500m; Eng spkn; adv bkg acc; quiet; ccard acc; CKE/CCI. "Interesting, friendly site; vg san facs; gd views; easy to find; some facs clsd LS; lovely area but site isolated in winter & poss heavy snow; conv NH fr m'way & for Madrid & Segovia; excel wooded site; v rural but well worth the sh drive fr the N1 E5; clean; lovely surroundings." ♦ € 41.70 2014*

GAVA *3C3* (5km S Coastal) *41.27245, 2.04250* **Camping Tres Estrellas, C31, Km 186.2, 08850 Gavà (Barcelona) [936-33 06 37; fax 936-33 15 25; info@camping3estrellas.com; www.camping3estrellas.com]** Fr S take C31 (Castelldefels to Barcelona), exit 13. Site at km 186.2 300m past rd bdge. Fr N foll Barcelona airport sp, then C31 junc 13 Gavà-Mar slip rd immed under rd bdge. Cross m'way, turn R then R again to join m'way heading N for 400m. Lge, mkd, pt shd, pt sl, wc (htd); chem disp; mv service pnt; fam bthrm; shwrs inc; EHU (5A) €6.56 (poss rev pol &/ or no earth); gas; lndry; shop; rest; snacks; bar; bbq; playgrnd; pool (htd); beach sand adj; entmnt; wifi; TV; 20% statics; dogs €4.90; phone; bus 400m; Eng spkn; adv bkg acc; ccard acc; tennis; CKE/CCI. "20 min by bus to Barcelona cent; poss smells fr stagnant stream in corner of site; poss mosquitoes." ♦ 15 Mar-15 Oct. € 52.50 2013*

GERONA *3B3* (8km S Rural) *41.9224, 2.82864* **Camping Can Toni Manescal, Ctra de la Barceloneta, 17458 Fornells de la Selva (Gerona) [972-47 61 17; fax 972-47 67 35; campinggirona@campinggirona.com; www.campinggirona.com]** Fr N leave AP7 at junc 7 onto N11 dir Barcelona. In 2km turn L to Fornells de la Selva; in vill turn L at church (sp); over rv; in 1km bear R & site on L in 400m. NB Narr rd in Fornells vill not poss lge o'fits. Sm, mkd, pt shd, pt sl, wc; chem disp; fam bthrm; shwrs inc; EHU (5A) inc (poss long lead req); gas; lndry; shop nr; rest nr; playgrnd; pool; dogs; train nr; Eng spkn; adv bkg acc; quiet; ccard acc; CKE/CCI. "Pleasant, open site on farm; gd base for lovely medieval city Gerona - foll bus stn sp for gd, secure m'van parking; welcoming & helpful owners; lge pitches; ltd san facs; excel cycle path into Gerona, along old rlwy line; Gerona mid-May flower festival rec; gd touring base away fr cr coastal sites." 1 Jun-30 Sep. € 23.00 2012*

GIJON *1A3* (10km NW Coastal) *43.58343, -5.75713* **Camping Perlora, Ctra Candás, Km 12, Perán, 33491 Candás (Asturias) [985-87 00 48; recepcion@campingperlora.com; www.campingperlora.com]** Exit A8 dir Candás; in 9km at rndabt turn R sp Perlora (AS118). At sea turn L sp Candás, site on R. Avoid Sat mkt day. Med, mkd, unshd, pt sl, terr, serviced pitches; wc; chem disp; shwrs inc; EHU (10A) €3.50; gas; lndry; shop; rest; bar; playgrnd; beach sand 1km; red long stay; wifi; 80% statics; dogs free; phone; bus adj; Eng spkn; quiet; ccard acc; watersports; fishing; tennis. "Excel; helpful staff; attractive, well-kept site on dramatic headland o'looking Candas Bay; ltd space for tourers; easy walk to Candás; gem of a site; train (5 mins); excel fish rests; no mob signal; facs gd & clean, ltd LS; rec." ♦ 19 Jan-9 Dec. € 27.50 2018*

GORLIZ *1A4* (1km N Coastal) *43.41782, -2.93626* **Camping Arrien, Uresarantze Bidea, 48630 Gorliz (Bizkaia) [946-77 19 11; fax 946-77 44 80; recepcion@campinggorliz.com or campingarrien@gmail.com; www.campinggorliz.com]** Fr Bilbao foll m'way to Getxo, then 637/634 thro Sopelana & Plentzia to Gorliz. In Gorliz turn L at 1st rndabt, foll sps for site, pass TO on R, then R at next rndabt, strt over next, site on L adj sports cent/running track. Not sp locally. Lge, pt shd, pt sl, wc; chem disp; shwrs inc; EHU (6A) inc; gas; lndry; shop; rest; snacks; bar; bbq; playgrnd; beach sand 700m; red long stay; wifi; 60% statics; dogs €1; phone; bus 150m; Eng spkn; ccard acc; CKE/CCI. "Useful base for Bilbao & ferry (approx 1hr); bus to Plentzia every 20 mins, fr there can get metro to Bilbao; friendly, helpful staff; poss shortage of hot water; very cramped; no mkd pitches." 1 Mar-31 Oct. € 36.00 2016*

⊞ **GRANADA** *2G4* (4km N Urban) *37.19832, -3.61166* **Camping Motel Sierra Nevada, Avda de Madrid 107, 18014 Granada [958-15 00 62; fax 958-15 09 54; campingmotel@terra.es; www.campingsierranevada.com]** App Granada S-bound on A44 & exit at junc 123, foll dir Granada. Site on R in 1.5km just beyond bus stn & opp El Campo supmkt, well sp. Lge, shd, wc; chem disp; mv service pnt; fam bthrm; shwrs inc; EHU (6A) €4.20; gas; lndry; shop nr; rest; snacks; bbq; playgrnd; wifi; dogs; bus to city cent 500m; Eng spkn; ccard acc; sports facs; CKE/CCI. "V helpful staff; excel san facs, but poss ltd LS; 2 pools adj; motel rms avail; can book Alhambra tickets at recep (24 hrs notice); conv city; excel site." ♦ € 40.00 2014*

⊞ **GRANADA** *2G4* (13km E Rural) *37.16085, -3.45388* **Camping Las Lomas, 11 Ctra de Güejar-Sierra, Km 6, 18160 Güejar-Sierra (Granada) [958-48 47 42; fax 958-48 40 00; info@campinglaslomas.com; www.campinglaslomas.com]** Fr A44 exit onto by-pass 'Ronda Sur', then exit onto A395 sp Sierra Nevada. In approx 4km exit sp Cenes, turn under A395 to T-junc & turn R sp Güejar-Sierra, Embalse de Canales. After approx 3km turn L at sp Güejar-Sierra & site. Site on R 6.5km up winding mountain rd. Med, mkd, hdg, pt shd, terr, wc (htd); chem disp; mv service pnt; fam bthrm; shwrs inc; EHU (10A) €4 (poss no earth/rev pol); gas; lndry; shop; rest; snacks; bar; playgrnd; pool; paddling pool; red long stay; wifi; dogs free; bus adj; Eng spkn; adv bkg req; quiet; ccard acc; CKE/CCI. "Helpful, friendly owners; well-run site; conv Granada (bus at gate); waterskiing nr; access poss diff for lge o'fits; excel san facs; gd shop & rest; beautiful mountain scenery; excel site." ♦ € 40.00 2014*

⊞ **GRANADA** *2G4* (10km SE Urban) *37.12444, -3.58611* **Camping Reina Isabel, Calle de Laurel de la Reina, 18140 La Zubia (Granada) [958-59 00 41; info@campingreinaisabel.com; www.campingreinaisabel.es]** Exit A44 nr Granada at junc sp Ronda Sur, dir Sierra Nevada, Alhambra, then exit 2 sp La Zubia. Foll site sp approx 1.2km on R; narr ent set back fr rd. Med, hdstg, hdg, pt shd, wc (htd); chem disp; mv service pnt; fam bthrm; shwrs inc; EHU (6A) poss rev pol €4.20; gas; lndry; shop; snacks; bar; pool; red long stay; wifi; TV; dogs free; phone; bus; Eng spkn; adv bkg rec; quiet; ccard acc; CKE/CCI. "Well-run, busy site; poss shwrs v hot/cold - warn children; ltd touring pitches & sm; poss student groups; conv Alhambra (order tickets at site), shwr block not htd; tight site; elec unrel." ♦ € 29.00 2017*

GUADALUPE *2E3* (2km S Rural) *39.44232, -5.31708* **Camping Las Villuercas, Ctra Villanueva-Huerta del Río, Km 2, 10140 Guadalupe (Cáceres) [927-36 71 39; fax 927-36 70 28; www.campinglasvilluercasguadalupe.es]** Exit A5/E90 at junc 178 onto EX118 to Guadalupe. Do not ent town. Site sp on R at rndabt at foot of hill. Med, shd, wc; shwrs; EHU €2.50 (poss no earth/rev pol); lndry; shop; rest; bar; playgrnd; pool; ccard acc; tennis. "Vg; helpful owners; ltd facs LS; some pitches sm & poss not avail in wet weather; nr famous monastery." 1 Mar-15 Dec. € 14.50 2012*

⊞ **GUARDAMAR DEL SEGURA** *4F2* (4km N Rural/Coastal) *38.10916, -0.65472* **Camping Marjal, Ctra N332, Km 73.4, 03140 Guardamar del Segura (Alicante) [965-48 49 45; camping@marjal.com; www.campingmarjal.com]** Fr N exit A7 junc 72 sp Aeropuerto/Santa Pola; in 5km turn R onto N332 sp Santa Pola/Cartagena, U-turn at km 73.4, site sp on R at km 73.5. Fr S exit AP7 at junc 740 onto CV91 twd Guardamar. In 9km join N332 twd Alicante, site on R at next rndabt. Lge, mkd, hdstg, hdg, pt shd, serviced pitches; wc; chem disp; fam bthrm; shwrs inc; EHU (16A) €3 or metered; gas; lndry; shop; rest; snacks; bar; bbq; playgrnd; pool (covrd, htd); beach sand 1km; red long stay; entmnt; wifi; TV; 50% statics; dogs €2.20; phone; Eng spkn; adv bkg rec; quiet; ccard acc; bike hire; tennis; sauna; CKE/CCI. "Fantastic facs; friendly, helpful staff; excel family entmnt & activities; recep 0800-2300; sports cent; tropical waterpark; excel; well sign-posted; lge shwr cubicles; highly rec." ♦ € 67.00 2018*

"I like to fill in the reports as I travel from site to site"

You'll find report forms at the back of this guide, or you can fill them in online at camc.com/europereport.

GUARDIOLA DE BERGUEDA *3B3* (4km SW Rural) *42.21602, 1.83705* **Camping El Berguedà, Ctra B400, Km 3.5, 08694 Guardiola de Berguedà (Barcelona) [938-22 74 32; info@campingbergueda.com; www.campingbergueda.com]** On C16 S take B400 W dir Saldes. Site is approx 10km S of Cadí Tunnel. Med, hdstg, mkd, pt shd, terr, wc; chem disp; fam bthrm; shwrs; EHU (6A) €3.90 (poss rev pol); gas; lndry; shop; rest; snacks; bar; bbq; playgrnd; pool; paddling pool; TV; 10% statics; dogs; phone; Eng spkn; quiet; games area; games rm; CKE/CCI. "Helpful staff; vg clean san facs; beautiful, remote situation; gd walking; poss open w/ends in winter; highly rec; spectacular mountains/scenery; gd touring area; rec Gaudi's Garden in La Pobla." ♦ 1 Apr-1 Nov. € 39.00 2014*

GUITIRIZ *1A2* (1km N Rural) *43.17850, -7.82640* **Camping El Mesón, 27305 Guitiriz (Lugo) [982-37 32 88; fax 626 50 91 40; elcampingelmeson@gmail.com; http://elmesoncamping.wix.com/campingelmeson.]** On A6 NW fr Lugo, exit km 535 sp Guitiriz, site sp. Sm, mkd, pt shd, pt sl, wc; chem disp; fam bthrm; shwrs inc; EHU (10A) €4.50; gas; lndry; shop nr; snacks; bar; bbq; playgrnd; sw nr; phone; bus adj; quiet. "Vg site." 15 Jun-15 Sep. € 17.00 2014*

HARO *1B4* (0km N Urban) *42.57824, -2.85421* **Camping de Haro, Avda Miranda 1, 26200 Haro (La Rioja) [941-31 27 37; fax 941-31 20 68; camping deharo@fer.es; www.campingdeharo.com]** Fr N or S on N124 take exit sp A68 Vitoria/Logrono & Haro. In 500m at rndabt take 1st exit, under rlwy bdge, cont to site on R immed bef rv bdge. Fr AP68 exit junc 9 to town; at 2nd rndabt turn L onto LR111 (sp Logroño). Immed after rv bdge turn sharp L & foll site sp. Avoid cont into town cent. Med, mkd, hdg, pt shd, wc (htd); chem disp; mv service pnt; shwrs inc; EHU (6A) inc; gas; lndry; shop; snacks; bar; bbq; playgrnd; pool (htd); wifi; 70% statics; dogs €3; phone; bus 800m; Eng spkn; adv bkg acc; ccard acc; site clsd 9 Dec-13 Jan; car wash; CKE/CCI. "Clean, tidy, well run, lovely site - peaceful LS; friendly owner; some sm pitches & diff turns; excel facs; statics busy at w/ends; conv Rioja 'bodegas' & Bilbao & Santander ferries; recep clsd 1300-1500 no entry then due to security barrier; excel & conv NH for Santander/ Bilbao; lge o'night area with electric; big, busy, vg site; easy walk to lovely town; san facs constantly cleaned!" ♦ 27 Jan-10 Dec. € 30.00 2017*

HARO *1B4* (10km SW Rural) *42.53017, -2.92173* **Camping De La Rioja, Ctra de Haro/Santo Domingo de la Calzada, Km 8.5, 26240 Castañares de la Rioja (La Rioja) [941 30 01 74; fax 941-30 01 56; info@ campingdelarioja.com]** Exit AP68 junc 9, take rd twd Santo Domingo de la Calzada. Foll by-pass round Casalarreina, site on R nr rvside just past vill on rd LR111. Lge, hdg, pt shd, wc (htd); chem disp; shwrs; EHU (4A) €3.90 (poss rev pol); gas; lndry; shop; rest; snacks; bar; pool; entmnt; 90% statics; dogs; bus adj; adv bkg acc; ccard acc; bike hire; tennis; clsd 10 Dec-8 Jan; site clsd 9 Dec-11 Jan. "Fair site but fairly isolated; basic san facs but clean; ltd facs in winter; sm pitches; conv for Rioja wine cents; Bilbao ferry." 10 Jan-10 Dec. € 36.00 2016*

⊞ **HECHO** *3B1* (1km S Rural) *42.73222, -0.75305* **Camping Valle de Hecho, Ctra Puente La Reina-Hecho s/n, 22720 Hecho (Huesca) [974-37 53 61; fax 976-27 78 42; campinghecho@campinghecho. com; www.campinghecho.com]** Leave Jaca W on N240. After 25km turn N on A176 at Puente La Reina de Jaca. Site on W of rd, o'skts of Hecho/Echo. Med, mkd, pt shd, pt sl, wc (htd); chem disp; mv service pnt; shwrs inc; EHU (5-15A) €4.20; gas; lndry; shop; rest; snacks; bar; playgrnd; pool; 40% statics; dogs; phone; bus 200m; quiet; ccard acc; games area; CKE/CCI. "Pleasant site in foothills of Pyrenees; excel, clean facs but poss inadequate hot water; gd birdwatching area; Hecho fascinating vill; shop & bar poss clsd LS except w/end; v ltd facs LS; not suitable lge o'fits; gd rest in vill." € 27.00 2016*

⊞ **HORCAJO DE LOS MONTES** *2E4* (0km E Rural) *39.32440, -4.6358* **Camping Mirador de Cabañeros, Calle Cañada Real Segoviana s/n, 13110 Horcajo de los Montes (Ciudad Real) [926 77 54 39; fax 926 77 50 03; info@campingcabaneros.com; www. campingcabaneros.com]** At km 53 off CM4103 Horcajo-Alcoba rd, 200m fr vill. CM4106 to Horcajo fr NW poor in parts. Med, mkd, hdstg, pt shd, terr, wc (htd); chem disp; mv service pnt; fam bthrm; shwrs; EHU (6A) €4.20; gas; shop nr; rest; bar; bbq; playgrnd; pool; red long stay; entmnt; TV; 10% statics; dogs €2; phone; adv bkg rec; quiet; ccard acc; tennis 500m; games area; bike hire; games rm; CKE/CCI. "Beside Cabañeros National Park; beautiful views; rd fr S much better." ♦ € 32.40 2013*

HORNOS *4F1* (9km SW Rural) *38.18666, -2.77277* **Camping Montillana Rural, Ctra Tranco-Hornos A319, km 78.5, 23292 Hornos de Segura (Jaén) [953-12 61 94 or 680-15 21 10; jrescalvor@hotmail. com; www.campingmontillana.es]** Fr N on N322 take A310 then A317 S then A319 dir Tranco & Cazorla. Site nr km 78.5, ent by 1st turning. Fr S on N322 take A6202 N of Villaneuva del Arzobispo. In 26km at Tranco turn L onto A319 & nr km 78.5 ent by 1st turning up slight hill. Sm, mkd, hdstg, pt shd, terr, wc, chem disp, shwrs inc; EHU (10A) €3.20; lndry; shop; rest; snacks; bar; pool; 5% statics; dogs; phone; Eng spkn; adv bkg acc; quiet; CKE/CCI. "Beautiful area; conv Segura de la Sierra, Cazorla National Park; much wildlife; lake adj; friendly, helpful staff; gd site." 19 Mar-30 Sep. € 16.00 2013*

"We must tell the Club about that great site we found"

Get your site reports in by mid-August and we'll do our best to get your updates into the next edition.

HOSPITAL DE ORBIGO *1B3* (1km N Urban) *42.4664, -5.8836* **Camp Municipal Don Suero, 24286 Hospital de Órbigo (León) [987-36 10 18; fax 987-38 82 36; camping@hospitaldeorbigo.com; www.hospital deorbigo.com]** N120 rd fr León to Astorga, km 30. Site well sp fr N120. Narr rds in Hospital. Med, hdg, pt shd, wc; shwrs; EHU (6A) €1.90; lndry; shop; rest nr; bar; bbq; 50% statics; dogs; phone; bus to León 1km; Eng spkn; ccard acc; poss open w/end only mid Apr-May; CKE/CCI. "Statics v busy w/ends, facs stretched; pool adj; phone ahead to check site open if travelling close to opening/closing dates." ♦ Easter-1 Oct. € 15.00 2016*

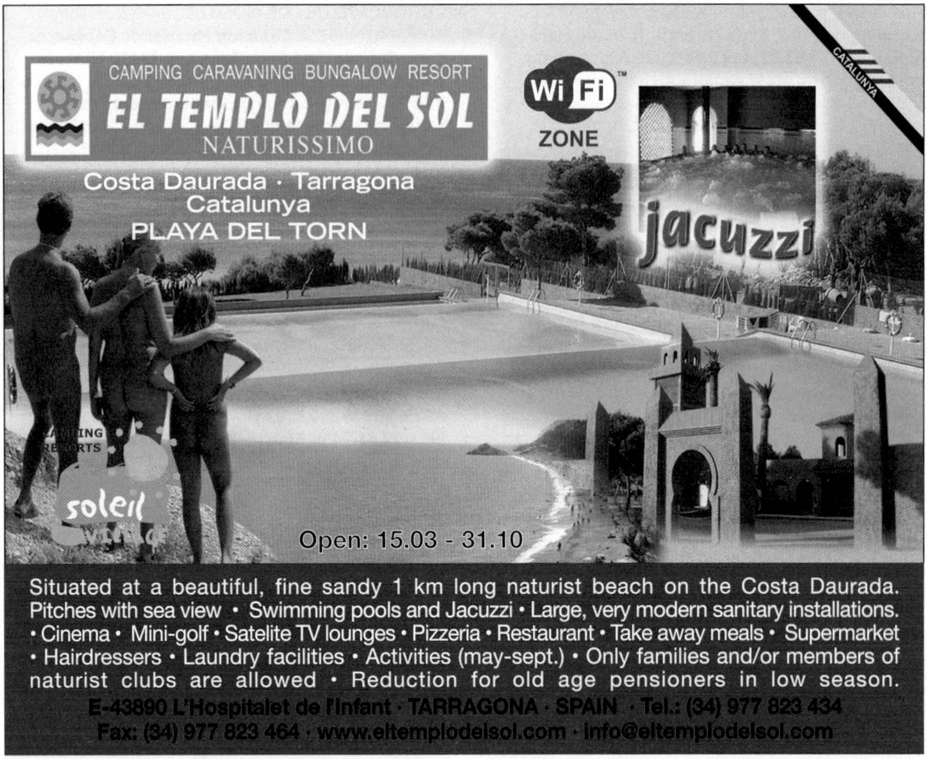

HOSPITALET DE L'INFANT, L' *3C2* (2km S Coastal) *40.97722, 0.90083* **Camping El Templo del Sol (Naturist), Polígon 14-15, Playa del Torn, 43890 L'Hospitalet de l'Infant (Tarragona) [977-82 34 34; fax 977-82 34 64; info@eltemplodelsol.com; www.eltemplodelsol.com]** Leave A7 at exit 38 or N340 twds town cent. Turn R (S) along coast rd for 2km. Ignore 1st camp sp on L, site 200m further on L. Lge, mkd, hdg, pt shd, pt sl, serviced pitches; wc; chem disp; shwrs inc; EHU (6A) inc; gas; lndry; shop; rest; snacks; bar; playgrnd; pool; red long stay; TV; 5% statics; Eng spkn; adv bkg req; ccard acc; INF card. "Excel naturist site; no dogs, radios or TV on pitches; cinema/theatre; solar-energy park; jacuzzi; official naturist sand/shgl beach adj; lge private wash/shwr rms; pitches v tight - take care o'hanging branches; conv Port Aventura; mosquito problem; poss strong winds - take care with awnings." ♦
1 Apr-22 Oct. € 43.00 2016*

See advertisement

"I need an on-site restaurant"

We do our best to make sure site information is correct, but it is always best to check any must-have facilities are still available or will be open during your visit.

HOYOS DEL ESPINO *1D3* (2km S Rural) *40.34055, -5.17527* **Camping Gredos, Ctra Plataforma, Km.1.8, 05634 Hoyos del Espino (Ávila) [920-20 75 85; campingredos@campingredos.com; www.campingredos.com]** Fr N110 turn E at El Barco onto AV941 for approx 41km; at Hoyos del Espino turn S twd Plataforma de Gredos. Site on R in 1.8km. Or fr N502 turn W onto AV941 dir Parador de Gredos to Hoyos del Espino, then as above. Sm, pt shd, pt sl, wc; chem disp; shwrs inc; EHU €2.90; gas; lndry; shop nr; snacks; playgrnd; sw nr; adv bkg acc; quiet; horseriding; bike hire; CKE/CCI. "Lovely mountain scenery; beautiful loc in forest nr rv, san facs basic, mountain walks." ♦
Holy Week & 1 May-1 Oct. € 16.00 2012*

⊞ **HUMILLADERO** *2G4* (1km S Rural) *37.10750, -4.69611* **Camping La Sierrecilla, Avda de Clara Campoamor s/n, 29531 Humilladero (Málaga) [951-19 90 90; fax 952-83 43 73; info@lasierrecilla.com; www.campinglasierrecilla.com]** Exit A92 junc 138 onto A7280 twd Humilladero. At vill ent turn L at 1st rndabt, site visible. Med, mkd, hdstg, pt shd, pt sl, terr, serviced pitches; wc (htd); chem disp; mv service pnt; fam bthrm; shwrs inc; EHU (10A) €3.50; lndry; shop nr; rest; snacks; bar; bbq; playgrnd; pool (htd); paddling pool; entmnt; wifi; 10% statics; dogs €1.50; Eng spkn; adv bkg acc; quiet; CKE/CCI. "Excel new site; gd modern, san facs; vg touring base; gd walking; horseriding, caving, archery high ssn; Fuentepiedra lagoon nrby; lots of facs clsd until Jun/Jul." ♦ € 24.00 2017*

IRUN *3A1* (2km N Rural) *43.36638, -1.80436* **Camping Jaizkibel, Ctra Guadalupe Km 22, 20280 Hondarribia (Guipúzkoa) [943-64 16 79; fax 943-64 26 53; recepcion@campingjaizkibel.com; www.campingjaizkibel.com]** Fr Hondarribia/Fuenterrabia inner ring rd foll sp to site below old town wall. Do not ent town. Med, hdg, hdstg, pt shd, terr, wc; fam bthrm; shwrs; EHU (6A) €4.35 (check earth); lndry; rest; bar; bbq; playgrnd; beach sand 1.5km; wifi; 90% statics; phone; bus 1km; Eng spkn; adv bkg acc; quiet; ccard acc; tennis; CKE/CCI. "Easy 20 mins walk to historic town; scenic area; gd walking; gd touring base but ltd turning space for tourers; clean facs; gd rest & bar; helpful staff; gd lndry facs." 28 Mar-15 Nov. € 34.00 2017*

⊞ **IRUN** *3A1* (6km W Coastal) *43.37629, -1.79939* **Camping Faro de Higuer, Ctra. Del Faro, 58, 20280 Hondarribia [943 64 10 08; fax 943 64 01 50; faro@campingseuskadi.com; www.campingseuskadi.com]** Fr AP8 exit at junc 2 Irun. Foll signs for airport. At rndabt after airport take 2nd exit, cross two more rndabts. 2nd exit at next 2 rndabts. Cont uphill to lighthouse & foll signs for Faro. Med, hdg, mkd, unshd, pt sl, terr, wc; chem disp; mv service pnt; fam bthrm; shwrs; EHU (10A) €5.20; lndry; shop; rest; snacks; bar; bbq; cooking facs; playgrnd; pool; paddling pool; beach adj; entmnt; wifi; TV; 50% statics; dogs €1.20; phone; Eng spkn; adv bkg acc; bike hire; games rm; waterslide; games area. "Vg site on top of winding rd, is v busy outside; sep ent & exit; exit has low stone arch, be careful when leaving." ♦ € 26.00 2014*

ISABA *3B1* (13km E Rural) *42.86618, -0.81247* **Camping Zuriza, Ctra Anso-Zuriza, Km 14, 22728 Ansó (Huesca) [974-37 01 96; contacto@campingzuriza.es; www.campingzuriza.es]** On NA1370 N fr Isaba, turn R in 4km onto NA2000 to Zuriza. Foll sp to site. Fr Ansó, take HUV2024 N to Zuriza. Foll sp to site; narr, rough rd not rec for underpowered o'fits. Lge, pt shd, pt sl, serviced pitches; wc; shwrs inc; EHU €6; lndry; shop; rest; bar; playgrnd; 50% statics; phone; quiet; ccard acc; CKE/CCI. "Beautiful, remote valley; no vill at Zuriza, nearest vills Isaba & Ansó; no direct rte to France; superb location for walking; best to call prior to journey to check opening dates." 1 Jul-31 Oct. € 18.00 2018*

ISLA *1A4* (1km NW Coastal) *43.50261, -3.54351* **Camping Playa de Isla, Calle Ardanal 1, 39195 Isla [942-67 93 61; consultas@playadeisla.com; www.playadeisla.com]** Turn off A8/E70 at km 185 Beranga sp Noja & Isla. Foll sp Isla. In town to beach, site sp to L. Then in 100m keep R along narr seafront lane (main rd bends L) for 1km (rd looks like dead end). Med, mkd, pt shd, pt sl, terr, wc; chem disp; shwrs inc; EHU (3A) €4.50; gas; lndry; shop; snacks; bar; playgrnd; beach sand adj; 90% statics; phone; bus 1km; quiet; ccard acc; CKE/CCI. "Beautiful situation; ltd touring pitches; busy at w/end." Easter-30 Sep. € 26.40 2013*

⊞ **ISLA CRISTINA** *2G2* (4km E Coastal) *37.20555, -7.26722* **Camping Playa Taray, Ctra La Antilla-Isla Cristina, Km 9, 21430 La Redondela (Huelva) [959-34 11 02; fax 959-34 11 96; info@campingplayataray.es; www.campingtaray.com]** Fr W exit A49 sp Isla Cristina & go thro town heading E. Fr E exit A49 at km 117 sp Lepe. In Lepe turn S on H4116 to La Antilla, then R on coast rd to Isla Cristina & site. Lge, pt shd, wc; mv service pnt; shwrs; EHU (10) €4.28; gas; lndry; shop; rest; bar; playgrnd; beach sand adj; red long stay; 10% statics; dogs; phone; bus; quiet; ccard acc; CKE/CCI. "Gd birdwatching, cycling; less cr than other sites in area in winter; poss untidy LS & ltd facs; poss diff for lge o'fits; friendly, helpful owner." ♦ € 19.00 2016*

ISLARES *1A4* (1km W Coastal) *43.40361, -3.31027* **Camping Playa Arenillas, Ctra Santander-Bilbao, Km 64, 39798 Islares (Cantabria) [942-86 31 52 or 609-44-21-67 (mob); cueva@mundivia.es; www.campingplayaarenillas.com]** Exit A8 at km 156 Islares. Turn W on N634. Site on R at W end of Islares. Steep ent & sharp turn into site, exit less steep. Lge, mkd, pt shd, wc; chem disp; mv service pnt; fam bthrm; shwrs inc; EHU (5A) €4.63 (poss no earth); gas; lndry; shop; rest nr; snacks; bar; bbq; playgrnd; beach sand 100m; TV; 40% statics; phone; bus 500m; adv bkg rec; ccard acc; bike hire; horseriding; games area; CKE/CCI. "Facs ltd LS & stretched in ssn; facs constantly cleaned; hot water to shwrs and washing up; rec arr early for choice of own pitch, and avoid Sat & Sun due to parked traff; conv Guggenheim Museum; excel NH for Bilbao ferry; beautiful setting." 24 Mar-30 Sep. € 30.60 2016*

⊞ **JACA** *3B2* (2km W Urban) *42.56416, -0.57027* **Camping Victoria, Avda de la Victoria 34, 22700 Jaca (Huesca) [974-35 70 08; fax 974-35 70 09; campingvictoria@eresmas.com; www.campingvictoria.es]** Fr Jaca cent take N240 dir Pamplona, site on R. Med, mkd, pt shd, wc; chem disp; mv service pnt; shwrs inc; EHU (10A) €5; lndry; snacks; bar; bbq; playgrnd; pool (htd); 80% statics; dogs; bus adj; quiet. "Basic facs, but clean & well-maintained; friendly staff; conv NH/sh stay Somport Pass." € 18.50 2017*

SPAIN

⊞ **JAVEA/XABIA** *4E2* (1km S Rural) *38.78333, 0.17294* **Camping Jávea, Camí de la Fontana 10, 03730 Jávea (Alicante) [965-79 10 70; fax 966-46 05 07; info@campingjavea.es; www.campingjavea.es]** Exit N332 for Jávea on A132, cont in dir Port on CV734. At rndabt & Lidl supmkt, take slip rd to R immed after rv bdge sp Arenal Platjas & Cap de la Nau. Strt on at next rndabt to site sp & slip rd 100m sp Autocine. If you miss slip rd go back fr next rndabt. Lge, mkd, pt shd, wc; chem disp; fam bthrm; shwrs inc; EHU (8A) €4.56 (long lead rec); gas; lndry; shop nr; rest; snacks; bar; bbq; playgrnd; pool; paddling pool; beach sand 1.5km; red long stay; wifi; 15% statics; dogs €2; adv bkg acc; quiet; ccard acc; games area; tennis; CKE/CCI. "Excel site & rest; variable pitch sizes/ prices; some lge pitches - lge o'fits rec phone ahead; gd, clean san facs; mountain views; helpful staff; m'vans beware low trees; gd cycling; site a bit tired but gd." ♦ € 35.00 2017*

"Satellite navigation makes touring much easier"

Remember most sat navs don't know if you're towing or in a larger vehicle – always use yours alongside maps and site directions.

⊞ **JAVEA/XABIA** *4E2* (3km S Coastal) *38.77058, 0.18207* **Camping El Naranjal, Cami dels Morers 15, 03730 Jávea (Alicante) [965-79 29 89; fax 966-46 02 56; info@campingelnaranjal.com; www.campingelnaranjal.com]** Exit A7 junc 62 or 63 onto N332 València/Alicante rd. Exit at Gata de Gorgos to Jávea. Foll sp Camping Jávea/Camping El Naranjal. Access rd by tennis club, foll sp. Med, hdstg, mkd, pt shd, wc (htd); chem disp; mv service pnt; fam bthrm; shwrs inc; EHU (10A) inc (poss rev pol); gas; lndry; shop; rest; snacks; bar; bbq; playgrnd; pool; paddling pool; beach sand 500m; red long stay; wifi; TV; 35% statics; dogs free; phone; bus 500m; Eng spkn; adv bkg acc; quiet; ccard acc; bike hire; tennis 300m; games rm; golf 3km; CKE/CCI. "Gd scenery & beach; pitches poss tight lge o'fits; excel rest; immac facs; tourist info - tickets sold; rec; ACSI rate." ♦ € 33.50 2017*

LAREDO *1A4* (1km W Urban/Coastal) *43.40888, -3.43277* **Camping Carlos V, Avnda Los Derechos Humanos 15, Ctra Residencial Playa, 39770 Laredo (Cantabria) [942-60 55 93]** Leave A8 at junc 172 to Laredo, foll yellow camping sp, site on W side of town. Med, mkd, pt shd, wc; mv service pnt; fam bthrm; shwrs inc; EHU €2.60; gas; lndry; shop; rest; bar; playgrnd; beach sand 200m; dogs €2.14; bus 100m; CKE/CCI. "Well sheltered & lively resort; sm area for tourers; gd, clean, modern facs." 6 May-30 Sep. € 25.00 2012*

LAREDO *1A4* (3km W Coastal) *43.41176, -3.45329* **Camping Playa del Regatón, El Sable 8, 39770 Laredo (Cantabria) [942-60 69 95; info@campingplayaregaton.com; www.campingplayaregaton.com]** Fr W leave A8 junc 172, under m'way to rndabt & take exit sp Calle Rep Colombia. In 800m turn L at traff lts, in further 800m turn L onto tarmac rd to end, passing other sites. Fr E leave at junc 172, at 1st rndabt take 2nd exit sp Centro Comercial N634 Colindres. At next rndabt take exit Calle Rep Colombia, then as above. Lge, mkd, pt shd, wc; chem disp; mv service pnt; shwrs inc; EHU (6-10A) €4.30; gas; lndry; shop; rest; bar; beach sand adj; red long stay; wifi; 75% statics; bus 600m; Eng spkn; adv bkg acc; quiet; ccard acc; horseriding nr; CKE/CCI. "Clean site; sep area for tourers; wash up facs (cold water) every pitch; gd, modern facs; gd NH/sh stay (check opening times of office for EHU release); gd bird watching." ♦ 18 Mar-25 Sep. € 26.70 2016*

LEKEITIO *3A1* (3km S Coastal) *43.35071, -2.49260* **Camping Leagi, Calle Barrio Leagi s/n, 48289 Mendexa (Vizcaya) [946-84 23 52; fax 946-24 34 20; leagi@campingleagi.com; www.campingleagi.com]** Fr San Sebastian leave A8/N634 at Deba twd Ondarroa. At Ondarroa do not turn into town, but cont on BI633 beyond Berriatua, then turn R onto BI3405 to Lekeitio. Fr Bilbao leave A8/N634 at Durango & foll B1633 twd Ondarroa. Turn L after Markina onto BI3405 to Lekeitio - do not go via Ondarroa, foll sp to Mendexa & site. Steep climb to site & v steep tarmac ent to site. Only suitable for o'fits with v high power/ weight ratio. Med, mkd, unshd, pt sl, serviced pitches; wc; chem disp; mv service pnt; shwrs inc; EHU (5A) €3.90 (rev pol); lndry; shop; rest; snacks; bar; playgrnd; beach sand 1km; 80% statics; dogs; ccard acc; CKE/ CCI. "Ltd facs LS; tractor tow avail up to site ent; beautiful scenery; excel local beach; lovely town; gd views; gd walking; san facs under pressure due to many tents; bus to Bilbao & Gurnika." 28 Feb-9 Nov. € 36.00 2014*

LEON *1B3* (7km SE Urban) *42.5900, -5.5331* **Camping Ciudad de León, Ctra N601, 24195 Golpejar de la Sobarriba [987-26 90 86; camping_leon@yahoo.es; www.vivaleon.com/campingleon.htm]** SE fr León on N601 twds Valladolid, L at top of hill at rndabt & Opel g'ge & foll site sp Golpejar de la Sobarriba; 500m after radio masts turn R at site sp. Narr track to site ent. Sm, shd, pt sl, wc; chem disp; shwrs inc; EHU inc (6A) €3.60; gas; lndry; shop; rest; snacks; bar; playgrnd; pool; paddling pool; wifi; dogs €1.50; bus 200m; Eng spkn; adv bkg acc; quiet; bike hire; tennis; CKE/CCI. "Clean, pleasant site; helpful, welcoming staff; access some sm pitches poss diff; easy access to León; shwrs need refurb (2015)." ♦ 1 Jun-20 Sep. € 19.00 2015*

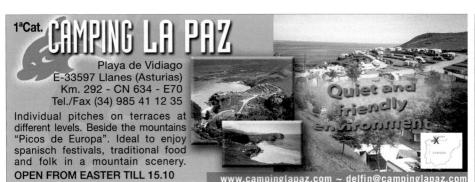

LEON *1B3* (12km SW Urban) *42.51250, -5.77472* **Camping Camino de Santiago, Ctra N120, Km 324.4, 24392 Villadangos del Páramo (León) [987-68 02 53; info@campingcaminodesantiago.com; www.campingcaminodesantiago.com]** Access fr N120 to W of vill, site sp on R (take care fast, o'taking traff). Fr E turn L in town & foll sp to site. Lge, mkd, pt shd, wc; chem disp; mv service pnt; fam bthrm; shwrs inc; EHU €3.90; gas; lndry; shop; rest; snacks; bar; pool; red long stay; wifi; 50% statics; dogs; phone; bus 300m; adv bkg acc, ccard acc; CKE/CCI. "Poss no hot water LS; facs tired; pleasant, helpful staff; mosquitoes; vill church worth visit; gd NH." ♦ 23 Mar-30 Sep, € 18.60 2013*

"There aren't many sites open at this time of year"

If you're travelling outside peak season remember to call ahead to check site opening dates – even if the entry says 'open all year'.

⊞ **LINEA DE LA CONCEPCION, LA** *2H3* (4km N Urban/Coastal) *36.19167, -5.3350* **Camping Sureuropa, Camino de Sobrevela s/n, 11300 La Línea de la Concepción (Cádiz) [956-64 35 87; fax 956-64 30 59; info@campingsureuropa.es; www.campingsureuropa.es]** Fr AP7, ext junc 124. Use junc124 fr both dirs. Just bef Gibraltar turn R up lane, in 200m turn L into site. Fr N on AP7, exit junc 124 onto A383 dir La Línea; foll sp Santa Margarita thro to beach. Foll rd to R along sea front, site in approx 1km - no advance sp. App rd to site off coast rd poss floods after heavy rain. Med, hdg, hdstg, mkd, pt shd, wc; chem disp; shwrs inc; EHU €4.30; lndry; bar; beach sand 500m; phone; Eng spkn; adv bkg acc; site clsd 20 Dec-7 Jan; CKE/CCI. "Clean, flat, pretty site; vg, modern san facs; sm pitches & tight site rds poss diff twin axles & l'ge o'fits; sports club adj; quiet but noise fr adj sports club; ideal for Gibraltar 4km; stay ltd to 4 days; wifi in recep only; sh stay only." ♦ € 20.00 2016*

LLANES *1A4* (8km E Coastal) *43.39948, -4.65350* **Camping La Paz, Ctra N634, Km 292, 33597 Playa de Vidiago (Asturias) [985-41 12 35; delfin@campinglapaz.com; www.campinglapaz.eu]** Take Fr A8/N634/E70 turn R at sp to site bet km stone 292 & 293 bef Vidiago.Site access via narr 1km lane. Stop bef bdge & park on R, staff will tow to pitch. Narr site ent & steep access to pitches. Lge, mkd, shd, terr, wc; chem disp; mv service pnt; fam bthrm; shwrs inc; EHU (9A) €4.82 (poss rev pol); gas; lndry; shop; rest; bar; bbq; playgrnd; beach sand adj; wifi; TV; dogs €2.50; phone; Eng spkn, adv bkg acc, quiet; ccard acc; fishing; golf 4km; watersports; games rm; horseriding; CKE/CCI. "Exceptionally helpful owner & staff; sm pitches; gd, modern san facs; mountain sports; excel views; cliff top rest; superb beaches in area." ♦ Easter-30 Sep. € 42.00 2017*

See advertisement

LLANES *1A4* (3km W Coastal) *43.42500, -4.78944* **Camping Las Conchas de Póo, Ctra General, 33509 Póo de Llanes (Asturias) [985-40 22 90 or 674-16 58 79 (mob); campinglasconchas@gmail.com; www.campinglasconchas.com]** Exit A8/E70 at Llanes West junc 307 & foll sp. Site on rd AS263. Med, pt shd, sl, terr, wc; chem disp; fam bthrm; shwrs; EHU (6A) €3.20; lndry; shop; rest; bar; playgrnd; beach sand adj; wifi; 50% statics; dogs; phone; bus adj; quiet. "Pleasant site; footpath to lovely beach; lovely coastal walk to Celorio; stn in Póo vill." 1 Jun-15 Sep. € 20.00 2017*

LLANES *1A4* (5km W Coastal) *43.43471, -4.81810* **Camping Playa de Troenzo, Ctra de Celerio-Barro, 33595 Celorio (Asturias) [985-40 16 72; fax 985-74 07 23; troenzo@telepolis.com]** Fr E take E70/A8 exit at junc 300 to Celorio. At T-junc with AS263 turn L dir Celorio & Llanes. Turn L on N9 (Celorio) thro vill & foll sp to Barro. Site on R after 500m (after Maria Elena site). Lge, pt shd, terr, wc; chem disp; mv service pnt; shwrs inc; EHU (6A) €2.51; gas; lndry; rest; snacks; bar; playgrnd; beach sand 400m; wifi; 90% statics; dogs; phone; Eng spkn; adv bkg acc; CKE/CCI. "Lovely, old town; most pitches sm; for pitches with sea views go thro statics to end of site; gd, modern facs; gd rests in town; nr harbour." 16 Feb-19 Dec. € 21.00 2014*

⊞ **LLORET DE MAR** *3B3* (1km SW Coastal) *41.6984, 2.8265* **Camping Santa Elena-Ciutat, Ctra Blanes/ Lloret, 17310 Lloret de Mar (Gerona) [972-36 40 09; fax 972-36 79 54; santaelana@betsa.es; www.betsa.es]** Exit A7 junc 9 dir Lloret. In Lloret take Blanes rd, site sp at km 10.5 on rd GI 682. V lge, pt sl, wc; fam bthrm; shwrs; EHU (5A) €3.90; gas; lndry; shop; rest; snacks; bar; playgrnd; pool; paddling pool; beach shgl 600m; phone; Eng spkn; quiet; games area; CKE/CCI. "Ideal for teenagers; cash machine." ♦ € 34.60 2016*

⊞ **LOGRONO** *3B1* (2km N Urban) *42.47122, -2.45493* **Camping La Playa, Avda de la Playa 6, 26006 Logroño (La Rioja) [941-25 22 53; fax 941-25 86 61; info@campinglaplaya.com; www.campinglaplaya.com]** Leave Logroño by bdge 'Puente de Piedra' on N111, then turn L at rndabt into Camino de las Norias. Site well sp in town & fr N111, adj sports cent Las Norias, on N side of Rv Ebro. Med, hdg, shd, wc; mv service pnt; shwrs inc; EHU (5A) €4.80; gas; lndry; shop; snacks; bar; playgrnd; pool; sw nr; 80% statics; dogs €2; tennis; CKE/CCI. "Sh walk to town cent; ltd facs LS & site poss clsd; vg; nr rest." ♦ € 25.00 2016*

LOGRONO *3B1* (14km SW Rural) *42.41613, -2.55169* **Camping Navarrete, Ctra La Navarrete-Entrena, Km 1.5, 26370 Navarrete (La Rioja) [941-44 01 69; fax 941-44 06 39; campingnavarrete@fer.es; www.campingnavarrete.com]** Fr AP68 exit junc 11, at end slip rd turn L onto LR137 to Navarette. Foll sp thro town for Entrena (S) on LR137 dir Entrena. Site 1km on R. Lge, mkd, pt shd, wc; chem disp; mv service pnt; fam bthrm; shwrs inc; EHU (6A) inc; gas; lndry; shop; snacks; bar; bbq; playgrnd; pool; paddling pool; red long stay; wifi; 95% statics; dogs €2.40; Eng spkn; ccard acc; golf 5km; tennis; car wash; horseriding. "Helpful staff; some sm pitches; Bilbao ferry 2 hrs via m'way; interesting area; under new management, site not clean; gd NH only." 13 Jan-10 Dec. € 31.00 2018*

⊞ **LORCA** *4G1* (8km W Rural) *37.62861, -1.74888* **Camping La Torrecilla, Ctra Granada-LaTorrecilla, 30817 Lorca (Murcia) [968-44 21 36; fax 968-44 21 96; campinglatorrecilla@hotmail.es]** Leave A7/E15 at junc 585. In 1km turn L, site well sp. Med, mkd, hdstg, pt shd, pt sl, wc (htd); chem disp; mv service pnt; shwrs inc; EHU (6A) €3.88; gas; lndry; shop nr; rest; snacks; bar; bbq; playgrnd; pool; red long stay; TV; 95% statics; dogs; phone; bus 1km; Eng spkn; quiet; ccard acc; games area; tennis. "Ltd touring pitches & EHU; friendly, helpful staff; excel pool; vg san facs." ♦ € 20.00 2013*

⊞ **LUARCA** *1A3* (1km NE Coastal) *43.54914, -6.52426* **Camping Los Cantiles, Ctra N634, Km 502.7, 33700 Luarca (Asturias) [985-64 09 38; fax 984-11 14 58; cantiles@campingloscantiles.com; www.campingloscantiles.com]** On A8 exit junc 467 (sp Luarca/Barcia/Almuña), At rndabt foll sp to Luarca, after petrol stn turn R, foll sp to site. Not rec to ent town fr W. Not rec to foll sat nav as may take you up v steep & narr rd. On leaving site, retrace to main rd - do not tow thro Luarca. Med, hdg, pt shd, wc; chem disp; mv service pnt; fam bthrm; shwrs inc; EHU inc (3-6A) €2-2.50; gas; lndry; shop; rest; snacks; bar; beach shgl; red long stay; wifi; dogs €1; phone; Eng spkn; adv bkg acc; quiet; CKE/CCI. "Site on cliff top; some narr site rds; pitches soft after rain; steep climb down to beach; 30 min walk to interesting town & port; pool 300m; wonderful setting; san fac's OK, water in shwrs v hot." ♦ € 23.00 2017*

LUARCA *1A3* (2km W Coastal) *43.55116, -6.55310* **Camping Playa de Taurán, 33700 Luarca (Asturias) [985-64 12 72 or 619-88 43 06 (mob); tauran@campingtauran.com; www.campingtauran.com]** Exit A8/N634 junc 471 sp Luarca, El Chano. Cont 3.5km on long, narr, rough access rd. Rd thro Luarca unsuitable for c'vans. Med, hdg, pt shd, pt sl, wc; chem disp; mv service pnt; fam bthrm; shwrs inc; EHU (10A) €3.50; gas; lndry; shop; rest; snacks; bar; bbq; pool; paddling pool; beach shgl 200m; red long stay; dogs €1; phone; quiet; bike hire. "Sea & mountain views; off beaten track; conv fishing & hill vills; peaceful, restful, attractive, well-kept site; beach sand 2km; steep access to beach; excel." 1 Apr-30 Sep. € 18.50 2013*

LUMBIER *3B1* (1km S Rural) *42.65111, -1.30222* **Camping Iturbero, Ctra N240 Pamplona-Huesca, 31440 Lumbier (Navarra) [948-88 04 05; fax 948-88 04 14; iturbero@campingiturbero.com; www.campingiturbero.com]** SE fr Pamplona on N240 twds Yesa Reservoir. In 30km L on NA150 twds Lumbier. In 3.5km immed bef Lumbier turn R at rndabt then over bdge, 1st L to site, adj sw pool. Well sp fr N240. Med, hdg, hdstg, mkd, pt shd, wc; chem disp; mv service pnt; shwrs inc; EHU (5A) €4.95; gas; lndry; shop nr; rest; snacks; bar; bbq; playgrnd; 25% statics; dogs; bus 1km; quiet; tennis; CKE/CCI. "Beautiful, well-kept site; clean, basic facs; pool 100m; excel touring base; open w/end only Dec-Easter (poss fr Sep) but clsd 19 Dec-19 Feb; eagles & vultures in gorge & seen fr site; hang-gliding; helpful staff; Lumbier lovely sm town." 15 Mar-15 Dec. € 26.50 2014*

"That's changed – Should I let the Club know?"

If you find something on site that's different from the site entry, fill in a report and let us know. See camc.com/europereport.

MACANET DE CABRENYS *3B3* (1km S Rural) *42.37314, 2.75419* **Camping Maçanet de Cabrenys, Mas Roquet s/n, 17720 Maçanet de Cabrenys (Gerona) [667-77 66 48 (mob); info@campingmassanet.com; www.campingmassanet.com]** Fr N, exit AP7 junc 2 at La Jonquera onto N-II dir Figueres; at km 767 turn R onto GI-502/GI-503 dir Maçanet de Cabrenys; turn L 500m bef vill; site sp. Or fr S, exit AP7 junc 4 at Figueres onto N-II dir France; at km 766 turn L onto GI-502 dir Maçanet de Cabrenys; then as above. Sm, mkd, hdstg, pt shd, sl, terr, wc (htd); chem disp; mv service pnt; fam bthrm; shwrs; EHU (10A) €6.50; gas; lndry; shop; rest; snacks; bar; bbq; playgrnd; pool; wifi; TV; 10% statics; dogs €5.50; Eng spkn; adv bkg acc; quiet; ccard acc; games rm; bike hire; CKE/CCI. "Cycle rtes fr site." ♦ 1 Mar-31 Dec. € 30.00 2013*

⊞ **MADRID** *1D4* (13km NE Urban) *40.45361, -3.60333* **Camping Osuna, Jardines de Aranjuez 1, 28042 Madrid [917-41 05 10; fax 913-20 63 65; camping.osuna.madrid@microgest.es]** Fr M40, travelling S clockwise (anti-clockwise fr N or E) exit junc 8 at Canillejas sp 'Avda de Logroño'. Turn L under m'way, then R under rlwy, immed after turn R at traff lts. Site on L corner - white painted wall. Travelling N, leave M40 at junc 7 (no turn off at junc 8) sp Avda 25 Sep, U-turn at km 7, head S to junc 8, then as above. Med, hdg, mkd, pt shd, pt sl, wc; chem disp; shwrs inc; EHU (6A) €4.85 (long lead rec); lndry; shop nr; playgrnd; 10% statics; dogs free; phone; Eng spkn; CKE/CCI. "Sm pitches poss diff lge o'fits; poss neglected LS & facs tired (June 2010); poss travellers; conv city cent; busy; gd basic facs; metro to town 600m; easy walk to metro with frequent svr into Madrid." ♦ € 33.00 2015*

⊞ **MADRID** *1D4* (13km S Urban) *40.31805, -3.68888* **Camping Alpha, Ctra de Andalucía N-IV, Km 12.4, 28906 Getafe (Madrid) [916-95 80 69; fax 916-83 16 59; info@campingalpha.com; www.campingalpha.com]** Fr S on A4/E5 twd Madrid, leave at km 12B to W dir Ocaña & foll sp. Fr N on A4/E5 at km 13B to change dir back onto A4; then exit 12B sp 'Polígono Industrial Los Olivos' to site. Lge, hdstg, hdg, pt shd, wc; chem disp; mv service pnt; shwrs inc; EHU (15A) €5.90 (poss no earth); lndry; shop; rest; snacks; bar; playgrnd; pool; 20% statics; dogs; phone; Eng spkn; adv bkg acc; ccard acc; games area; tennis; CKE/CCI. "Lorry depot adj; poss vehicle movements 24 hrs but minimal noise; bus & metro to Madrid 30-40 mins; sm pitches poss tight for space; vg, clean facs; helpful staff; NH or sh stay." € 28.00 2016*

⊞ **MADRIGAL DE LA VERA** *1D3* (1km E Rural) *40.14864, -5.35769* **Camping Alardos, Ctra Madrigal-Candeleda, 10480 Madrigal de la Vera (Cáceres) [927-56 50 66; info@campingalardos.com; www.campingalardos.com]** Fr Jarandilla take EX203 E to Madrigal. Site sp in vill nr rv bdge. Med, mkd, pt shd, wc; chem disp; shwrs inc; EHU (6-10A) €4; lndry; shop; rest; snacks; bar; bbq; playgrnd; sw nr; TV; 95% statics; phone; Eng spkn; adv bkg acc. "Friendly owners; ltd facs LS; beautiful scenery; excel touring area; ancient Celtic settlement at El Raso 5km." € 24.60 2013*

⊞ **MANGA DEL MAR MENOR, LA** *4G2* (4km W Urban) *37.6244, -0.7447* **Caravaning La Manga, Autovia Cartagena - La Manga, exit 11, E-30370 La Manga del Mar Menor [968-56 30 19; fax 968-543426; lamanga@caravaning.es; www.caravaning.es]** Take Autovia CT-32 fr Cartagena to La Manga; take exit 800B twds El Algar/Murcia; keep L, merge onto Autovia MU312; cont to foll MU-312; cont onto Ctra a La Manga & cont onto Av Gran Via; site clearly sp. Lge, hdg, hdstg, pt shd, serviced pitches; wc; chem disp; mv service pnt; fam bthrm; shwrs inc; EHU (10A) inc; lndry; shop; rest; snacks; bar; bbq; playgrnd; pool; beach; entmnt; wifi; 10% statics; dogs €1.45; bus to Murcia and Cartagena; Eng spkn; adv bkg acc; ccard acc; watersports; tennis; games rm. "Immac, busy, popular site; Mar Menor shallow & warm lagoon; open air cinema & children's programme high ssn; lovely location & gd for golfers; horseridng nrby; recep open 24 hrs; Mar Menor well worth visiting; vg rest; gd for families; outdoor fitness; some narr site rds & trees - rec park in car park on arr & walk to find pitch; gd walking; gym; sauna; jacuzzi; mountain biking; bird sanctuary; poss lge rallies on site Dec-Mar; excel; friendly, helpful staff; immac facs." € 38.00 2016*

See advertisement opposite

SPAIN

⊞ **MANZANARES EL REAL** *1D4* (8km NE Rural) *40.74278, -3.81583* **Camping La Fresneda, Ctra M608, Km 19.5, 28791 Soto del Real (Madrid) [918-47 72 13; fresnedacamp.com]** Fr AP6/NV1 turn NE at Collado-Villalba onto M608 to Cerceda & Manzanares el Real. Foll rd round lake to Soto del Real, site sp at km 19.5. Med, shd, wc; chem disp; fam bthrm; shwrs; EHU (6A) €3.50; gas; lndry; shop; snacks; bar; playgrnd; pool; dogs €3; phone; ccard acc; tennis. ♦ € 27.50 2014*

⊞ **MARBELLA** *2H3* (14km E Coastal) *36.48881, -4.74294* **Kawan Village Cabopino, Ctra N340/A7, Km 194.7, 29600 Marbella (Málaga) [952-83 43 73; info@campingcabopino.com; www.campingcabopino.com]** Fr E site is on N side of N340/A7; turn R at km 195 'Salida Cabopino' past petrol stn, site on R at rndabt. Fr W on A7 turn R at 'Salida Cabopino' km 194.7, go over bdge to rndabt, site strt over. NB Do not take sm exit fr A7 immed at 1st Cabopino sp. Lge, mkd, pt shd, pt sl, wc; chem disp; mv service pnt; fam bthrm; shwrs inc; EHU (6-10A) inc (poss long lead req); lndry; shop; rest; snacks; bar; bbq (elec, gas); playgrnd; pool (covrd); beach sand 200m; wifi; TV; 50% statics; dogs €2; bus 100m; Eng spkn; ccard acc; golf driving range; archery; games area; games rm; watersports; CKE/CCI. "V pleasant site set in pine woodland; rd noise & lge groups w/enders; marina 300m; busy, particularly w/end; varied pitch size, poss diff access lge o'fits, no o'fits over 11m high ssn; blocks req some pitches; gd, clean san facs; feral cats on site (2009)." ♦ € 36.00 2016*

⊞ **MARBELLA** *2H3* (7km E Coastal) *36.50259, -4.80413* **Camping La Buganvilla, Ctra N340, Km 188.8, 29600 Marbella (Málaga) [952-83 56 21 or 952-83 19 73; info@campingbuganvilla.com; www.campingbuganvilla.com]** E fr Marbella for 6km on N340/E15 twds Málaga. Pass site & cross over m'way at Elviria & foll site sp. Fr Málaga exit R off autovia immed after 189km marker. Lge, shd, pt sl, terr, wc; chem disp; shwrs; EHU (16A) inc; gas; lndry; shop; rest; bar; bbq; playgrnd; pool; beach sand 350m; red long stay; wifi; TV; 40% statics; dogs €4 (not acc Jul/Aug); phone; Eng spkn; adv bkg acc; ccard acc; fishing; tennis; games rm; CKE/CCI. "Relaxed, conv site; helpful staff; excel beach; facs being upgraded (2014); bus stop to Marbella at ent; 20min walk to nice beach; bigger & more shaded pitches at top of site." ♦ € 25.00 2016*

"I like to fill in the reports as I travel from site to site"

You'll find report forms at the back of this guide, or you can fill them in online at camc.com/europereport.

⊞ **MARIA** *4G1* (8km W Rural) *37.70823, -2.23609* **Camping Sierra de María, Ctra María a Orce, Km 7, Paraje La Piza, 04838 María (Málaga) [620-23 22 23; fax 950-48 54 16; info@campingsierrademaria.com; www.campingsierrademaria.es]** Exit A92 at junc 408 to Vélez Rubio, Vélez Blanco & María. Foll A317 to María & cont dir Huéscar & Orce. Site on R. Med, mkd, pt shd, pt sl, wc; chem disp; shwrs; EHU (16A) €3.75; shop; rest; bar; 10% statics; dogs; adv bkg acc; quiet; ccard acc; bike hire; horseriding; CKE/CCI. "Lovely, peaceful, ecological site in mountains; much wildlife; variable pitch sizes; facs poss stretched high ssn; v cold in winter; v helpful mgrs, gd food in rest." ♦ € 23.00 2016*

"We must tell the Club about that great site we found"

Get your site reports in by mid-August and we'll do our best to get your updates into the next edition.

⊞ **MARINA, LA** *4F2* (2km S Coastal) *38.12972, -0.65000* **Camping Internacional La Marina, Ctra N332a, Km 76, 03194 La Marina (Alicante) [965-41 92 00; fax 965-41 91 10; info@campinglamarina.com; www.lamarinaresort.com]** Fr N332 S of La Marina turn E twd sea at rndabt onto Camino del Cementerio. At next rndabt turn S onto N332a & foll site sp along Avda de l'Alegría. V lge, hdg, hdstg, mkd, shd, terr, serviced pitches; wc (htd); chem disp; mv service pnt; fam bthrm; shwrs inc; EHU (10A) €3.21; gas; lndry; shop; rest; snacks; bar; playgrnd; pool (covrd, htd); beach sand 500m; red long stay; entmnt; wifi; TV; 10% statics; dogs €2.14; phone; bus 50m; Eng spkn; adv bkg acc; ccard acc; watersports; games rm; solarium; fishing; sauna; tennis; games area; waterslide; CKE/CCI. "Popular winter site - almost full late Feb; v busy w/end; disco; clean, high quality facs; various pitch sizes/prices; fitness cent; bus fr gate; gd security; car wash; security; excel rest; gd site; v helpful; hire cars avail; fantastic site." ♦ € 65.00 2017*

See advertisement opposite

⊞ **MASNOU, EL** *3C3* (1km W Coastal) *41.4753, 2.3033* **Camping Masnou, Ctra NII, Km 633, Carrer de Camil Fabra 33, 08320 El Masnou (Barcelona) [935-55 15 03; masnou@campingsonline.es; www.campingmasnoubarcelona.com]** App site fr N on N11. Pass El Masnou rlwy stn on L & go strt on at traff lts. Site on R on N11 after km 633. Not clearly sp. Med, shd, pt sl, wc; mv service pnt; shwrs inc; EHU €5.88; shop; snacks; bar; bbq; playgrnd; pool; beach sand adj; wifi; dogs; phone; bus 300m, train to Barcelona nr; Eng spkn; ccard acc; CKE/CCI. "Gd pitches, no awnings; some sm pitches, poss shared; facs vg, though poss stretched when site busy; no restriction on NH vehicle movements; well-run, friendly site but v tired; gd service LS; rlwy line bet site & excel beach - subway avail; excel train service to Barcelona; excel pool." ♦ € 35.00 2017*

MATARO *3C3* (3km E Coastal) *41.55060, 2.48330* **Camping Barcelona, Ctra NII, Km 650, 08304 Mataró (Barcelona) [937-90 47 20; fax 937-41 02 82; info@campingbarcelona.com; www.campingbarcelona.com]** Exit AP7 onto C60 sp Mataró. Turn N onto NII dir Gerona, site sp on L after rndabt. Lge, hdstg, mkd, shd, wc; chem disp; mv service pnt; fam bthrm; shwrs inc; EHU (6A) €5.50; gas; lndry; shop; rest; snacks; bar; playgrnd; pool; paddling pool; beach sand 1.5km; red long stay; entmnt; wifi; TV; 5% statics; dogs €4; Eng spkn; adv bkg acc; ccard acc; games area; games rm; CKE/CCI. "Conv Barcelona 28km; pleasant site; shuttle bus to beach; animal farm; friendly, welcoming staff." ♦ 4 Mar-1 Nov. € 45.70 2013*

"I need an on-site restaurant"

We do our best to make sure site information is correct, but it is always best to check any must-have facilities are still available or will be open during your visit.

⊞ **MAZAGON** *2G2* (10km E Coastal) *37.09855, -6.72650* **Camping Doñana Playa, Ctra San Juan del Puerto-Matalascañas, Km 34.6, 21130 Mazagón (Huelva) [959-53 62 81; fax 959-53 63 13; info@campingdonana.com; www.campingdonana.com]** Fr A49 exit junc 48 at Bullullos del Condado onto A483 sp El Rocio, Matalascañas. At coast turn R sp Mazagón, site on L in 16km. V lge, hdstg, mkd, pt shd, wc; chem disp; shwrs inc; EHU (6A) €5.20; shop; rest; snacks; bar; playgrnd; pool; beach sand 300m; entmnt; 10% statics; dogs €4.10; bus 500m; adv bkg acc; games area; site clsd 14 Dec-14 Jan; watersports; bike hire; tennis; CKE/CCI. "Pleasant site amongst pine trees but lack of site care LS; ltd LS; lge pitches but poss soft sand; quiet but v noisy Fri/Sat nights; new (2014) lge shwr block on lower pt of site." ♦ € 48.00 2014*

MENDIGORRIA *3B1* (1km SW Rural) *42.62416, -1.84277* **Camping El Molino, Ctra Larraga, 31150 Mendigorría (Navarra) [948-34 06 04; fax 948-34 00 82; info@campingelmolino.com; www.campingelmolino.com]** Fr Pamplona on N111 turn L at 25km in Puente la Reina onto NA601 sp Mendigorría. Site sp thro vill dir Larraga. Med, mkd, pt shd, serviced pitches; wc; chem disp; fam bthrm; shwrs inc; EHU (6A) inc; gas; lndry; shop; rest; snacks; bar; bbq; playgrnd; pool; paddling pool; wifi; TV; dogs; phone; adv bkg acc; ccard acc; games area; canoe hire; waterslide; tennis; clsd 23 Dec-14 Jan & poss Mon-Thurs fr Nov to Feb, phone ahead to check; CKE/CCI. "Gd clean san facs; solar water heating - water poss only warm; vg leisure facs; v ltd facs LS; for early am dep LS, pay night bef & obtain barrier key; friendly, helpful staff; statics (sep area); lovely medieval vill." ♦ 1 Feb-15 Dec. € 29.60 2013*

See advertisement

⊞ **MERIDA** *2E3* (4km NE Urban) *38.93558, -6.30426* **Camping Mérida, Avda de la Reina Sofia s/n, 06800 Mérida (Badajoz) [924-30 34 53; fax 924-30 03 98; proexcam@jet.es]** Fr E on A5/E90 exit junc 333/334 to Mérida, site on L in 2km. Fr W on A5/E90 exit junc 346, site sp. Fr N exit A66/E803 at junc 617 onto A5 E. Leave at junc 334, site on L in 1km twd Mérida. Fr S on A66-E803 app Mérida, foll Cáceres sp onto bypass to E; at lge rndabt turn R sp Madrid; site on R after 2km. Med, mkd, pt shd, pt sl, wc; chem disp; shwrs inc; EHU (6A) €3.30 (long lead poss req & poss rev pol); gas; lndry; shop; rest; snacks; bar; pool; paddling pool; wifi; TV; 10% statics; dogs €1.65; phone; quiet; CKE/CCI. "Roman remains & National Museum of Roman Art worth visit; poss diff lge o'fits manoeuvring onto pitch due trees & soft grnd after rain; ltd facs & run down in LS; conv NH; taxi to town costs 5-9 euros; grass pitches; bread can be ordered fr rest; poss nightclub noise at w/end." ♦ € 21.00 2017*

⊞ **MIAJADAS** *2E3* (14km SW Rural) *39.09599, -6.01333* **Camping-Restaurant El 301, Ctra Madrid-Lisbon, Km 301, 10100 Miajadas (Cáceres) [927-34 79 14; camping301@hotmail.com; www.camping301.com]** Leave A5/E90 just bef km stone 301 & foll sp 'Via de Servicio' with rest & camping symbols; site in 500m. Med, pt shd, wc; chem disp; shwrs inc; EHU (8A) €5.50 (poss no earth); gas; lndry; shop; rest; snacks; bar; playgrnd; pool; TV; phone; ccard acc; CKE/CCI. "Well-maintained, clean site; grass pitches; OK wheelchair users but steps to pool; gd NH; gd bird life." € 24.00 2016*

⊞ **MOJACAR** *4G1* (4km S Coastal) *37.12656, -1.83250* **Camping El Cantal di Mojácar, Ctra Garrucha-Carboneras, 04638 Mojácar (Almería) [950-47 82 04; fax 950-47 23 93; campingelcantal@hotmail.com]** Fr N on coast rd AL5105, site on R 800m after Parador, opp 25km sp.Or exit A7 junc 520 sp Mojácar Parador. Foll Parador sps by-passing Mojácar, to site. Med, hdstg, pt shd, wc; chem disp; mv service pnt; shwrs inc; EHU (15A) €3; gas; lndry; shop nr; rest; snacks; bar nr; bbq; beach sand adj; red long stay; 5% statics; dogs; phone; bus; CKE/CCI. "Pitches quite lge, not mkd; lge o'fits rec use pitches at front of site; busy site; staff unhelpful, quite expensive; bar adj; facs run down." ♦ € 30.00 2014*

⊞ **MOJACAR** *4G1* (9km S Rural) *37.06536, -1.86864* **Camping Sopalmo, Sopalmo, 04638 Mojácar (Almería) [950-47 84 13; fax 950-47 30 02; info@campingsopalmoelcortijillo.com; www.campingsopalmoelcortijillo.com]** Exit A7/E15 at junc 520 onto AL6111 sp Mojácar. Fr Mojácar turn S onto A1203/AL5105 dir Carboneras, site sp on W of rd about 1km S of El Agua del Medio. Sm, mkd, hdstg, pt shd, wc; chem disp; shwrs; EHU (15A) €3; gas; lndry; shop; rest nr; bar; beach shgl 1.7km; wifi; 10% statics; dogs €1; Eng spkn; adv bkg acc; CKE/CCI. "Clean, pleasant, popular site; remote & peaceful; friendly owner; gd walking in National Park; lovely san facs." ♦ € 29.00 2012*

⊞ **MOJACAR** *4G1* (1km W Rural) *37.14083, -1.85916* **Camping El Quinto, Ctra Mojácar-Turre, 04638 Mojácar (Almería) [950-47 87 04; fax 950-47 21 48; campingelquinto@hotmail.com]** Fr A7/E15 exit 520 sp Turre & Mojácar. Site on R in approx 13km at bottom of Mojácar vill. Sm, hdg, hdstg, mkd, pt shd, wc; chem disp; mv service pnt; shwrs inc; EHU (6-10A) €3.21; gas; lndry; shop; rest nr; snacks; bar; bbq; playgrnd; pool; beach sand 3km; red long stay; dogs €1; phone; Eng spkn; adv bkg acc; quiet; CKE/CCI. "Neat, tidy site; mkt Wed; close National Park; excel beaches; metered 6A elect for long winter stay; popular in winter, poss cr & facs stretched; security barrier; poss mosquitoes; drinking water ltd to 5L a time." ♦ € 20.50 2012*

⊞ **MONCOFA** *3D2* (2km E Urban/Coastal) *39.80861, -0.12805* **Camping Mon Mar, Camino Serratelles s/n, 12593 Platja de Moncófa (Castellón) [964-58 85 92; campingmonmar@hotmail.com]** Exit 49 fr A7 or N340, foll sp Moncófa Platja passing thro Moncófa & foll sp beach & tourist info thro 1-way system. Site sp, adj Aqua Park. Lge, hdg, hdstg, pt shd, serviced pitches; wc (htd); chem disp; fam bthrm; shwrs inc; EHU (6A) inc; gas; lndry; shop; rest; snacks; bar; bbq; playgrnd; pool; beach shgl 200m; red long stay; entmnt; wifi; 80% statics; phone; bus 300m; Eng spkn; adv bkg acc; quiet; ccard acc; CKE/CCI. "Helpful owner & staff; rallies on site Dec-Apr; mini-bus to stn & excursions; sunshades over pitches poss diff high o'fits; excel clean, tidy site." ♦ € 27.00 2014*

MONESTERIO *2F3* (3km S Rural) *38.06276, -6.24689* **Camping Tentudia, CN 630 Km 727, 06260 Monesterio (Badajoz) [924-51 63 16; fax 924-51 63 52; ctentudia@turiex.com; www.camping-extremadura.com]** On E803/A66 N Mérida-Sevilla take exit 722 and foll dir to Monesterio then Santa Olallio - don't take rd into Nature Park. Med, hdstg, mkd, shd, terr, serviced pitches; wc; shwrs; EHU (15A); lndry; shop; rest; snacks; bar; playgrnd; pool; red long stay; Eng spkn; adv bkg acc; quiet; ccard acc; horseriding nr; bike hire; CKE/CCI. "Gd stopping place in quiet area; Sep '99 future of site uncertain, contact in advance; site a little run down; poss rd noise fr A66." ♦ Easter-15 Sep. € 20.00 2013*

⊞ **MONTBLANC** *3C2* (2km NE Rural) *41.37743, 1.18511* **Camping Montblanc Park, Ctra Prenafeta, Km 1.8, 43400 Montblanc [977-86 25 44; fax 977-86 05 39; montblancpark@franceloc.fr; www.montblancpark.com]** Exit AP2 junc 9 sp Montblanc; foll sp Montblanc/Prenafeta/TV2421; site on L on TV2421. Med, hdg, pt shd, pt sl, terr, wc (htd); chem disp; mv service pnt; fam bthrm; shwrs inc; EHU (10A) inc; lndry; shop; rest; snacks; bar; bbq; playgrnd; pool; paddling pool; red long stay; entmnt; wifi; 50% statics; dogs €4.50; phone; Eng spkn; adv bkg acc; ccard acc; CKE/CCI. "Excel site; excel facs; lovely area; many static pitches only suitable for o'fits up to 7m; Cistercian monestaries nrby; conv NH Andorra." ♦ € 39.50 2014*

MONTERROSO *1B2* (1km S Rural) *42.78720, -7.84414* **Camp Municipal de Monterroso, A Peneda, 27560 Monterroso (Lugo) [982-37 75 01; fax 982-37 74 16; campingmonterroso@aged-sl.com; www.campingmonterroso.com]** Fr N540 turn W onto N640 to Monterroso. Fr town cent turn S on LU212. In 100m turn sharp R then downhill for 1km; 2 sharp bends to site. Sm, mkd, hdg, pt shd, pt sl, wc; chem disp; shwrs inc; EHU (10A) €3.50; shop; rest nr; bar nr; wifi; dogs; Eng spkn; quiet; games area; CKE/CCI. "Helpful staff; v quiet & ltd facs LS; pool adj; vg." ♦ 30 Mar-24 Sep. € 21.00 2017*

⊞ **MONZON** *3B2* (6km NE Rural) *41.93673, 0.24146* **Camping Almunia, Calle del Nao 10, 22420 Almunia de san Juan [696-77 18 51; camping-almunia@hotmail.com; www.camping-almunia.es]** Foll the A-22 to Monzón. Turn R onto the A-1237 to campsite. Sm, hdstg, pt shd, terr, wc; chem disp; mv service pnt; fam bthrm; shwrs; EHU (6A) €4; lndry; bbq; playgrnd; pool; twin axles; wifi; adv bkg acc. "Friendly German couple; Monzon splendid medieval castle to visit; conv NH." ♦ € 13.00 2013*

"Satellite navigation makes touring much easier"

Remember most sat navs don't know if you're towing or in a larger vehicle – always use yours alongside maps and site directions.

⊞ **MORELLA** *3D2* (2km NE Rural) *40.62401, -0.09141* **Motor Caravan Parking, 12300 Morella (Castellón)** Exit N232 at sp (m'van emptying). Sm, hdstg, pt shd, chem disp; mv service pnt; quiet. "Free of charge; stay up to 72 hrs; clean; superb location; lge m'vans acc; excel Aire with fine views of hilltop town of Morella (floodlit at night), clean, well maintained." 2013*

⊞ **MOTILLA DEL PALANCAR** *4E1* (10km NW Rural) *39.61241, -2.10185* **Camping Pantapino, Paraje de Hontanar, s/n, 16115 Olmedilla de Alarcón (Cuenca) [969-33 92 33 or 676-47 86 11 (mob); fax 969-33 92 44; pantapina@hotmail.com]** Fr cent of Motilla foll NIII; turn NW onto rd CM2100 at sp for Valverde de Júcar; site on L just bef 12km marker. Med, mkd, pt shd, pt sl, serviced pitches; wc; chem disp; mv service pnt; fam bthrm; shwrs inc; EHU (6A) €4; gas; lndry; shop; rest; bar; bbq; playgrnd; pool; 40% statics; dogs €1.50; adv bkg acc; quiet; ccard acc; tennis; bike hire; horseriding; games area; CKE/CCI. "Clean, attractive site but tatty statics; poor facs; gd size pitches; resident owners hospitable; poss clsd in winter - phone ahead to check; vg; san facs old but clean; ltd facs LS; gd NH; poss problem with earth on elec." ♦ € 18.00 2014*

⊞ **MOTRIL** *2H4* (3km SW Urban/Coastal) *36.71833, -3.54616* **Camping Playa de Poniente de Motril, 18600 Motril (Granada) [958-82 03 03; fax 958-60 41 91; info@campingplayadeponiente.com; www.campingplayadeponiente.com]** Turn off coast rd N340 to port bef flyover; at rndabt take rd for Motril. Turn R in town, site sp. Lge, mkd, hdstg, pt shd, wc (htd); chem disp; mv service pnt; fam bthrm; shwrs; EHU (6-10) €3.35; gas; lndry; shop; rest; snacks; bar; bbq (elec, gas); playgrnd; pool; beach adj; red long stay; wifi; 40% statics; dogs €1.50; bus; Eng spkn; adv bkg acc; ccard acc; bike hire; games rm; horseriding; tennis; golf; games area. "Well-appointed site but surrounded by blocks of flats; gd, clean facs; helpful recep; gd shop; access diff for lge o'fits; poss lge flying beetles; excel long stay winter; lovely promenade with dedicated cycle track." ♦ € 31.70 2017*

⊞ **MUNDAKA** *3A1* (1km S Coastal) *43.39915, -2.69620* **Camping Portuondo, Ctra Amorebieta-Bermeo, Km 43, 48360 Mundaka (Bilbao) [946-87 77 01; fax 946-87 78 28; recepcion@campingportuondo.com; www.campingportuondo.com]** Fr Bermeo pass Mundaka staying on main rd, do not enter Mundaka. Stay on BI-2235 sp Gernika. After approx 1km site on L down steep slip rd. Med, pt shd, terr, wc; shwrs inc; EHU (6A) €4.20; lndry; rest; snacks; bar; playgrnd; pool; paddling pool; beach 500m; 30% statics; dogs; train 800m; adv bkg rec; ccard acc; site clsd end Jan-mid Feb. "Excel clean, modern facs; pitches tight not suitable for lge o'fits; popular with surfers; conv Bilbao by train; site suitable sm m'vans only; v ltd touring space; ent is very steep single track." € 35.00 2016*

MUROS *1B1* (3km SSW Coastal) *42.76072, -9.06222* **A'Vouga, Ctra Mouros-Finisterre, km 3 15291 Louro [34 98 18 26 115; avouga@hotmail.es]** On Coast rd fr Muros (3km) on L side. Med, mkd, unshd, pt sl, wc (htd); chem disp; mv service pnt; fam bthrm; shwrs inc; EHU (6A); lndry; shop nr; rest; snacks; bar; bbq; beach adj; twin axles; entmnt; wifi; TV; 10% statics; dogs; phone; Eng spkn; adv bkg acc; quiet; ccard acc; games rm. "Excel site & rest; seaviews; friendly & helpful staff; rec; guided walking tours." 1 Mar-31 Oct. € 38.00 2014*

MUROS *1B1* (7km W Coastal) *42.76100, 9.11100* **Camping Ancoradoiro, Ctra Corcubión-Muros, Km.7.2, 15250 Louro (La Coruña) [981-87 88 97; fax 981-87 85 50; wolfgang@mundo-r.com; www.rc-ancoradoiro.com/camping]** Foll AC550 W fr Muros. Site on L (S), well sp. Immed inside ent arch, to thro gate on L. Med, hdg, mkd, pt shd, terr, wc; chem disp; shwrs inc; EHU (6-15A) €3.50; lndry; shop nr; rest; bar nr; playgrnd; beach sand 500m; entmnt; phone; bus 500m; adv bkg acc; quiet; watersports; CKE/CCI. "Excel, lovely, well-run, well-kept site; superb friendly site on headland bet 2 sandy beaches; welcoming owner; excel rest; excel san facs; poss diff for lge o'fits; beautiful beaches; scenic area." 15 Mar-15 Sep. € 22.00 2014*

⊞ **MUXIA** *1A1* (10km E Coastal) *43.1164, -9.1583* **Camping Playa Barreira Leis, Playa Berreira, Leis, 15124 Camariñas-Muxia (La Coruña) [981-73 03 04; playaleis@yahoo.es; http://campingplayaleis.es/]** Fr Ponte do Porto turn L sp Muxia; foll camp sp. Site is 1st after Leis vill on R. Med, mkd, pt shd, terr, wc; chem disp; shwrs inc; EHU €3.50; lndry; shop; rest; bar; bbq; playgrnd; beach sand 100m; TV; dogs €1; quiet; ccard acc; CKE/CCI. "Beautiful situation on wooded hillside; dir acces to gd beach; ltd, poorly maintained facs LS; mkt in Muxia Thurs; scruffy & rundown; basic san facs." € 16.00 2012*

NAJERA *3B1* (1km S Urban) *42.41310, -2.73145* **Camping El Ruedo, San Julián 24, 26300 Nájera (La Rioja) [941-36 01 02; www.campingslarioja.es]** Take Nájera town dirs off N120. In town turn L bef x-ing bdge. Site sp. Sm, pt shd, wc (htd); chem disp; shwrs inc; EHU (10-16A) €3 (rev pol & poss no earth); gas; lndry; shop; rest; snacks; bar; playgrnd; entmnt; TV; phone; bus 200m; adv bkg acc; quiet; ccard acc; CKE/CCI. "Pleasant site in quiet location, don't be put off by 1st impression of town; monastery worth visit, some pitches in former bullring; san facs tired; many trees, could be diff for lge o'fits." 1 Apr-10 Sep. € 26.00 2016*

⊞ **NAVAJAS** *3D2* (1km W Rural) *39.87489, -0.51034* **Camping Altomira, Carretera, CV-213 Navajas Km. 1, E-12470 Navajas (Castellón) [964-71 32 11; fax 964-71 35 12; reservas@campingaltomira.com; www.campingaltomira.com]** Exit A23/N234 at junc 33 to rndabt & take CV214 dir Navajas. In approx 2km turn L onto CV213, site on L just past R turn into vill, sp. Med, hdstg, pt shd, terr, serviced pitches; wc (htd); chem disp; fam bthrm; shwrs; EHU (6A) inc; gas; lndry; shop; rest; snacks; bar; bbq; playgrnd; pool; paddling pool; red long stay; wifi; TV; 70% statics; dogs; phone; bus 500m; Eng spkn; adv bkg acc; ccard acc; games rm; tennis; bike hire; fishing; CKE/CCI. "Friendly welcome; panoramic views fr upper level (steep app) but not rec for lge o'fits due tight bends & ramped access/kerb to some pitches; gd birdwatching, walking, cycling; excel san facs; some sm pitches poss diff for lge o'fits without motor mover; poss clsd LS - phone ahead to check; vg, useful NH & longer; excel; waterfall in walking dist." ♦ € 28.50 2017*

⊞ **NERJA** *2H4* (4km E Rural) *36.76035, -3.83490* **Nerja Camping, Ctra Vieja Almeria, Km 296.5, Camp de Maro, 29787 Nerja (Málaga) [952-52 97 14; fax 952-52 96 96; info@nerjacamping.com; www.nerjacamping.com]** On N340, cont past sp on L for 200m around RH corner, bef turning round over broken white line. Foll partly surfaced rd to site on hillside. Fr Almuñécar on N340, site on R approx 20km. Med, pt shd, pt sl, terr, wc; chem disp; shwrs inc; EHU (5A) €3.75 (check earth); gas; lndry; shop; rest; snacks; bar; playgrnd; pool; beach sand 2km; red long stay; Eng spkn; adv bkg rec; site clsd Oct; bike hire; CKE/CCI. "5 mins to Nerja caves; mkt Tue; annual carnival 15 May; diff access lge o'fits; gd horseriding; site rds steep but gd surface; gd views; friendly owners." ♦ € 24.00 2016*

"There aren't many sites open at this time of year"

If you're travelling outside peak season remember to call ahead to check site opening dates – even if the entry says 'open all year'.

⊞ **NOIA** *1B2* (5km SW Coastal) *42.77198, -8.93761* **Camping Punta Batuda, Playa Hornanda, 15970 Porto do Son (La Coruña) [981-76 65 42; camping@puntabatuda.com; www.puntabatuda.com]** Fr Santiago take C543 twd Noia, then AC550 5km SW to Porto do Son. Site on R approx 1km after Boa. Lge, mkd, pt shd, terr, wc (htd); chem disp; shwrs inc; EHU (3A) €3.74 (poss rev pol); gas; lndry; shop; rest nr; snacks; bar; playgrnd; beach sand adj; red long stay; 50% statics; Eng spkn; adv bkg acc; quiet; tennis; CKE/CCI. "Wonderful views; htd pool w/end only; exposed to elements & poss windy; ltd facs LS; hot water to shwrs only; some pitches very steep &/or sm; gd facs; naturist beach 5km S." ♦ € 23.60 2012*

⊞ **NOJA** *1A4* (20km W Coastal) *43.46306, -3.72379* **Camping Derby Loredo, Calle Bajada a lay Playa, 19 39160 Loredo [942 504106; info@camping loredo.com; campingloredo.com]** Fr Santander on S10 twrd Bilbao. L at J12, foll CA141 to Pedrena/Somo. After Somo L onto CA440 Loredo. On ent Loredo L sp @400m Playa Deloredo.' Med, mkd, pt shd, wc; chem disp; shwrs inc; EHU; gas; lndry; shop nr; rest; snacks; bar; playgrnd; beach sand adj; twin axles; wifi; 70% statics; dogs €2; Eng spkn; ccard acc. "Gd loc by beach; watersports; not suitable for lge o'fits; surf board hire; busy; friendly site with fams and surfers; gd." ♦ € 27.00 2017*

NOJA *1A4* (1km NW Coastal) *43.49011, -3.53636* **Camping Playa Joyel, Playa del Ris, 39180 Noja (Cantabria) [942-63 00 81; fax 942-63 12 94; info@playayoyel.com; www.playajoyel.com]** Fr Santander or Bilbao foll sp A8/E70 (toll-free). Approx 15km E of Solares exit m'way junc 184 at Beranga onto CA147 N twd Noja & coast. On o'skirts of Noja turn L sp Playa del Ris, (sm brown sp) foll rd approx 1.5km to rndabt, site sp to L, 500m fr rndabt. Fr Santander take S10 for approx 8km, then join A8/E70. V lge, mkd, pt shd, pt sl, wc; chem disp; mv service pnt; fam bthrm; shwrs inc; EHU (6A) €4.30; gas; lndry; shop, rest, snacks; bar; bbq; playgrnd; pool; paddling pool; beach sand adj; entmnt; wifi; TV; 40% statics; phone; Eng spkn; adv bkg acc; ccard acc; sailing; windsurfing; jacuzzi; tennis; games rm; CKE/CCI. "Well-organised site on sheltered bay; cash dispenser; very busy high ssn; pleasant staff; hairdresser; car wash; no o'fits over 8m high ssn; gd, clean facs; superb pool & beach; recep 0800-2200; some narr site rds with kerbs; midnight silence enforced; highly rec." ♦ 27 Mar-25 Sep. € 49.00 2015*

⊞ **NUEVALOS** *3C1* (1km N Rural) *41.21846, -1.79211* **Camping Lago Park, Ctra De Alhama de Aragón a Cillas, Km 39, 50210 Nuévalos (Zaragoza) [976-84 90 38; lagoresort@gmail.com; www.lagoresort.com]** Fr E on A2/E90 exit junc 231 to Nuévalos, turn R sp Madrid. Site 1.5km on L when ent Nuévalos. Fr W exit junc 204, site well sp. Steep ent fr rd. V lge, hdg, mkd, pt shd, terr, wc; chem disp; fam bthrm; shwrs inc; EHU (10A) €5.40; gas; lndry; shop nr; rest; snacks; bar nr; bbq; playgrnd; pool; red long stay; 10% statics; dogs free; bus 500m; adv bkg acc; fishing; games area; boating; CKE/CCI. "Nr Monasterio de Piedra & Tranquera Lake; excel facs on top terr, but stretched high ssn & poss long, steepish walk; lake nrby; ltd facs LS; gd birdwatching; bar 500m; only site in area; gd; very friendly owner; an oasis en rte to Madrid in picturesque setting; vg rest; gd welcome; pool not open yet (2016); rec." € 19.00 2016*

⊞ **OCHAGAVIA** *3A1* (1km S Rural) *42.90777, -1.08750* **Camping Osate, Ctra Salazar s/n, 31680 Ochagavia (Navarra) [948-89 01 84; info@campingosate.net; www.campingosate.net]** On N135 SE fr Auritz, turn L onto NA140 & cont for 24km bef turning L twd Ochagavia on NA140. Site sp in 2km on R, 500m bef vill. Med, mkd, pt shd, serviced pitches; wc; chem disp; shwrs inc; EHU (4A) €5.50; gas; lndry; shop; rest; snacks; bar; bbq; 50% statics; dogs €2; Eng spkn; quiet. "Attractive, remote vill; gd, well-maintained site; touring pitches under trees, sep fr statics; facs ltd & poss stretched high ssn; site clsd 3 Nov-15 Dec & rec phone ahead LS; facs require maintenance (2018); TO v helpful; gd walks fr site."
€ 22.00 2018*

"That's changed – Should I let the Club know?"

If you find something on site that's different from the site entry, fill in a report and let us know. See camc.com/europereport.

OCHAGAVIA *3A1* (7km S Rural) *42.85486, -1.09766* **Camping Murkuzuria, 31453 Esparza de Salazar [948-89 01 90 or 661-08 87 35; campingesparza@gmail.com; www.campingmurkuzuria.com]** Fr N or S on Pic d'Orhy rte thro Pyrenees, NA178(Spain)/D26(France). Situated in vill. Pamplona approx 80km. Med, mkd, pt shd, wc; chem disp; shwrs; EHU; lndry; shop; rest; bar; bbq; playgrnd; pool; twin axles; TV; phone; bus adj; Eng spkn; adv bkg acc; quiet. "Discounted forest passes avail for Foret d'Iraty; excel." 15 May-30 Oct. € 19.00 2015*

⊞ **OLIVA** *4E2* (2km E Coastal) *38.93278, -0.09778* **Camping Kiko Park, Calle Assagador de Carro 2, 46780 Playa de Oliva (València) [962-85 09 05; fax 962-85 43 20; kikopark@kikopark.com; www.kikopark.com]** Exit AP7/E15 junc 61; fr toll turn R at T-junc onto N332.At rndabt turn L foll sp Platjas; next rdbt take 1st exit sp Platja; next rndabt foll sp Kiko Park. Do not drive thro Oliva. Access poss diff on app rds due humps. Lge, mkd, hdstg, hdg, shd, serviced pitches; wc (htd); chem disp; mv service pnt; fam bthrm; shwrs inc; EHU (16A) inc; gas; lndry; shop; rest; snacks; bar; bbq; playgrnd; pool (covrd); paddling pool; beach sand adj; red long stay; entmnt; wifi; dogs €3.10; phone; Eng spkn; adv bkg acc; quiet; ccard acc; horseriding nr; watersports; games rm; bike hire; fishing; games area; tennis; windsurfing school; golf nr; CKE/CCI. "Gd, family-run site; whirlpool; spa; very helpful staff; vg, clean san facs; excel rest in Michelin Guide; pitch price variable (lge pitches avail); cash machine; beauty cent; access tight to some pitches." ♦ € 66.00 2014*

See advertisement above

⊞ **OLIVA** *4E2* (3km SE Coastal) *38.90555, -0.06666* **Eurocamping, Ctra València-Oliva, Partida Rabdells s/n, 46780 Playa de Oliva (València) [962-85 40 98; fax 962-85 17 53; info@eurocamping-es.com; www.eurocamping-es.com]** Fr N exit AP7/E15 junc 61 onto N332 dir Alicante. Drive S thro Oliva & exit N332 km 209.9 sp 'urbanización'. At v lge hotel Oliva Nova Golf take 3rd exit at rndabt sp Oliva & foll camping sp to site. Fr S exit AP7 junc 62 onto N332 dir València, exit at km 209 sp 'urbanización', then as above. Lge, mkd, hdstg, hdg, pt shd, wc (htd); chem disp; mv service pnt; fam bthrm; shwrs inc; EHU (6-10A) €4.64-6.70; gas; lndry; shop; rest; snacks; bar; bbq; playgrnd; beach sand adj; red long stay; entmnt; wifi; TV; dogs €2.16; phone; quiet; ccard acc; bike hire; CKE/CCI. "Gd facs; busy, well-maintained, clean site adj housing development; helpful British owners; beautiful clean beach; gd rest; gd beach walks; cycle rte thro orange groves to town; pitch far fr recep if poss, night noise fr generators 1700-2400; recep clsd 1400-1600; highly rec; rest stretched; busy site." ♦
€ 44.40 2012*

See advertisement opposite

⊞ **OLIVA** *4E2* (3km S Coastal) *38.89444, -0.05361* **Camping Olé, Partida Aigua Morta s/n, 46780 Playa de Oliva (València) [962-85 75 17; fax 962-85 75 16; campingole@hotmail.com; www.camping-ole.com]** Exit AP7/E15 junc 61 onto N332 dir Valencia/Oliva. At km 209 (bef bdge) turn R sp 'Urbanización. At 1st rndabt, take 2nd exit past golf club ent, then 1st exit at next rndabt, turn L sp ' Camping Olé' & others. Site down narr rd on L. Lge, hdg, mkd, hdstg, pt shd, wc (htd); chem disp; fam bthrm; shwrs inc; EHU (6-10A) €5.74; gas; lndry; shop; rest; snacks; bar; bbq; playgrnd; pool; beach sand adj; red long stay; entmnt; wifi; 15% statics; dogs €3.15; phone; Eng spkn; adv bkg acc; quiet; ccard acc; bike hire; tennis 600m; fishing; golf adj; games rm; horseriding 2km; CKE/CCI. "Many sports & activities; direct access to beach; excel site; rest across rd very nice; gd value; pool only opens 1st July." ♦ € 50.00 2014*

⊞ **OLVERA** *2H3* (4km E Rural) *36.93905, -5.21719* **Camping Pueblo Blanco, Ctra N384, Km 69, 11690 Olvera (Cadiz) [619 45 35 34; fax 952 83 43 73; info@campingpuebloblanco.com; www.camping puebloblanco.com]** Bet Antequera and Jerez de la Frontera, on the A384, at 69km marker. About 3km bef Olvera on the R. Wide driveway 600m to the top. Lge, unshd, pt sl, terr, wc; chem disp; mv service pnt; fam bthrm; shwrs; EHU (16A) €4; shop; rest; bar; bbq; playgrnd; red long stay; entmnt; wifi; TV; dogs €1.50; Eng spkn; adv bkg acc; games rm. "Site has 360 degree mountain views; ideal for walking; 12 bungalows; pool games area; bird watching and Pueblo Blanco; vg site, but not quite finished." ♦ € 27.50 2013*

⊞ **OROPESA** *3D2* (4km N Coastal) *40.12125, 0.15848* **Camping Didota, Avenida de la Didota s/n, 12594 Oropesa del Mar (Castellón) [964 31 95 51; fax 964 31 98 47; info@campingdidota.es; www.campingdidota.es]** N on rd E-15 fr València to Barcelona, bear L at exit 45 sp Oropesa del Mar. Turn L onto N-340. Turn R at next exit, then cont strt at rndabt onto on Avenida La Ratlla. Foll camping signs. Med, pt shd, wc; chem disp; fam bthrm; shwrs inc; EHU (6-10A) €4.30; gas; lndry; shop; rest; snacks; playgrnd; pool; beach sand; 10% statics; dogs; adv bkg acc; ccard acc. "Gd site, helpful friendly staff; excel pool." ♦ € 33.70 2014*

⊞ **OROPESA** *3D2* (4km NE Coastal) *40.1275, 0.15972* **Camping Torre La Sal 2, Cami L'Atall s/n, 12595 Ribera de Cabanes (Castellón) [964-31 95 67; fax 964-31 97 44; camping@torrelasal2.com; www.torrelasal2.com]** Leave AP7 at exit 45 & take N340 twd Tarragona. Foll camp sp fr km 1000 stone. Site adj Torre La Sal 1. Lge, mkd, hdstg, hdg, pt shd, serviced pitches; wc (htd); chem disp; shwrs inc; EHU (10A) inc; gas; lndry; shop; rest; snacks; bar; playgrnd; pool (covrd, htd); beach shgl adj; red long stay; entmnt; wifi; TV; 10% statics; dogs free; Eng spkn; adv bkg acc; quiet; games area; tennis; sauna; CKE/CCI. "Vg, clean, peaceful, well-run site; lger pitches nr pool; library; more mature c'vanners very welcome; many dogs; poss diff for lge o'fits & m'vans; excel rest; excel beach with dunes; excel site, spotless facs, highly rec." ♦ € 51.00 2013*

"I like to fill in the reports as I travel from site to site"

You'll find report forms at the back of this guide, or you can fill them in online at camc.com/europereport.

PALAFRUGELL *3B3* (5km E Coastal) *41.9005, 3.1893* **Kim's Camping, Calle Font d'en Xeco s/n, 17211 Llafranc (Gerona) [972-30 11 56; fax 972-61 08 94; info@campingkims.com; www.campingkims.com]** Exit AP7 at junc 6 Gerona Nord if coming fr France, or junc 9 fr S dir Palamós. Foll sp for Palafrugell, Playa Llafranc. Site is 500m N of Llafranc. Lge, hdg, mkd, hdstg, shd, sl, terr, wc; chem disp; fam bthrm; shwrs inc; EHU (5A) inc; gas; lndry; shop; rest; snacks; bar; bbq (gas); playgrnd; pool; beach sand 500m; red long stay; entmnt; wifi; TV; 10% statics; dogs; phone; Eng spkn; adv bkg acc; quiet; ccard acc; watersports; golf 10km; games area; excursions; tennis 500m; games rm; CKE/CCI. "Excel, well-organised, friendly, fam run site; steep site rds, new 2nd ent fr dual c'way fr Palafrugell to llafranc for lge o'fits & steps to rd to beach; bike hire 500m; guarded; discount in high ssn for stays over 1 wk; excel, modern san facs; beautiful coastal area; mostly gd size pitches." ♦ 14 Apr-24 Sep. € 51.00 2017*

PALAFRUGELL *3B3* (5km S Coastal) *41.88879, 3.17928* **Camping Moby Dick, Carrer de la Costa Verda 16-28, 17210 Calella de Palafrugell (Gerona) [972-61 43 07; fax 972-61 49 40; info@camping mobydick.com; www.campingmobydick.com]** Fr Palafrugell foll sps to Calella. At rndabt just bef Calella turn R, then 4th L, site clearly sp on R. Med, hdstg, pt shd, sl, terr, wc; chem disp; mv service pnt; fam bthrm; shwrs inc; EHU (10A); lndry; shop; rest nr; snacks; bar; playgrnd; pool; beach shgl 100m; wifi; TV; 15% statics; dogs €3.30; phone; bus 100m; Eng spkn; adv bkg acc; quiet; ccard acc; CKE/CCI. "Nice views fr upper terraces; gd rest; very friendly; very pretty sm resort; lovely coastal walks; gd value; excel; lovely sea views; great site." ♦ 25 Mar-30 Sep. € 35.00 2016*

PALAMOS *3B3* (1km N Coastal) *41.85044, 3.13873* **Camping Palamós, Ctra La Fosca 12, 17230 Palamós (Gerona) [972-31 42 96; fax 972-60 11 00; campingpal@grn.es; www.campingpalamos.com]** App Palamós on C66/C31 fr Gerona & Palafrugell turn L 16m after overhead sp Sant Feliu-Palamós at sm sp La Fosca & campsites. Lge, pt shd, pt sl, terr, wc; fam bthrm; shwrs; EHU (4A) €2.70; gas; lndry; shop; rest nr; playgrnd; pool (htd); beach shgl adj; wifi; 30% statics; dogs €2; phone; ccard acc; golf; tennis. ♦ 27 Mar-30 Sep. € 40.00 2013*

PALAMOS *3B3* (2km NE Coastal) *41.87277, 3.15055* **Camping Benelux, Paratge Torre Mirona s/n, 17230 Palamós (Gerona) [972-31 55 75; fax 972-60 19 01; www.cbenelux.com]** Turn E off Palamós-La Bisbal rd (C66/C31) at junc 328. Site in 800m on minor metalled rd, twd sea at Playa del Castell. Lge, hdstg, mkd, pt shd, terr, wc; chem disp; mv service pnt; shwrs inc; EHU (10A) €6.90; gas; lndry; shop; rest; snacks; bar; bbq; playgrnd; pool; beach sand 1km; red long stay; wifi; TV; 30% statics; dogs; phone; Eng spkn; adv bkg acc; ccard acc; CKE/CCI. "In pine woods; many long stay British/Dutch; friendly owner; safe dep; clean facs poss ltd LS; car wash; currency exchange; poss flooding in heavy rain; rough grnd; marvellous walking/cycling area; many little coves." ♦ 24 Mar-25 Sep. € 36.70 2016*

PALAMOS *3B3* (3km W Coastal) *41.84700, 3.09861* **Eurocamping, Avda de Catalunya 15, 17252 Sant Antoni de Calonge (Gerona) [972-65 08 79; fax 972-66 19 87; info@euro-camping.com; www.euro-camping.com]** Exit A7 junc 6 dir Palamós on C66 & Sant Feliu C31. Take exit Sant Antoni; on ent Sant Antoni turn R at 1st rndabt. Visible fr main rd at cent of Sant Antoni. V lge, hdg, mkd, shd, serviced pitches; wc; chem disp; mv service pnt; fam bthrm; shwrs inc; EHU (6A) inc; lndry; shop; rest; snacks; bar; bbq; playgrnd; pool; paddling pool; beach sand 300m; red long stay; entmnt; wifi; TV; 15% statics; dogs €4; phone; Eng spkn; adv bkg acc; quiet; ccard acc; games area; games rm; tennis; golf 7km; waterpark. "Excel facs for families; fitness rm; doctor Jul & Aug; car wash; lots to do in area; excel; lots of ssnal pitches; immac san facs; generous flat pitches; waterpark 5km; helpful, friendly staff." ♦ 28 Apr-17 Sep. € 54.00 2017*

PALS *3B3* (6km NE Coastal) *41.98132, 3.20125* **Camping Inter Pals, Avda Mediterránea s/n, Km 45, 17256 Playa de Pals (Gerona) [972-63 61 79; fax 972-66 74 76; interpals@interpals.com; www.interpals.com]** Exit A7 junc 6 dir Palamós onto C66. Turn N sp Pals & foll sp Playa/Platja de Pals, pass Camping Neptune sp on L then Golf Aparthotel on R. At rndbt take 2nd exit, site clearly sp on L approx 500m. Lge, shd, pt sl, terr, wc (htd); chem disp; mv service pnt; fam bthrm; shwrs inc; EHU (5-10A) inc; lndry; shop nr; rest; snacks; bar; playgrnd; pool; paddling pool; beach sand 600m; entmnt; wifi; TV; 20% statics; dogs €3.50; phone; adv bkg acc; quiet; golf 1km; tennis; bike hire; watersports; games area; CKE/CCI. "Lovely, well-maintained site in pine forest; poss diff lge o'fits - lge pitches at lower end of site; naturist beach 1km; modern, well-maintained facs." ♦ 1 Apr-25 Sep. € 55.60 2014*

PALS *3B3* (6km NE Coastal) *42.00120, 3.19388* **Camping Playa Brava, Playa Pals, 17256 Pals (Gerona) [972-63 68 94; fax 972-63 69 52; info@playabrava.com; www.playabrava.com]** App Pals on rd 650 fr N or S. Avoid Pals cent. Fr by-pass take rd E sp Playa de Pals at rndabt. In 4km turn L opp shops, in 400m turn L past golf course, site on L bef beach. Avoid Begur & coast rd. Lge, shd, wc; chem disp; fam bthrm; shwrs; EHU inc; gas; lndry; shop nr; rest; snacks; bar; playgrnd; pool; beach sand adj; adv bkg acc; quiet; tennis; car wash; bike hire. "Guarded; gd facs for children & families." ♦ 15 May-12 Sep. € 34.00 2016*

See advertisement opposite

⊞ **PALS** *3B3* (1km E Rural) *41.95541, 3.15780* **Camping Resort Mas Patoxas, Ctra Torroella-Palafrugell, Km 339, 17256 Pals (Gerona) [972-63 69 28; fax 972-66 73 49; info@campingmaspatoxas.com; www.campingmaspatoxas.com]** AP7 exit 6 onto C66 Palamós/La Bisbal, turn L via Torrent to Pals. Turn R & site on R almost opp old town of Pals on rd to Torroella de Montgri. Or fr Palafrugell on C31 turn at km 339. Lge, mkd, shd, terr, serviced pitches; wc (htd); chem disp; mv service pnt; fam bthrm; shwrs inc; EHU (5A) inc; gas; lndry; shop; rest; snacks; bar; playgrnd; pool; beach sand 1km; red long stay; entmnt; TV; dogs €3.60; phone; Eng spkn; adv bkg req; quiet; ccard acc; bike hire; games area; tennis; site clsd 14 Dec-16 Jan; golf 4km; CKE/CCI. "Excel; recep clsd Monday LS; gd security." ♦ € 47.00 2015*

PALS *3B3* (4km E Rural) *41.98555, 3.18194* **Camping Cypsela, Rodors 7, 17256 Playa de Pals (Gerona) [972-66 76 96; fax 972-66 73 00; info@cypsela.com; www.cypsela.com]** Exit AP7 junc 6, rd C66 dir Palamós. 7km fr La Bisbal take dir Pals & foll sp Playa/Platja de Pals, site sp. V lge, hdg, mkd, hdstg, shd, serviced pitches; wc; chem disp; mv service pnt; fam bthrm; shwrs inc; EHU (6-10A) inc; gas; lndry; shop; rest; snacks; bar; bbq; playgrnd; pool; beach sand 1.5km; red long stay; entmnt; wifi; TV; 60% statics; Eng spkn; adv bkg acc; ccard acc; tennis; golf 1km; games rm; bike hire; CKE/CCI. "Noise levels controlled after midnight; excel san facs; mini golf & other sports; free bus to beach; private bthrms avail; 4 grades of pitch/price (highest price shown); vg site." ♦ 4 May-10 Sep. € 56.00 2016*

"We must tell the Club about that great site we found"

Get your site reports in by mid-August and we'll do our best to get your updates into the next edition.

SPAIN

⊞ **PAMPLONA** *3B1* (7km N Rural) *42.85776, -1.62250* **Camping Ezcaba, Ctra a Francia, km 2,5, 31194 Eusa-Oricain (Navarre) [948-33 03 15; fax 948-33 13 16; info@campingezcaba.com; www.campingezcaba.com]** Fr N leave AP15 onto NA30 (N ring rd) to N121A sp Francia/Iruña. Pass Arre & Oricáin, turn L foll site sp 500m on R dir Berriosuso. Site on R in 500m - fairly steep ent. Or fr S leave AP15 onto NA32 (E by-pass) to N121A sp Francia/Iruña, then as above. Med, mkd, pt shd, pt sl, wc; shwrs inc; EHU (10A) €5.50; gas; lndry; shop; rest; snacks; bar; playgrnd; pool; wifi; dogs €2.95; phone; bus 1km; adv bkg acc; horseriding; tennis. "Helpful, friendly staff; sm pitches unsuitable lge o'fits & poss diff due trees, esp when site full; attractive setting; gd pool, bar & rest; ltd facs LS & poss long walk to san facs; in winter use as NH only; phone to check open LS; excel cycle track to Pamplona; quiet rural site; gd facs not htd." ♦ € 30.00 2018*

"I need an on-site restaurant"

We do our best to make sure site information is correct, but it is always best to check any must-have facilities are still available or will be open during your visit.

⊞ **PARADA DE SIL** *1B2* (3km NW Rural) *42.38941, -7.58885* **Camping Cañón do Sil, Lugar de Castro s/n, 32740 Parada de Sil [608 537 017; info@canondosilcamping.com; www.canondosilcamping.com]** OU-0604 to Ctra De Castro, turn R on OU-0605. Turn L to Ctra da Castro, site on L. Med, pt shd, terr, wc; chem disp; shwrs; EHU €4; shop nr; snacks; bar. "Amazing location on the edge of Rv Sil Gorge; gd walks fr site; vg." € 23.00 2016*

PENAFIEL *1C4* (1km SW Rural) *41.59538, -4.12811* **Camping Riberduero, Avda Polideportivo 51, 47300 Peñafiel [983-88 16 37; camping@campingpenafiel.com; www.campingpenafiel.com]** Fr Valladolid 56km or Aranda de Duero 38km on N122. In Peñafiel take VA223 dir Cuéllar, foll sp to sports cent/camping. Med, hdstg, mkd, shd, wc (htd); chem disp; mv service pnt; fam bthrm; shwrs inc; EHU (5A) €5; gas; lndry; shop; rest; snacks; bar; playgrnd; pool; red long stay; TV; 20% statics; dogs €1.50; phone; bus 1km; Eng spkn; adv bkg acc; quiet; ccard acc; site open w/end only LS; bike hire. "Excel, well-kept site; interesting, historical area; ideal for wheelchair users; sm pitches and access diff due to trees." ♦ Holy Week & 1 Apr-30 Sep. € 16.50 2016*

⊞ **PENISCOLA** *3D2* (2km N Coastal) *40.37916, 0.38833* **Camping Los Pinos, Calle Abellars s/n, 12598 Peñíscola (Castellón) [964-48 03 79; info@campinglospinos.com; www.campinglospinos.com]** Exit A7 junc 43 or N340 sp Peñíscola. Site sp on L. Med, mkd, hdg, pt shd, wc; chem disp; mv service pnt; fam bthrm; shwrs; EHU (10A) €5.95; gas; lndry; shop; rest; snacks; bar; bbq; playgrnd; pool; paddling pool; beach 1.5km; wifi; TV; 10% statics; dogs; phone; bus fr site; Eng spkn; adv bkg acc; games rm. "Narr site rds, lots of trees; poss diff access some pitches; vg." € 30.00 2016*

⊞ **PENISCOLA** *3D2* (3km NE Coastal) *40.37152, 0.40269* **Camping El Edén, Ctra CS501 Benicarló-Peñíscola Km 6, 12598 Peñíscola (Castellón) [964-48 05 62; fax 964-48 98 28; camping@camping-eden.com; www.camping-eden.com]** Exit AP7 junc 43 onto N340 then CV141 to Peñíscola ctr. Take 3rd exit off rndabt at seafront, after 1km turn L after Hotel del Mar. Rec avoid sat nav rte across marshes fr Peñíscola. Lge, mkd, hdg, pt shd, wc (htd); chem disp; mv service pnt; fam bthrm; shwrs inc; EHU (10A) inc; gas; lndry; shop nr; rest; snacks; bar; playgrnd; pool; paddling pool; beach shgl adj; red long stay; wifi; 40% statics; dogs €0.75; bus adj; ccard acc; ACSI acc. "San facs refurbished & v clean; beach adj cleaned daily; gd security; excel pool; easy access to sandy/gravel pitches but many sm trees poss diff for awnings or high m'vans; cash dispenser; poss vicious mosquitoes at dusk; easy walk/cycle to town; 4 diff sizes of pitch (some with tap, sink & drain) with different prices; ltd facs LS; excel." ♦ € 53.00 2015*

⊞ **PENISCOLA** *3D2* (2km NW Rural) *40.40158, 0.38116* **Camping Spa Natura Resort, Partida Villarroyos s/n, Playa Montana, 12598 Peñíscola-Benicarló (Castellón) [964-47 54 80; fax 964-78 50 51; info@spanaturaresort.com; www.spanaturaresort.com]** Exit AP7 junc 43, within 50m of toll booths turn R immed then immed L & foll site sp twd Benicarló (NB R turn is on slip rd). Fr N340 take CV141 to Peñíscola. Cross m'way bdge & immed turn L; site sp. Med, hdstg, mkd, shd, serviced pitches; wc (htd); chem disp; mv service pnt; fam bthrm; shwrs inc; EHU (6A) inc; gas; lndry; rest; snacks; bar; bbq; playgrnd; pool (htd); paddling pool; beach sand 2.5km; red long stay; twin axles; entmnt; wifi; TV; 50% statics; dogs €3 (free LS); phone; bus 600m; Eng spkn; adv bkg acc; ccard acc; games rm; tennis; waterslide; car wash; sauna; games area; gym; bike hire; CKE/CCI. "Vg site; helpful, enthusiastic staff; c'van storage; spa; wellness cent; jacuzzi; gd clean san facs; wide range of facs; gd cycling." ♦ € 40.00 2013*

⊞ **PILAR DE LA HORADADA** *4F2* (4km NE Coastal) *37.87916, -0.76555* **Lo Monte Camping & Caravaning, Avenida Comunidada Valenciana No 157 CP 03190 [00 34 966 766 782; fax 00 34 966 746 536; info@camplinglomonte-alicante.es; www.campinglomonte-alicante.es]** Exit 770 of AP7 dir Pilar de la Horadada; take the 1st L. Med, mkd, hdg, serviced pitches; wc (htd); chem disp; fam bthrm; shwrs inc; EHU (16A) €0.40; lndry; shop; rest; snacks; bar; bbq; playgrnd; pool (covrd, htd); beach 1km; entmnt; wifi; dogs €1; Eng spkn; adv bkg acc; quiet; ccard acc; bike hire; games rm; CKE/CCI. "New site; superb facs, exceptionally clean; great location, lots of golf & gd for walks & cycling; rec; excel; gym/wellness cent; beautifully laid out; neat; gd pool; v gd rest; isolated, nothing around site." ♦ € 36.00 2018*

PINEDA DE MAR *3C3* (1km SW Coastal) *41.61827, 2.67891* **Camping Bellsol, Passeig Maritim 46, 08397 Pineda de Mar [937-67 17 78; fax 937-65 55 51; info@campingbellsol.com; www.campingbellsol.com]** Fr N, take exit AP7 Junc 9 & immed turn R onto N11 dir Barcelona. Foll sp Pineda de Mar and turn L twd Paseo Maritim at exit at rv x-ing. Fr S on C32, exit 122 dir Pineda de Mar & foll dir Paseo Maritim & Campings fr same rndabt. Beware narr rds at other turnings. Lge, hdstg, shd, wc; chem disp; fam bthrm; shwrs; EHU (4A); lndry; rest; snacks; playgrnd; pool; beach; twin axles; red long stay; wifi; 15% statics; dogs; bus nrby, train 800m; Eng spkn; adv bkg acc; CKE/CCI. "V friendly & helpful staff; walking; bike & moped hire; sea fishing trips; vg." ♦ 19 Mar-31 Dec. € 34.60 2016*

"Satellite navigation makes touring much easier"

Remember most sat navs don't know if you're towing or in a larger vehicle – always use yours alongside maps and site directions.

⊞ **PLASENCIA** *1D3* (4km NE Urban) *40.04348, -6.05751* **Camping La Chopera, Ctra N110, Km 401.3, Valle del Jerte, 10600 Plasencia (Caceras) [927-41 66 60; lachopera@campinglachopera.com; www.campinglachopera.com]** In Plasencia on N630 turn E on N110 sp Ávila & foll sp indus est & sp to site. Med, shd, serviced pitches; wc; chem disp; fam bthrm; shwrs inc; EHU (6A) inc; gas; lndry; shop nr; rest; bar; bbq; playgrnd; pool; paddling pool; wifi; dogs; ccard acc; tennis; bike hire; CKE/CCI. "Peaceful & spacious; much birdsong; conv Manfragüe National Park (breeding of black/Egyptian vultures, black storks, imperial eagles); excel pool & modern facs; helpful owners; shop (Jul & Aug); Carrefour in town; 35 min walk to town." ♦ € 22.00 2016*

⊞ **PLASENCIA** *1D3* (14km S Rural) *39.94361, -6.08444* **Camping Parque Natural Monfragüe, Ctra Plasencia-Trujillo, Km 10, 10680 Malpartida de Plasencia (Cáceres) [927- 45 92 33 or 605 94 08 78 (mob); campingmonfrague@hotmail.com; www.campingmonfrague.com]** Fr N on A66/N630 by-pass town, 5km S of town at flyover junc take EXA1 (EX108) sp Navalmoral de la Mata. In 6km turn R onto EX208 dir Trujillo, site on L in 5km. Med, hdg, pt shd, pt sl, terr, wc (htd); chem disp; mv service pnt; fam bthrm; shwrs inc; EHU (10A) €4; gas; lndry; shop; rest; snacks; bar; bbq; playgrnd; pool; wifi; TV; 10% statics; dogs; phone; Eng spkn; quiet; ccard acc; tennis; archery; bike hire; horseriding; games area. "Friendly, helpful staff; red ACSI; vg, gd rest; clean, tidy, busy site but poss dusty, hoses avail; 10km to National Park (birdwatching trips); rambling; 4x4 off-rd; many birds on site; excel year round base; new excel san facs; discounted fees must be paid in cash; pitches muddy after heavy rain; peaceful." ♦ € 20.40 2018*

"There aren't many sites open at this time of year"

If you're travelling outside peak season remember to call ahead to check site opening dates – even if the entry says 'open all year'.

PLAYA DE ARO *3B3* (2km N Coastal) *41.83116, 3.08366* **Camping Cala Gogo, Avda Andorra 13, 17251 Calonge (Gerona) [972-65 15 64; fax 972-65 05 53; calagogo@calagogo.es; www.calagogo.es]** Exit AP7 junc 6 dir Palamós/Sant Feliu. Fr Palamós take C253 coast rd S twd Sant Antoni, site on R 2km fr Playa de Aro, sp. Lge, pt shd, pt sl, terr, serviced pitches; wc; chem disp; mv service pnt; fam bthrm; shwrs inc; EHU (10A) inc; gas; lndry; shop; rest; snacks; bar; bbq; playgrnd; pool (htd); paddling pool; beach sand adj; red long stay; entmnt; wifi; TV; dogs (except 3/7-21/8, otherwise €2); Eng spkn; adv bkg acc; quiet; boat hire; tennis; games area; golf 4km; bike hire; games rm. "Clean & recently upgraded san facs; rest/bar with terr; diving school; site terraced into pinewood on steep hillside; excel family site." ♦ 16 Apr-18 Sep. € 52.00 2016*

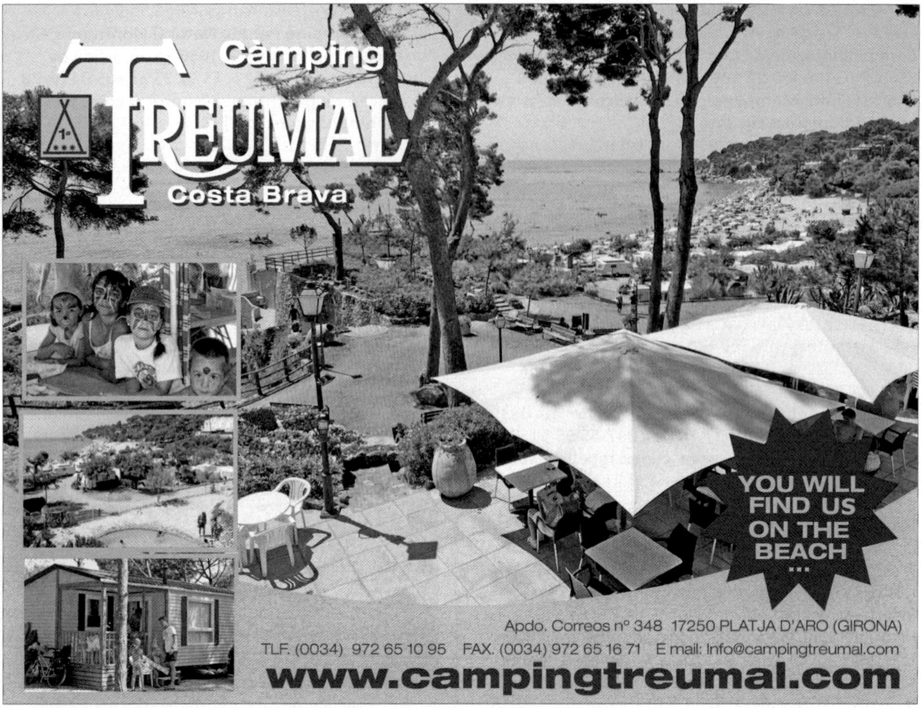

PLAYA DE ARO *3B3* (2km N Coastal) *41.83666, 3.08722* **Camping Treumal, Ctra Playa de Aro/ Palamós, C253, Km 47.5, 17250 Playa de Arro (Gerona) [972-65 10 95; fax 972-65 16 71; info@ campingtreumal.com; www.campingtreumal.com]** Exit m'way at junc 6, 7 or 9 dir Sant Feliu de Guixols to Playa de Aro; site is sp at km 47.5 fr C253 coast rd SW of Palamós. Lge, mkd, shd, terr, wc; chem disp; mv service pnt; fam bthrm; shwrs inc; EHU (10A) inc; gas; lndry; shop; rest; snacks; bar; playgrnd; pool; beach sand adj; entmnt; wifi; 25% statics; phone; Eng spkn; adv bkg acc; quiet; ccard acc; car wash; sports facs; games rm; tennis 1km; bike hire; golf 5km; fishing; CKE/CCI. "Peaceful site in pine trees; excel san facs; manhandling poss req onto terr pitches; gd beach." ♦ 31 Mar-30 Sep. € 49.00 2016*

See advertisement

POLA DE SOMIEDO *1A3* (0km E Rural) *43.09222, -6.25222* **Camping La Pomerada de Somiedo, 33840 Pola de Somiedo (Asturias) [985-76 34 04; csomiedo@infonegocio.com]** W fr Oviedo on A63, turn S onto AS15/AS227 to Augasmestas & Pola de Somiedo. Site adj Hotel Alba, sp fr vill. Route on steep, winding, mountain rd - suitable sm, powerful o'fits only. Sm, mkd, pt shd, wc; chem disp; mv service pnt; shwrs inc; EHU €4.20; shop nr; rest nr; quiet. "Mountain views; nr national park." 1 Apr-31 Dec. € 19.00 2016*

⊞ **PONFERRADA** *1B3* (16km W Rural) *42.56160, -6.74590* **Camping El Bierzo, 24550 Villamartín de la Abadia (León) [987-56 25 15; info@campingbierzo.com; www.campingbierzo.com]** Exit A6 junc 399 dir Carracedelo; after rndabt turn onto NV1 & foll sp Villamartín. Bef ent Villamartín turn L & foll site sp. Med, pt shd, wc; chem disp; mv service pnt; shwrs inc; EHU (5A) €4.20; shop nr; rest; bar; playgrnd; phone; bus 1km; adv bkg acc; quiet; CKE/CCI. "Attractive, rvside site in pleasant area; gd facs; friendly, helpful owner takes pride in his site; Roman & medieval attractions nr." ♦ € 22.00 2016*

PONT DE SUERT *3B2* (3km N Rural) *42.43083, 0.73861* **Camping Can Roig, Ctra Boí, Km 0.5, 25520 El Pont de Suert (Lleida) [973-69 05 02; fax 973-69 12 06; info@campingcanroig.com; www.campingcanroig.com]** N of Pont de Suert on N230 turn NE onto L500 dir Caldes de Boí. Site in 1km. App narr for 100m. Med, mkd, hdstg, pt shd, pt sl, wc; chem disp; shwrs inc; EHU (5A) €5.15; gas; lndry; shop; snacks; bar; playgrnd; paddling pool; 5% statics; dogs €3.60; adv bkg acc; quiet; ccard acc. "NH en rte S; beautiful valley; informal, friendly, quirky site (free range poultry); v relaxed atmosphere; helpful owner; fabulous valley & national park with thermal springs; poss scruffy (2015)." 1 Mar-31 Oct. € 27.00 2015*

⊞ **PONT DE SUERT** *3B2* (16km NE Rural) *42.51900, 8.84600* **Camping Taüll, Ctra Taüll s/n, 25528 Taüll (Lleida) [973 69 61 74; www.campingtaull.com]** Fr Pont de Suert 3km N on N230 then NE on L500 dir Caldes de Boí. In 13km turn R into Taüll. Site sp on R. Sm, pt shd, pt sl, terr, wc (htd); chem disp; fam bthrm; shwrs inc; EHU €6; lndry; shop nr; rest nr; bar nr; 30% statics; dogs €3; quiet; clsd 15 Oct-15 Nov; CKE/CCI. "Excel facs; taxis into National Park avail; ltd touring pitches; suitable sm m'vans only; many bars & rest in pretty vill." € 28.50 2014*

⊞ **PONT DE SUERT** *3B2* (5km NW Rural) *42.43944, 0.69860* **Camping Baliera, Ctra N260, Km 355.5, Castejón de Sos, 22523 Bonansa (Huesca) [974-55 40 16; fax 974-55 40 99; info@baliera.com; www.baliera.com]** N fr Pont de Suert on N230 turn L opp petrol stn onto N260 sp Castejón de Sos. In 1km turn L onto A1605 sp Bonansa, site on L immed over rv bdge. Site sp fr N230. Lge, mkd, shd, pt sl, terr, wc (htd); chem disp; mv service pnt; fam bthrm; shwrs inc; EHU (5-10A) €5; gas; lndry; shop; rest; snacks; bar; bbq; playgrnd; pool; paddling pool; wifi; TV (pitch); 50% statics; dogs €3.80; phone; Eng spkn; quiet; ccard acc; rv fishing; horseriding 4km; site clsd Nov & Xmas; bike hire; golf 4km; CKE/CCI. "Excel, well-run, peaceful site in parkland setting; walking in summer, skiing in winter; excel cent for touring; conv Vielha tunnel; weights rm; all facs up steps; pt of site v sl; helpful owner proud of his site; clean facs, some rvside pitches." ♦ € 32.40 2014*

PORT DE LA SELVA, EL *3B3* (3km W Coastal) *42.34222, 3.18333* **Camping Port de la Vall, Ctra Port de Llançà, 17489 El Port de la Selva (Gerona) [972-38 71 86; fax 972-12 63 08; portdelavall@terra.es]** On coast rd fr French border at Llançà take GI612 twd El Port de la Selva. Site on L, easily seen. Lge, pt shd, wc; shwrs; EHU (3-5A) €6; gas; lndry; shop; rest; snacks; bar; playgrnd; beach shgl adj; wifi; 10% statics; dogs €2.95; phone; adv bkg acc; ccard acc. "Easy 1/2 hr walk to harbour; gd site; sm pitches & low branches poss diff - check bef siting; san facs v clean." 1 Mar-15 Oct. € 29.00 2016*

POTES *1A4* (1km W Rural) *43.15527, -4.63694* **Camping La Viorna, Ctra Santo Toribio, Km 1, Mieses, 39570 Potes (Cantabria) [942-73 20 21; info@campinglaviorna.com; www.campinglaviorna.com]** Exit N634 at junc 272 onto N621 dir Panes & Potes - narr, winding rd (passable for c'vans). Fr Potes take rd to Fuente Dé sp Espinama; in 1km turn L sp Toribio. Site on R in 1km, sp fr Potes. Do not use Sat Nav. Med, mkd, pt shd, terr, wc (htd); chem disp; mv service pnt; fam bthrm; shwrs inc; EHU (6A) €3.40 (poss rev pol); lndry; shop; rest; snacks; bar; bbq; playgrnd; pool; paddling pool; wifi; bus 1km; Eng spkn; adv bkg acc; quiet; ccard acc; bike hire; CKE/CCI. "Lovely views; gd walks; friendly, family-run, clean, tidy site; gd pool; ideal Picos de Europa; conv cable car, 4x4 tours, trekking; mkt on Mon; festival mid-Sep v noisy; some pitches diff in wet & diff lge o'fits; excel san facs; voted 8th best camp in Spain; excel." ♦ 1 Apr-1 Nov. € 27.70 2017*

POTES *1A4* (3km W Rural) *43.15742, -4.65617* **Camping La Isla-Picos de Europa, Ctra Potes-Fuente Dé, 39586 Turieno (Cantabria) [942-73 08 96; campinglaislapicosdeeuropa@gmail.com; www.campinglaislapicosdeeuropa.com]** Take N521 W fr Potes twd Espinama, site on R in 3km thro vill of Turieno (app Potes fr N). Med, mkd, shd, pt sl, wc; chem disp; mv service pnt; shwrs inc; EHU (6A) €4 (poss rev pol); gas; lndry; shop; rest; snacks; bar; bbq; playgrnd; pool; red long stay; wifi; 10% statics; dogs; phone; Eng spkn; adv bkg acc; ccard acc; horseriding; cycling; CKE/CCI. "Delightful, family-run site; friendly, helpful owners; gd san facs; conv cable car & mountain walks (map fr recep); many trees & low branches; 4x4 touring; walking; mountain treks in area; hang-gliding; rec early am dep to avoid coaches on gorge rd; highly rec; lovely loc, gd facs." 1 Apr-15 Oct. € 25.00 2017*

"That's changed – Should I let the Club know?"

If you find something on site that's different from the site entry, fill in a report and let us know. See camc.com/europereport.

PUEBLA DE SANABRIA *1B3* (10km NW Rural) *42.13111, -6.70111* **Camping El Folgoso, Ctra Puebla de Sanabria-San Martin de Castañeda, Km 13, 49361 Vigo de Sanabria (Zamora) [980-62 67 74; fax 980-62 68 00; info@.campingelfolgoso.com; www.campingelfolgoso.com]** Exit A52 sp Puebla de Sanabria & foll sp for Lago/Vigo de Sanabria thro Puente de Sanabria & Galende; site 2km beyond vill of Galende; sp. Med, shd, pt sl, terr, wc; chem disp; shwrs; EHU (5A) €2.50; gas; lndry; shop; rest; snacks; bar; playgrnd; phone; ccard acc; bike hire. "Lovely setting beside lake, and woods, v cold in winter." ♦ 1 Apr-31 Oct. € 18.00 2012*

⊞ **PUERTO DE MAZARRON** *4G1* (5km NE Coastal) *37.5800, -1.1950* **Camping Los Madriles, Ctra a la Azohía 60, Km 4.5, 30868 Isla Plana (Murcia) [968-15 21 51; fax 968-15 20 92; info@campinglosmadriles.com; www.campinglosmadriles.com]** Fr Cartegena on N332 dir Puerto de Mazarrón. Turn L at rd junc sp La Azohía (32km). Site in 4km sp. Fr Murcia on E15/N340 dir Lorca exit junc 627 onto MU603 to Mazarrón, then foll sp. (Do not use rd fr Cartegena unless powerful tow vehicle/gd weight differential - use rte fr m'way thro Mazarrón). Lge, mkd, hdstg, hdg, pt shd, pt sl, serviced pitches; wc; chem disp; mv service pnt; shwrs inc; EHU (10A) €5; gas; lndry; shop; rest; bar; playgrnd; pool (htd); beach shgl 500m; red long stay; wifi; bus; Eng spkn; adv bkg req; quiet; ccard acc; games area; jacuzzi; CKE/CCI. "Clean, well-run, v popular winter site; adv bkg req; some sm pitches, some with sea views; sl bet terrs; 3 days min stay high ssn; v helpful staff; excel." ♦ € 48.50 2014*

SPAIN

⊞ **PUERTO DE SANTA MARIA, EL** *2H3* (2km SW Coastal) *36.58768, -6.24092* **Camping Playa Las Dunas de San Antón, Paseo Maritimo La Puntilla s/n, 11500 El Puerto de Santa María (Cádiz) [956-87 22 10; fax 956-86 01 17; info@lasdunascamping.com; www.lasdunascamping.com]** Fr N or S exit A4 at El Puerto de Sta María. Foll site sp carefully to avoid narr rds of town cent. Site 2-3km S of marina & leisure complex of Puerto Sherry. Alt, fr A4 take Rota rd & look for sp to site & Hotel Playa Las Dunas. Site better sp fr this dir & avoids town. Lge, pt shd, pt sl, wc; chem disp; mv service pnt; fam bthrm; shwrs inc; EHU (10A) inc; gas; lndry; shop; snacks; bar; playgrnd; beach sand 50m; wifi; 30% statics; dogs; phone; Eng spkn; adv bkg rec; ccard acc; sports facs; CKE/CCI. "Friendly staff; conv Cádiz & Jerez sherry region, birdwatching areas & beaches; conv ferry or catamaran to Cádiz; facs poss stretched high ssn; pitches quiet away fr rd; take care caterpillars in spring, poss dangerous to dogs; dusty site but staff water rds; gd; excel facs; busy; guarded; pitches not defined & soft sand in places; excel new shwr block (2016); pool adj; old elec conns." ♦ € 26.00 2017*

⊞ **RIAZA** *1C4* (2km W Rural) *41.26995, -3.49750* **Camping Riaza, Ctra de la Estación s/n, 40500 Riaza (Segovia) [921-55 05 80; info@camping-riaza.com; www.camping-riaza.com]** Fr N exit A1/E5 junc 104, fr S exit 103 onto N110 N. In 12km turn R at rndabt on ent to town, site on L. Lge, hdg, unshd, wc (htd); chem disp; mv service pnt; fam bthrm; shwrs inc; EHU (10A) €4.70 (rev pol); lndry; shop; rest; snacks; bar; bbq; playgrnd; pool; paddling pool; wifi; 30% statics; dogs free; phone; bus 900m; Eng spkn; adv bkg acc; quiet; games rm; games area. "Vg site; various pitch sizes - some lge; excel san facs; easy access to/fr Santander or Bilbao; beautiful little town." ♦ € 31.00 2017*

⊞ **RIBADEO** *1A2* (12km N Rural/Coastal) *43.554004, -7.111085* **Rinlo Costa Camping, Rua Campo Maria Mendez, s/n 27715 Rinlo [679-25 52 81; info@rinlocosta.es; www.rinlocosta.es]** Fr N634 take LU141 twrds Rinlo. Over rly bdge (0.5km) take 1st L and foll rd round for another 0.5km. Turn L and site ahead on R. Sm, EHU (6A) €4.50; shop; bbq; cooking facs; playgrnd; pool; entmnt; dogs; bike hire. € 24.00 2018*

RIBADEO *1A2* (4km E Coastal) *43.55097, -6.99699* **Camping Playa Peñarronda, Playa de Peñarronda-Barres, 33794 Castropol (Asturias) [985-62 30 22; campingpenarrondacb@hotmail.com; www.campingplayapenarronda.com]** Exit A8 at km 498 onto N640 dir Lugo/Barres; turn R approx 500m and then foll site sp for 2km. Med, mkd, pt shd, wc; chem disp; mv service pnt; shwrs inc; EHU (6A) €4 (poss rev pol); gas; lndry; shop; rest; snacks; bar; bbq; playgrnd; beach sand adj; red long stay; 10% statics; phone; Eng spkn; quiet; games area; bike hire; CKE/CCI. "Beautifully-kept, delightful, clean, friendly, family-run site on 'Blue Flag' beach; rec arr early to get pitch; facs clean; gd cycling along coastal paths & to Ribadeo; ltd facs LS, sm pitches; gd sized pitches, little shd." Holy Week-25 Sep. € 23.60 2012*

RIBADEO *1A2* (18km W Coastal) *43.56237, -7.20762* **Camping Gaivota, Playa de Barreiros, 27792 Barreiros [982 12 44 51; campinggaivota@gmail.com; www.campingpobladogaivota.com]** Fr Berreiros take N634, turn L at KM 567, betRibadeo & Foz. Foll camping sp. Med, hdg, pt shd, wc; chem disp; mv service pnt; shwrs inc; EHU (6A); gas; lndry; shop; rest; snacks; bar; bbq; beach adj; twin axles; wifi; TV; 5% statics; dogs; phone; quiet. "V well cared for; superb beaches; family run; pleasant bar & rest; excel." ♦ 28 Mar-15 Oct. € 30.00 2015*

RIBADESELLA *1A3* (3km W Rural) *43.46258, -5.08725* **Camping Ribadesella, Sebreño s/n, 33560 Ribadesella (Asturias) [985 858293 or 985 857721; info@camping-ribadesella.com; www.camping-ribadesella.com]** W fr Ribadesella take N632. After 2km fork L up hill. Site on L after 2km. Poss diff for lge o'fits & alt rte fr Ribadesella vill to site to avoid steep uphill turn can be used. Lge, mkd, pt shd, pt sl, terr, wc; chem disp; fam bthrm; shwrs inc; EHU (5A) €4.80; gas; lndry; shop; rest; snacks; bar; bbq; playgrnd; pool (covrd, htd); beach sand 4km; red long stay; dogs €2.50; Eng spkn; adv bkg acc; quiet; ccard acc; tennis; games area; games rm; CKE/CCI. "Clean san facs; some sm pitches; attractive fishing vill; prehistoric cave paintings nrby; excel; not much shd; steps or slopes to walk to top rate facs; 35min easy downhill walk to town, shorter walk down steep lane to beach; rec." ♦ 19 Mar-25 Sep. € 30.00 2016*

RIBADESELLA *1A3* (8km W Rural/Coastal) *43.47472, -5.13416* **Camping Playa de Vega, Vega, 33345 Ribadesella (Asturias) [985-86 04 06; info@campingplayadevega.com; www.campingplayadevega.com]** Fr A8 exit junc 336 sp Ribadesella W, thro Bones. At rndabt cont W dir Caravia, turn R opp quarry sp Playa de Vega. Fr cent of Ribadesella (poss congestion) W on N632. Cont for 5km past turning to autovia. Turn R at sp Vega & site. Med, hdg, pt shd, terr, serviced pitches; wc; chem disp; shwrs inc; EHU €4.15; lndry; shop; rest; snacks; bar; bbq; beach sand 400m; wifi; TV; dogs; phone; bus 700m; quiet; ccard acc; CKE/CCI. "Sh walk to vg beach thro orchards; beach rest; sm pitches not suitable lge o'fits; poss overgrown LS; immac san facs; a gem of a site; beware very narr bdge on ent rd." 15 Jun-15 Sep. € 24.50 2017*

⊞ **RIBEIRA** *1B2* (10km N Rural) *42.62100, -8.98600* **Camping Ría de Arosa II, Oleiros, 15993 Santa Eugenia (Uxía) de Ribeira (La Coruña) [981- 86 59 11; fax 981-86 55 55; rural@campingriadearosa.com; www.campingriadearosa.com]** Exit AP9 junc 93 Padrón & take N550 then AC305/VG11 to Ribeira. Then take AC550 to Oleiros to site, well sp. V lge, mkd, hdg, shd, wc (htd); chem disp; mv service pnt; fam bthrm; shwrs inc; EHU (6A) inc; gas; lndry; shop; rest; snacks; bar; bbq; playgrnd; pool; wifi; TV; 10% statics; dogs €2.50; phone; Eng spkn; adv bkg acc; quiet; ccard acc; fishing; tennis; games area; games rm; CKE/CCI. "Beautiful area; helpful, friendly staff; excel; lots to do; excel pool; great facs." ♦ € 28.50 2015*

⊞ **RIBES DE FRESER** *3B3* (1km NE Rural) *42.31260, 2.17570* **Camping Vall de Ribes, Ctra de Pardines, Km 0.5, 17534 Ribes de Freser (Girona) [972-72 88 20 or 620-78 39 20; fax 972 93 12 96; info@campingvallderibes.com; www.campingvallderibes.com]** N fr Ripoll on N152; turn E at Ribes de Freser; site beyond town dir Pardines. Site nr town but 1km by rd. App rd narr. Med, mkd, pt shd, terr, wc (htd); chem disp; shwrs inc; EHU (6A) €4.30; lndry; shop nr; rest; bar; playgrnd; pool; 50% statics; dogs €4.60; train 500m; quiet; CKE/CCI. "Gd, basic site; steep footpath fr site to town; 10-20 min walk to stn; cog rlwy train to Núria a 'must' - spectacular gorge, gd walking & interesting exhibitions; sm/med o'fits only; poss unkempt statics LS; spectacular walk down fr the Vall de Nuria to Queralbs." € 23.00 2013*

"I like to fill in the reports as I travel from site to site"

You'll find report forms at the back of this guide, or you can fill them in online at camc.com/europereport.

⊞ **ROCIO, EL** *2G3* (2km N Rural) *37.14194, -6.49250* **Camping La Aldea, Ctra del Rocío, Km 25, 21750 El Rocío,Almonte (Huelva) [959-44 26 77; fax 959-44 25 82; info@campinglaaldea.com; www.campinglaaldea.com]** Fr A49 turn S at junc 48 onto A483 by-passing Almonte, site sp just bef El Rocío rndabt. Fr W (Portugal) turn off at junc 60 to A484 to Almonte, then A483. Lge, mkd, hdstg, hdg, pt shd, wc (htd); chem disp; mv service pnt; fam bthrm; shwrs inc; EHU (10A) €6.50; gas; lndry; shop; rest; snacks; bar; bbq; playgrnd; pool (htd); red long stay; wifi; 30% statics; dogs €3; phone; bus 500m; Eng spkn; adv bkg acc; ccard acc; horseriding nr; CKE/CCI. "Well-appointed & maintained site; winter rallies; excel san facs; friendly, helpful staff; tight turns on site; most pitches have kerb or gully; van washing facs; pitches soft after rain; rd noise; easy walk to interesting town; avoid festival (in May-7 weeks after Easter) when town cr & site charges higher; poss windy; excel birdwatching nrby (lagoon 1km); beautiufl site; gd pool & rest." ♦ € 30.00 2017*

⊞ **RONDA** *2H3* (1km S Rural) *36.72111, -5.17166* **Camping El Sur, Ctra Ronda-Algeciras Km 1.5, 29400 Ronda (Málaga) [952-87 59 39; fax 952-87 70 54; info@campingelsur.com; www.campingelsur.com]** Site on W side of A369 dir Algeciras. Do not tow thro Ronda. Med, hdstg, mkd, pt shd, sl, terr, wc (htd); chem disp; mv service pnt; fam bthrm; shwrs inc; EHU (5-10A) €4.30-5.35 (poss rev pol &/or no earth); lndry; shop; rest nr; snacks; bar; playgrnd; pool; red long stay; wifi; dogs €1.70; phone; Eng spkn; adv bkg acc; quiet; CKE/CCI. "Gd rd fr coast with spectacular views; long haul for lge o'fits; busy family-run site in lovely setting; conv National Parks & Pileta Caves; poss diff access some pitches due trees & high kerbs; hard, rocky grnd; san facs poss stretched high ssn; easy walk to town; friendly staff; vg rest; excel." ♦ € 24.40 2014*

⊞ **ROSES** *3B3* (1km W Urban/Coastal) *42.26638, 3.16305* **Camping Joncar Mar, Ctra Figueres s/n, 17480 Roses (Gerona) [972-25 67 02; info@campingjoncarmar.com; www.campingjoncarmar.com]** At Figueres take C260 W for Roses. On ent Roses turn sharp R at last rndabt at end of dual c'way. Site on both sides or rd - go to R (better) side, park & report to recep on L. Lge, pt shd, pt sl, wc (htd); chem disp; fam bthrm; shwrs; EHU (6-10A) poss no earth; gas; lndry; shop; rest; bar; playgrnd; pool; beach sand 150m; red long stay; entmnt; wifi; 15% statics; dogs €2.40; phone; bus 500m; Eng spkn; adv bkg acc; ccard acc; golf 15km; games rm. "Conv walk into Roses; hotels & apartment blocks bet site & beach; poss cramped/tight pitches; narr rds; vg value LS; new san facs 2015." € 32.00 2015*

ROSES *3B3* (3km W Coastal) *42.26638, 3.15611* **Camping Salatà, Port Reig s/n, 17480 Roses (Gerona) [972-25 60 86; fax 972-15 02 33; info@campingsalata.com; www.campingsalata.com]** App Roses on rd C260. On ent Roses take 1st R after Roses sp & Caprabo supmkt. Lge, hdstg, mkd, pt shd, wc (htd); chem disp; fam bthrm; shwrs inc; EHU (6-10A) inc; gas; lndry; shop; rest; snacks; bar; playgrnd; pool (htd); beach sand 200m; red long stay; wifi; 10% statics; dogs €2.80 (not acc Jul/Aug); phone; Eng spkn; adv bkg acc; ccard acc; CKE/CCI. "Vg area for sub-aqua sports; vg clean facs, but not enough; red facs LS; pleasant walk/cycle to town; overpriced." ♦ 12 Mar-31 Oct. € 50.70 2016*

⊞ **SABINANIGO** *3B2* (6km N Rural) *42.55694, -0.33722* **Camping Valle de Tena, Ctra N260, Km 512.6, 22600 Senegüe (Huesca). [974-48 09 77 or 974-48 03 02; correo@campingvalledetena.com; www.campingvalledetena.com]** Fr Jaca take N330, in 12km turn L onto N260 dir Biescas. In 5km ignore site sp Sorripas, cont for 500m to site on L - new ent at far end. Lge, mkd, unshd, terr, serviced pitches; wc (htd); chem disp; mv service pnt; fam bthrm; shwrs inc; EHU (6A) €6; lndry; shop; rest; snacks; bar; playgrnd; pool; paddling pool; entmnt; wifi; TV; 60% statics; dogs €2.70; phone; Eng spkn; adv bkg acc; sports facs. "Helpful staff; steep, narr site rd; sm pitches; hiking nrby; excel, busy NH to/fr France; rv rafting nr; rd noise during day but quiet at night; beautiful area; in reach of ski runs; v busy." € 21.00 2013*

⊞ **SACEDON** *3D1* (1km E Rural) *40.48148, -2.72700* **Camp Municipal Ecomillans, Camino Sacedón 15, 19120 Sacedón (Guadalajara) [949-35 10 18 or 949-35 17 80; fax 949-35 10 73; ecomillans63@hotmail.com; www.campingsacedon.com]** Fr E on N320 exit km220, foll Sacedón & site sp; site on L. Fr W exit km 222. Med, hdstg, mkd, shd, pt sl, wc cont; shwrs; EHU €5; lndry; shop nr; sw nr; quiet. "NH only in area of few sites; sm pitches; san facs v poor; LS phone to check open." ♦ € 15.50 2013*

SPAIN

SAHAGUN *1B3* (1km W Rural) *42.37188, -5.04280* **Camping Pedro Ponce, Avda Tineo, s/n 24326 Sahagun [987 78 04 15; fax 987 78 00 84; camping sahagun@hotmail.com; www.villadesahagun.es]** Leave A231 at junc 46 onto N120. Foll sp Shagun. Site on in 1km. Lge, unshd, wc; chem disp; shwrs inc; EHU (6A); lndry; shop nr; rest; bar; playgrnd; pool; twin axles; entmnt; wifi; 60% statics; dogs; phone; bus adj; Eng spkn; ccard acc. "Excel municipal site; modern facs; sep area for tourers; interesting town; vg." ♦ 1 Mar-31 Oct. € 17.00 2015*

⊞ **SALAMANCA** *1C3* (17km NE Rural) *41.05805, -5.54611* **Camping Olimpia, Ctra de Gomecello, Km 3.150, 37427 Pedrosillo el Ralo (Salamanca) [923-08 08 54 or 620-46 12 07; fax 923-35 44 26; info@ campingolimpia.com; www.campingolimpia.com]** Exit A62 junc 225 dir Pedrosillo el Ralo & La Vellés, strt over rndabt, site sp. Sm, hdg, pt shd, wc (htd); chem disp; shwrs inc; EHU €3; lndry; rest; snacks; bar; dogs €1; phone; bus 300m; Eng spkn; adv bkg acc; site clsd 8-16 Sep; CKE/CCI. "Helpful, friendly & pleasant owner; really gd 2 course meal for €10 (2014); handy fr rd with little noise & easy to park; poss open w/ends only LS; excel; grass pitches; clean facs; perfect; some pitches tight." € 20.00 2018*

"We must tell the Club about that great site we found"

Get your site reports in by mid-August and we'll do our best to get your updates into the next edition.

SALAMANCA *1C3* (5km E Rural) *40.97611, -5.60472* **Camping Don Quijote, Ctra Aldealengua, Km 1930, 37193 Cabrerizos (Salamanca) [923-20 90 52; fax 923-20 97 87; info@campingdonquijote.com; www. campingdonquijote.com]** Fr Madrid or fr S cross Rv Tormes by most easterly bdge to join inner ring rd. Foll Paseo de Canalejas for 800m to Plaza España. Turn R onto SA804 Avda de los Comuneros & strt on for 5km. Site ent 2km after town boundary sp. Fr other dirs, head into city & foll inner ring rd to Plaza España. Site well sp fr rv & ring rd. Med, hdg, mkd, hdstg, pt shd, wc; chem disp; mv service pnt; fam bthrm; shwrs inc; EHU (10A) inc; lndry; shop; rest; snacks; bar; bbq; playgrnd; pool; paddling pool; beach sand 200m; twin axles; wifi; 10% statics; dogs; phone; bus; Eng spkn; adv bkg acc; quiet; ccard acc; rv fishing; CKE/CCI. "Gd rv walks; conv city cent; 45 mins easy cycle ride 6km to town along rv; rv Tormes flows alongside site with pleasant walks; friendly owner; highly rec; new excel san facs (2016); v friendly; 12 min walk to bus stop." ♦ 1 Mar-31 Oct. € 24.00 2017*

⊞ **SALAMANCA** *1C3* (7km E Urban) *40.94722, -5.6150* **Camping Regio, Ctra Ávila-Madrid, Km 4, 37900 Santa Marta de Tormes (Salamanca) [923-13 88 88; fax 923-13 80 44; recepcion@campingregio. com; www.campingregio.com]** Fr E on SA20/N501 outer ring rd, pass hotel/camping sp visible on L & exit Sta Marta de Tormes, site directly behind Hotel Regio. Foll sp to hotel. Lge, mkd, pt shd, pt sl, wc; chem disp; mv service pnt; fam bthrm; shwrs inc; EHU (10A) €3.95 (no earth); gas; lndry; shop; rest; snacks; bar; playgrnd; wifi; TV; 5% statics; dogs; phone; bus to Salamanca; Eng spkn; quiet; ccard acc; car wash; bike hire; CKE/CCI. "In LS stop at 24hr hotel recep; poss no hdstg in wet conditions; conv en rte Portugal; refurbished facs to excel standard; site poss untidy, & ltd security in LS; hotel pool high ssn; hypmkt 3km; spacious pitches but some poss tight for lge o'fits; take care lge brick markers when reversing; hourly bus in and out of city; excel pool; vg; facs up to gd standard & htd; highly rec." ♦ € 23.00 2018*

"I need an on-site restaurant"

We do our best to make sure site information is correct, but it is always best to check any must-have facilities are still available or will be open during your visit.

⊞ **SALAMANCA** *1C3* (3km NW Rural) *40.99945, -5.67916* **Camping Ruta de la Plata, Ctra de Villamayor, 37184 Villares de la Reina (Salamanca) [923-28 95 74; recepcion@campingrutadelaplata. com; www.campingrutadelaplata.com]** Fr N on A62/E80 Salamanca by-pass, exit junc 238 & foll sp Villamayor. Site on R about 800m after rndabt at stadium 'Helmántico'. Avoid SA300 - speed bumps. Fr S exit junc 240. Med, mkd, hdg, pt shd, pt sl, terr, wc (htd); chem disp; mv service pnt; shwrs inc; EHU (6A) €2.90; gas; lndry; shop; snacks; bar; playgrnd; pool; TV; dogs €1.50; bus to city at gate; golf 3km; CKE/ CCI. "Family-owned site; some gd, san facs tired (poss unhtd in winter); ltd facs LS; less site care LS & probs with EHU; conv NH; helpful owners; bus stop at camp gate, every hr." ♦ € 27.00 2014*

SALOU *3C2* (1km S Urban/Coastal) *41.0752, 1.1176* **Camping Sanguli, Paseo Miramar-Plaza Venus, 43840 Salou (Tarragona) [977-38 16 41; fax 977-38 46 16; mail@sanguli.es; www.sanguli.es]** Exit AP7/E15 junc 35. At 1st rndbt take dir to Salou (Plaça Europa), at 2nd rndabt foll site sp. V lge, hdstg, mkd, shd, pt sl, serviced pitches; wc (htd); chem disp; mv service pnt; fam bthrm; shwrs inc; EHU (10A) inc; gas; lndry; shop; rest; snacks; bar; bbq; playgrnd; pool; paddling pool; beach sand 50m; red long stay; entmnt; wifi; TV; 35% statics; dogs; phone; bus; Eng spkn; adv bkg rec; ccard acc; games area; waterslide; games rm; jacuzzi; car wash; tennis; CKE/CCI. "Quiet end of Salou nr Cambrils & 3km Port Aventura; site facs recently updated/upgraded; fitness rm; excursions; cinema; youth club; mini club; amphitheatre; excel, well-maintained site." ♦ 4 Apr-2 Nov. € 70.00 2014*

"Satellite navigation makes touring much easier"

Remember most sat navs don't know if you're towing or in a larger vehicle – always use yours alongside maps and site directions.

⊞ **SAN FULGENCIO** *4F2* (7km ENE Coastal) *38.12094, -0.65982* **Camper Park San Fulgencio, Calle Mar Cantábrico 7 Centro Comercial las Dunas (Alicante) [966-72 53 17 or 679-62 26 93; infosol@camperparksanfulgencio.com; www.camperparksanfulgencio.com]** Fr N on AP7, take exit 740 twd Guardamar; At rndabt exit onto CV-91. 2nd exit at next rndabt then 3rd exit at another rndabt and stay on CV91. Take 1st exit at rndabt onto N-332, merge onto N332 and go thro 1 rndabt. At next rndabt take 3rd exit onto Calle Mar Cantabrico, go thro next rndabt and site is on the L. Sm, hdstg, unshd, wc; chem disp; mv service pnt; shwrs; EHU (5A) inc; lndry; wifi; bus to Alicante 150m; Eng spkn. "Sm m'van only site; gd facs; 1.75km to wooden area nr beach & sea; friendly helpful owner; gd atmosphere; Sat mkt; vg." € 14.00 2013*

⊞ **SAN ROQUE** *2H3* (7km NE Rural) *36.25031, -5.33808* **Camping La Casita, Ctra N340, Km 126.2, 11360 San Roque (Cádiz) [956-78 00 31]** Site sp 'Via de Servicio' parallel to AP7/E15. Access at km 119 fr S, km 127 fr N. Site visible fr rd. Lge, pt shd, pt sl, terr, wc; chem disp; mv service pnt; shwrs inc; EHU (10A) €4.54; shop; rest; playgrnd; pool; beach sand 3km; red long stay; entmnt; 90% statics; dogs €2.67; phone; bus 100m; Eng spkn; adv bkg req; ccard acc; horseriding; CKE/CCI. "Shwrs solar htd - water temp depends on weather (poss cold); san facs poss unclean; friendly staff; conv Gibraltar & Morocco; daily buses to La Línea & Algeciras; ferries to N Africa (secure parking at port); golf course next to site; great rest; poor site." ♦ € 39.50 2016*

⊞ **SAN SEBASTIAN/DONOSTIA** *3A1* (7km W Rural) *43.30458, -2.04588* **Camping Igueldo, Paseo Padre Orkolaga 69, 20008 San Sebastián (Guipúzkoa) [943-21 45 02; fax 943-28 04 11; info@campingigueldo.com; www.campingigueldo.com]** Fr W on A8, leave m'way at junc 9 twd city cent, take 1st R & R at rndabt onto Avda de Tolosa sp Ondarreta. At sea front turn hard L at rndabt sp to site (Avda Satrústegui) & foll sp up steep hill 4km to site. Fr E exit junc 8 then as above. Site sp as Garoa Camping Bungalows. Steep app poss diff for lge o'fits. Lge, hdg, mkd, pt shd, terr, serviced pitches; wc; chem disp; fam bthrm; shwrs inc; EHU (10A) inc; gas; lndry; shop; rest; bar; playgrnd; beach sand 5km; red long stay; wifi; TV; phone; bus to city adj; Eng spkn; CKE/CCI. "Vg, clean facs; sm pitches poss diff; spectacular views; pitches muddy when wet; excel rest 1km (open in winter); pool 5km; frequent bus to beautiful, interesting town; new pool (2017)." ♦ € 37.00 2017*

SAN TIRSO DE ABRES *1A2* (0km N Rural) *43.41352, -7.14141* **Amaido, El Llano, 33774 San Tirso de Abres [985-47 63 94; amaido@amaido.com; www.amaido.com]** Head N on A6 twds Lugo & exit 497 for N-640 twds Oviedo/Lugo Centro cidade. At rndabt take 4th exit onto N-640, turn R at LU-P-6104, turn R onto Vegas, then take 2nd L. Site at end of rd. Med, hdg, pt shd, terr, wc; chem disp; mv service pnt; fam bthrm; shwrs; EHU (6A); lndry; shop; bbq; playgrnd; twin axles; wifi; TV; dogs; adv bkg acc; bike hire; games area. "Lovely wooded site set in a circle around facs; farm animals; vg site." ♦ 10 Apr-15 Sep. € 28.00 2014*

"There aren't many sites open at this time of year"

If you're travelling outside peak season remember to call ahead to check site opening dates – even if the entry says 'open all year'.

SAN VICENTE DE LA BARQUERA *1A4* (1km E Coastal) *43.38901, -4.3853* **Camping El Rosal, Ctra de la Playa s/n, 39540 San Vicente de la Barquera (Cantabria) [942-71 01 65; fax 942-71 00 11; info@campingelrosal.com; www.campingelrosal.com]** Fr A8 km 264, foll sp San Vicente. Turn R over bdge then 1st L (site sp) immed at end of bdge; keep L & foll sp to site. Barier height 3.1m. Med, mkd, pt shd, pt sl, terr, wc; chem disp; shwrs; EHU (6A) €4.80; gas; lndry; shop; rest; snacks; bar; beach sand adj; wifi; phone; Eng spkn; adv bkg acc; quiet; ccard acc; CKE/CCI. "Lovely site in pine wood o'looking bay; surfing beach; some modern, clean facs; helpful staff; vg rest; easy walk or cycle ride to interesting town; Sat mkt; no hot water at sinks." ♦ 1 Apr-30 Sep. € 27.00 2015*

SAN VICENTE DE LA BARQUERA *1A4* (6km E Coastal) *43.38529, -4.33831* **Camping Playa de Oyambre, Finca Peña Gerra, 39540 San Vicente de la Barquera (Cantabria) [942-71 14 61; fax 942-71 15 30; camping@oyambre.com; www.oyambre.com]** E70/A8 Santander-Oviedo, exit sp 264 S. Vicente de la Barquera, then N634 for 3 km to Comillas exit on the Ctra La Revilla-Comillas (CA 131) bet km posts 27 and 28. Lge, mkd, pt shd, pt sl, terr, wc; chem disp; mv service pnt; shwrs inc; EHU (10A) €4.55; gas; lndry; shop; rest; snacks; bar; playgrnd; pool; beach 800m; wifi; 40% statics; dogs; bus 200m; Eng spkn; adv bkg acc; ccard acc; CKE/CCI. "V well-kept site; clean, helpful owner; quiet week days LS; gd base for N coast & Picos de Europa; 4x4 avail to tow to pitch if wet; some sm pitches & rd noise some pitches; conv Santander ferry; immac san facs; excel site; staff speak gd english; rest rec; gd base to tour N coast & Picos de Europa." 4 Mar-30 Oct. € 27.50 2015*

"That's changed – Should I let the Club know?"

If you find something on site that's different from the site entry, fill in a report and let us know. See camc.com/europereport.

⊞ **SANT JORDI** *3D2* (2km SW Rural) *40.49318, 0.31806* **Camping Maestrat Park, 12320 Sant Jordi (Castellón) [964-86 08 89 or 679-29 87 95 (mob); info@maestratpark.es; www.maestratpark.es]** Exit AP7 junc 42 onto CV11 to Sant Rafel del Riu. At rndabt with fuel stn take CV11 to Traiguera; then at rndabt take 1st exit onto N232 dir Vinarós & at next rndabt take 2nd exit sp Calig. Site 2km on L. Sm, mkd, hdstg, hdg, pt shd, pt sl, wc; chem disp; mv service pnt; fam bthrm; shwrs inc; EHU (10-16A) €4.50; lndry; shop; rest; snacks; bar; bbq; playgrnd; pool; paddling pool; twin axles; red long stay; wifi; TV; 25% statics; dogs; phone; bus; Eng spkn; adv bkg acc; quiet; games rm; bike hire; CKE/CCI. "Excel; club memb owner; excel for v lge o'fits." ♦ € 25.00 2015*

SANT PERE PESCADOR *3B3* (1km SE Coastal) *42.18180, 3.10403* **Camping L'Àmfora, Avda Josep Tarradellas 2, 17470 Sant Pere Pescador (Gerona) [972-52 05 40; fax 972-52 05 39; info@campingamfora.com; www.campingamfora.com]** Fr N exit junc 3 fr AP7 onto N11 fro Figueres/Roses. At junc with C260 foll sp Castelló d'Empúries & Roses. At Castelló turn R at rndabt sp Sant Pere Pescador then foll sp to L'Amfora. Fr S exit junc 5 fr AP7 onto GI 623/GI 624 to Sant Pere Pescador. V lge, mkd, hdg, pt shd, serviced pitches; wc (htd); chem disp; mv service pnt; fam bthrm; shwrs inc; EHU (10A) inc; gas; lndry; shop; rest; snacks; bar; bbq (charcoal, elec); playgrnd; pool; paddling pool; beach sand adj; red long stay; entmnt; wifi; TV; 15% statics; dogs €4.95; phone; Eng spkn; adv bkg acc; quiet; fishing; horseriding 5km; bike hire; games rm; windsurfing school; tennis; waterslide; ice; CKE/CCI. "Excel, well-run, clean site; no o'fits over 10m Apr-Sep; helpful staff; immac san facs; gd rest; private san facs avail; poss flooding on some pitches when wet; Parque Acuatico 18km." ♦ 14 Apr-27 Sep. € 60.40 2014*

See advertisement opposite

SANT PERE PESCADOR *3B3* (2km SE Coastal) *42.16194, 3.10888* **Camping Las Dunas, 17470 Sant Pere Pescador (Gerona) (Postal Address: Aptdo Correos 23, 17130 L'Escala) [972-52 17 17 or 01205 366856 (UK); fax 972-55 00 46; info@campinglasdunas.com; www.campinglasdunas.com]** Exit AP7 junc 5 dir Viladamat & L'Escala; 2km bef L'Escala turn L for Sant Martí d'Empúries, turn L bef ent vill for 2km, camp sp. V lge, mkd, pt shd, pt sl, serviced pitches; wc; chem disp; mv service pnt; fam bthrm; shwrs inc; EHU (6A) inc; gas; lndry; shop; rest; snacks; bar; bbq; playgrnd; pool; paddling pool; beach sand adj; entmnt; wifi; TV; 5% statics; dogs €4.50; phone; Eng spkn; adv bkg req; quiet; games area; games rm; tennis; watersports; CKE/CCI. "Greco-Roman ruins in Empúries; gd sized pitches - extra for serviced; busy, popular site; souvenir shop; money exchange; cash machine; doctor; excel, clean facs; vg site." ♦ 17 May-19 Sep. € 56.50 2014*

See advertisement below

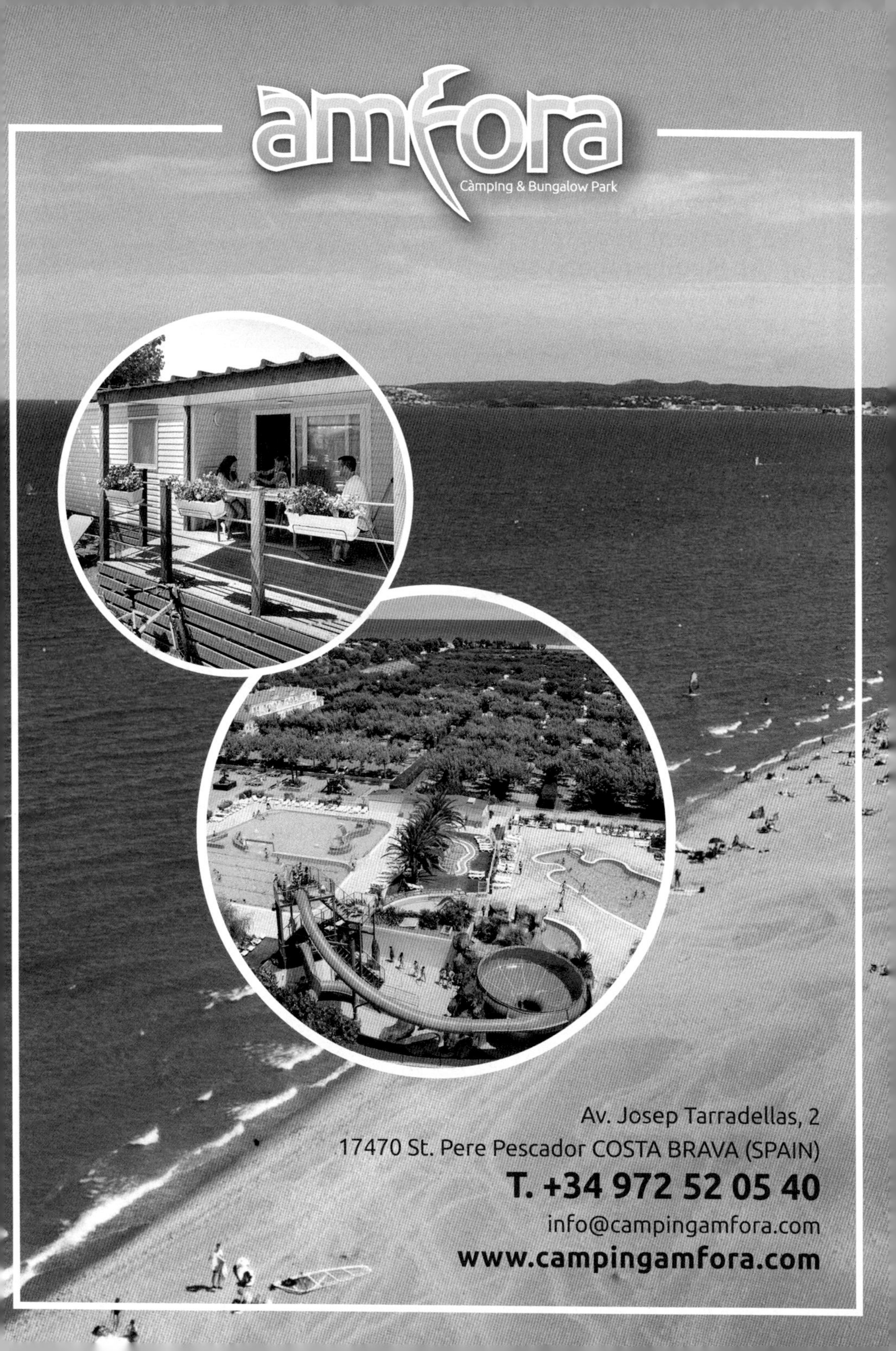
amfora
Càmping & Bungalow Park
Av. Josep Tarradellas, 2
17470 St. Pere Pescador COSTA BRAVA (SPAIN)
T. +34 972 52 05 40
info@campingamfora.com
www.campingamfora.com

SANT PERE PESCADOR *3B3* (3km SE Coastal) *42.17701, 3.10833* **Camping Aquarius, Camí Sant Martí d'Empúries, 17470 Sant Pere Pescador (Gerona) [972-52 00 03; fax 972-55 02 16; camping@aquarius.es; www.aquarius.es]** Fr AP7 m'way exit 3 on N11, foll sp to Figueres. Join C260, after 7km at rndabt at Castello d'Empúries turn R to Sant Pere Pescador. Cross rv bdge in vill, L at 1st rndabt & foll camp sp. Turn R at next rndabt, then 2nd L to site. Lge, pt shd, serviced pitches; wc; chem disp; mv service pnt; fam bthrm; shwrs; EHU (6-15A) €4-8; gas; lndry; shop; rest; snacks; bar; playgrnd; beach sand adj; red long stay; wifi; 10% statics; dogs €4.10; phone; Eng spkn; adv bkg rec; quiet; games rm; car wash; games area; CKE/CCI. "Immac, well-run site; helpful staff; nursery in ssn; cash point; vg rest; windsurfing; vast beach; recycling facs; excel site, highly rec, gd value ACSI site; wind gets v high." ♦ 15 Mar-2 Nov. € 56.70 2013*

See advertisement above

"I like to fill in the reports as I travel from site to site"

You'll find report forms at the back of this guide, or you can fill them in online at camc.com/europereport.

SANT PERE PESCADOR *3B3* (1km S Coastal) *42.18816, 3.10265* **Camping Las Palmeras, Ctra de la Platja 9, 17470 Sant Pere Pescador (Gerona) [972-52 05 06; fax 972-55 02 85; info@campinglaspalmeras.com; www.campinglas palmeras.com]** Exit AP7 junc 3 or 4 at Figueras onto C260 dir Roses/Cadaqués rd. After 8km at Castelló d'Empúries turn S for Sant Pere Pescador & cont twd beach. Site on R of rd. Lge, mkd, shd, serviced pitches; wc; chem disp; mv service pnt; fam bthrm; shwrs inc; EHU (5-16A) €3.90; gas; lndry; shop; rest; snacks; bar; playgrnd; pool (htd); paddling pool; beach sand 200m; entmnt; wifi; TV; dogs €4.50; phone; Eng spkn; adv bkg acc; quiet; games area; bike hire; games rm; tennis; CKE/CCI. "Pleasant site; helpful, friendly staff; superb, clean san facs; cash point; gd cycle tracks; nature reserve nrby; excel." ♦ 15 Apr-5 Nov. € 47.70 2016*

See advertisement below

⊞ **SANTA ELENA** *2F4* (0km E Rural) *38.34305, -3.53611* **Camping Despeñaperros, Calle Infanta Elena s/n, Junto a Autovia de Andulucia, Km 257, 23213 Santa Elena (Jaén) [953-66 41 92; info@campingdespenaperros.com; www.campingdespenaperros.com]** Leave A4/E5 at junc 257 or 259, site well sp to N side of vill nr municipal leisure complex. Med, hdstg, mkd, pt shd, serviced pitches; wc; chem disp; mv service pnt; shwrs inc; EHU (10A) €4.25 (poss rev pol); gas; lndry; shop; rest; snacks; bar; playgrnd; pool; red long stay; wifi; TV (pitch); 80% statics; dogs free; phone; bus 500m; adv bkg acc; ccard acc; CKE/CCI. "Gd winter NH in wooded location; gd size pitches but muddy if wet; gd walking area, perfect for dogs; friendly, helpful staff; clean san facs; disabled facs wc only; conv national park & m'way; gd rest; sh walk to vill & shops; site v rural; beautiful area." ♦ € 24.00 2017*

SANTA MARINA DE VALDEON *1A3* (1km N Rural) *43.13638, -4.89472* **Camping El Cares, El Cardo, 24915 Santa Marina de Valdeón (León) [987-74 26 76; campingelcares@hotmail.com]** Fr S take N621 to Portilla de la Reina. Turn L onto LE243 to Santa Marina. Avoid vill (narr rd), go to Northern end of vill bypass. Site is sp on L. Med, pt shd, terr, wc; chem disp; shwrs; EHU (5A) €3.20; lndry; shop; rest; bar; 10% statics; dogs €2.10; phone; bus 1km; quiet; ccard acc; CKE/CCI. "Lovely, scenic site high in mountains; gd base for Cares Gorge, friendly, helpful staff; gd views; tight access - not rec if towing or lge m'van." ♦ 1 Jun-15 Oct. € 26.50 2016*

"We must tell the Club about that great site we found"

Get your site reports in by mid-August and we'll do our best to get your updates into the next edition.

⊞ **SANTA POLA** *4F2* (1km NW Urban/Coastal) *38.20105, -0.56983* **Camping Bahía de Santa Pola, Ctra de Elche s/n, Km. 11, 03130 Santa Pola (Alicante) [965-41 10 12; fax 965-41 67 90; campingbahia@gmail.com; www.campingbahia.com]** Exit A7 junc 72 dir airport, cont to N332 & turn R dir Cartagena. At rndabt take exit sp Elx/Elche onto CV865, site 100m on R. Lge, mkd, hdstg, pt shd, wc (htd); chem disp; mv service pnt; fam bthrm; shwrs inc; EHU (10A) €3; gas; lndry; shop; rest; playgrnd; pool; beach sand 1km; red long stay; TV (pitch); 50% statics; dogs; phone; bus adj; Eng spkn; adv bkg acc; ccard acc; CKE/CCI. "Helpful, friendly manager; well-organised site; sm pitches; recep in red building facing ent; excel san facs; site rds steep; attractive coastal cycle path." ♦ € 25.00 2014*

⊞ **SANTAELLA** *2G3* (5km N Rural) *37.62263, -4.85950* **Camping La Campiña, La Guijarrosa-Santaella, 14547 Santaella (Córdoba) [957-31 53 03; fax 957-31 51 58; info@campinglacampina.com; www.campinglacampina.com]** Fr A4/E5 leave at km 441 onto A386 rd dir La Rambla to Santaella for 11km, turn L onto A379 for 5km & foll sp. Sm, mkd, hdstg, pt shd, pt sl, wc; chem disp; fam bthrm; shwrs inc; EHU (10A) €4; gas; lndry; shop; rest; snacks; bar; bbq; playgrnd; pool; red long stay; TV; dogs €2; bus at gate to Córdoba; Eng spkn; adv bkg acc; ccard acc; CKE/CCI. "Fine views; friendly, warm welcome; popular, family-run site; many pitches sm for lge o'fits; guided walks; poss clsd winter - phone to check; helpful & knowledgable owner; great site." ♦ € 29.00 2017*

SANTANDER *1A4* (5km NE Coastal) *43.48916, -3.79361* **Camping Cabo Mayor, Avda. del Faro s/n, 39012 Santander (Cantabria) [942-39 15 42; info@cabomayor.com; www.cabomayor.com]** Sp thro town but not v clearly. On waterfront (turn R if arr by ferry). At lge junc do not foll quayside, take uphill rd (resort type prom) & foll sp for Faro de Cabo Mayor. Site 200m bef lighthouse on L. Lge, mkd, unshd, terr, wc; chem disp; fam bthrm; shwrs inc; EHU (10A) inc; gas; lndry; shop; rest; snacks; bar; playgrnd; pool; beach adj; wifi; TV; 10% statics; phone; bus to Santander nrby; Eng spkn; quiet; CKE/CCI. "Med to lge pitches; site popular with lge youth groups hg ssn; shwrs clsd 2230-0800; conv ferry; pitches priced by size, pleasant coastal walk to Sardinero beachs; gd NH; well organised & clean; gd facs but dated; no hot water for washing up; excel; gd welcome." ♦ 27 Mar-12 Oct. € 20.00 2018*

⊞ **SANTANDER** *1A4* (12km E Rural) *43.44777, -3.72861* **Camping Somo Parque, Ctra Somo-Suesa s/n, 39150 Suesa-Ribamontán al Mar (Cantabria) [942-51 03 09; somoparque@somoparque.com; www.somoparque.com]** Fr car ferry foll sp Bilbao. After approx 8km turn L over bdge sp Pontejos & Somo. After Pedreña climb hill at Somo Playa & take 1st R sp Suesa. Foll site sp. Med, pt shd, wc; chem disp; shwrs; EHU (6A) €3 (poss rev pol); gas; shop; snacks; bar; playgrnd; beach 1.5km; 99% statics; Eng spkn; quiet; site clsd 16 Dec-31 Jan; CKE/CCI. "Friendly owners; peaceful rural setting; sm ferry bet Somo & Santander; poss unkempt LS & poss clsd; NH only." € 23.00 2012*

⊞ **SANTANDER** *1A4* (8km W Coastal) *43.47678, -3.87303* **Camping Virgen del Mar, Ctra Santander-Liencres, San Román-Corbán s/n, 39000 Santander (Cantabria) [942-34 24 25; fax 942-32 24 90; cvirdmar @ceoecant.es; www.campingvirgendelmar.com]** Fr ferry turn R, then L up to football stadium, L again leads strt into San Román. If app fr W, take A67 (El Sardinero) then S20, leave at junc 2 dir Liencres, strt on. Site well sp. Lge, mkd, pt shd, wc; chem disp; mv service pnt; shwrs; EHU (4-10A) €4; lndry; shop; rest; snacks; bar; playgrnd; pool; beach sand 300m; red long stay; bus 500m; adv bkg acc; quiet; CKE/CCI. "Basic facs, poss ltd hot water; some sm pitches not suitable lge o'fits; site adj cemetary; phone in LS to check site open; expensive LS; gd for ferry; new owner 2016." ♦ € 26.00 2017*

⊞ **SANTANDER** *1A4* (2km NW Coastal) *43.46762, -3.89925* **Camping Costa San Juan, Avda San Juan de la Canal s/n, 39110 Soto de la Marina (Cantabria) [942-57 95 80 or 629-30 36 86; info@hotelcostasanjuan.com; www.hotelcostasanjuan.com]** Fr A67 take S20 twds Bilbao, exit junc 2 & foll sp Liencres. In 2km at Irish pub turn 1st R to Playa San Juan de la Canal, site behind hotel on L. Sm, pt shd, wc; shwrs inc; EHU (3-6A) €3.20 (poss rev pol); rest; bar; beach sand 400m; wifi; TV; 90% statics; bus 600m; quiet. "NH for ferry; muddy in wet; poss diff lge o'fits; gd coastal walks; 2 pin adaptor needed for elec conn." € 28.70 2014*

⊞ **SANTIAGO DE COMPOSTELA** *1A2* (4km NE Urban) *42.88972, -8.52444* **Camping As Cancelas, Rua do Xullo 25, 35, 15704 Santiago de Compostela (La Coruña) [981-58 02 66 or 981-58 04 76; fax 981-57 55 53; info@campingascancelas.com; www.campingascancelas.com]** Exit AP9 junc 67 & foll sp Santiago. At rndabt with lge service stn turn L sp 'camping' & foll sp to site turning L at McDonalds. Site adj Guardia Civil barracks. NB-Do not use sat nav if app fr Lugo on N547. Lge, mkd, shd, pt sl, terr, wc; chem disp; fam bthrm; shwrs inc; EHU (5A) inc; gas; lndry; shop; rest; snacks; bar; bbq; playgrnd; pool; paddling pool; entmnt; wifi; TV; dogs; phone; bus 100m; Eng spkn; quiet; CKE/CCI. "Busy site-conv for pilgrims; rec arr early high ssn; some sm pitches poss diff c'vans & steep ascent; gd clean san facs, stretched when busy; gd rest; bus to city 100m fr gate avoids steep 15 min walk back fr town (LS adequate car parks in town); poss interference with car/c'van electrics fr local transmitter, if problems report to site recep; LS recep in bar; arr in sq by Cathedral at 1100 for Thanksgiving service at 1200; helpful owner; excel site; facs v clean; wifi vg." ♦ € 33.60 2017*

See advertisement below

SANTILLANA DEL MAR *1A4* (3km E Rural) *43.38222, -4.08305* **Camping Altamira, Barrio Las Quintas s/n, 39330 Queveda (Cantabria) [942-84 01 81; fax 942-26 01 55; nfo@campingaltamira.es; www.campingaltamira.es]** Clear sp to Santillana fr A67; site on R 3km bef vill. Med, mkd, pt shd, pt sl, terr, wc; shwrs; EHU (3A)- (5A) €2 (poss rev pol); gas; lndry; shop; rest; bar; pool; TV; 30% statics; bus 100m; Eng spkn; adv bkg req; ccard acc; horseriding; CKE/CCI. "Pleasant site; ltd facs LS; nr Altimira cave paintings; easy access Santander ferry on m'way; gd coastal walks; open w/end only Nov-Mar - rec phone ahead; excel; san facs v clean & modern; local rests nrby." 10 Mar-7 Dec. € 21.00 2017*

"I need an on-site restaurant"

We do our best to make sure site information is correct, but it is always best to check any must-have facilities are still available or will be open during your visit.

⊞ **SANTILLANA DEL MAR** *1A4* (1km NW Rural) *43.39333, -4.11222* **Camping Santillana del Mar, Ctra de Comillas s/n, 39330 Santillana del Mar (Cantabria) [942-81 82 50; fax 942-84 01 83; www.campingsantillana.com]** Fr W exit A8 junc 230 Santillana-Comillas, then foll sp Santillana & site on rd CA131. Fr E exit A67 junc 187 & foll sp Santillana. Turn R onto CA131, site on R up hill after vill. Lge, pt shd, sl, terr, wc; chem disp; mv service pnt; fam bthrm; shwrs inc; EHU (6A) inc (poss rev pol); gas; lndry; shop; rest; snacks; bar; playgrnd; pool; paddling pool; beach 5km; entmnt; wifi; 20% statics; dogs; phone; bus 300m; Eng spkn; horseriding; bike hire; tennis; golf 15km; CKE/CCI. "Useful site in beautiful historic vill; hot water only in shwrs; diff access to fresh water & to mv disposal point; narr, winding access rds, projecting trees & kerbs to some pitches - not rec lge o'fits or twin axles; car wash; cash machine; poss muddy LS & pitches rutted; poss travellers; gd views; lovely walk to town; poor facs (2014); NH." ♦ € 24.00 2017*

⊞ **SANTO DOMINGO DE LA CALZADA** *1B4* (4km E Rural) *42.44083, -2.91506* **Camping Banares, Ctra N120, Km 42.2, 26250 Santo Domingo de la Calzada [941-34 01 31; info@campingbanares.es; www.campingbanares.es]** Fr N120 Burgos-Logroño rd, turn N at Santo Domingo, foll sp Banares & site. Sm, unshd, pt sl, wc; chem disp; mv service pnt; fam bthrm; shwrs inc; EHU (5-10A) €5.50; gas; lndry; shop; rest; bar; playgrnd; pool; paddling pool; 90% statics; Eng spkn; adv bkg acc; ccard acc; games area; tennis; CKE/CCI. "Interesting, historic town; shops, bars, rests 3km; NH only." € 32.60 2016*

SANXENXO *1B2* (5km SW Coastal) *42.39254, -8.84517* **Camping Playa Paxariñas, Ctra C550, Km 2.3 Lanzada-Portonovo, 36960 Sanxenxo (Pontevedra) [986-72 30 55; fax 986-72 13 56; info@campingpaxarinas.com; www.campingpaxarinas.com]** Fr Pontevedra W on P0308 coast rd; 3km after Sanxenxo. Site thro hotel on L at bend. Site poorly sp. Fr AP9 fr N exit junc 119 onto VRG41 & exit for Sanxenxo. Turn R at 3rd rndabt for Portonovo to site in dir O Grove. Do not turn L to port area on ent Portonovo. Lge, mkd, shd, pt sl, terr, wc; chem disp; fam bthrm; shwrs inc; EHU (5A) €4.75; gas; lndry; shop; snacks; bar; bbq; playgrnd; beach sand adj; red long stay; wifi; TV; 75% statics; dogs; phone; bus adj; Eng spkn; adv bkg acc; quiet; ccard acc; CKE/CCI. "Site in gd position; secluded beaches, views over estuary; take care high kerbs on pitches; excel san facs - ltd facs LS & poss clsd; lovely unspoilt site; plenty of shd." ♦ 17 Mar-15 Oct. € 32.00 2016*

⊞ **SANXENXO** *1B2* (3km NW Coastal) *42.41777, -8.87555* **Camping Monte Cabo, Soutullo 174, 36990 Noalla (Pontevedra) [986-74 41 41; info@montecabo.com; www.montecabo.com]** Fr AP9 exit junc 119 onto upgraded VRG4.1 dir Sanxenxo. Ignore sp for Sanxenxo until rndabt sp A Toxa/La Toja, where turn L onto P308. Cont to Fontenla supmkt on R - minor rd to site just bef supmkt. Rd P308 fr AP9 junc 129 best avoided. Sm, mkd, pt shd, terr, wc; chem disp; mv service pnt; shwrs inc; EHU €4.25; lndry; shop; rest; snacks; bar; playgrnd; beach sand 250m; red long stay; wifi; TV; 10% statics; phone; bus 600m; Eng spkn; adv bkg acc; quiet; ccard acc; CKE/CCI. "Peaceful, friendly site set above sm beach (access via steep path) with views; sm pitches; beautiful coastline & interesting historical sites; vg." € 30.00 2017*

SARRIA *1B2* (2km E Rural) *42.77625, -7.39552* **Camping Vila de Sarria, Ctra. De Pintín, Km 1 Sarria 27600 [982 53 54 67; info@campingviladesarria.com; www.campingviladesarria.com]** Leave Sarria on LU5602 twds Pintin. Site on L in 1km. Med, pt shd, pt sl, wc; chem disp; shwrs inc; EHU; lndry; shop nr; rest; snacks; bar; twin axles; wifi; dogs; Eng spkn; adv bkg acc; ccard acc. "Quiet site on Camino de Santiago Rte; excel rest; v pleasant, welcoming staff; busy w/ends; vg." ♦ Easter-30 Sep. € 22.50 2015*

SAX *4F2* (6km NW Rural) *38.56875, -0.84913* **Camping Gwen & Michael, Colonia de Santa Eulalia 1, 03630 Sax (Alicante) [965-47 44 19 or 7718 18 58 05(UK)]** Exit A31 at junc km 191 & foll sp for Santa Eulalia, site on R just bef vill sq. Rec phone prior to arr. Sm, hdg, hdstg, unshd, wc; chem disp; fam bthrm; shwrs inc; EHU (3A) €1; lndry; shop nr; rest nr; dogs; quiet. "Vg CL-type site; friendly British owners; beautiful area; gd NH & touring base; c'van storage avail; 3 rest nrby." 15 Mar-30 Nov. € 15.00 2016*

SEGOVIA *1C4* (2km SE Urban) *40.93138, -4.09250* **Camping El Acueducto, Ctra de la Granja, 40004 Segovia [921-42 50 00; informacion@campingacueducto.com; www.campingacueducto.com]** Turn off Segovia by-pass N110/SG20 at La Granja exit, but head twd Segovia on DL601. Site in approx 500m off dual c'way just bef Restaurante Lago. Lge, mkd, pt shd, pt sl, wc; chem disp; mv service pnt; shwrs inc; EHU (6-10A) €5; gas; lndry; shop; rest nr; bar; bbq; playgrnd; pool; paddling pool; wifi; 10% statics; dogs; phone; bus 150m; bike hire; CKE/CCI. "Excel; helpful staff; lovely views; clean facs; gates locked 0000-0800; gd bus service; some pitches sm & diff for lge o'fits; city a 'must' to visit; new (unhtd) facs for women (2016); bus stop and gd spmkt 10min walk; site muddy after rain; unfriendly new owner (2018)." ♦ 26 Mar-15 Oct. € 32.50 2018*

SEO DE URGEL *3B3* (9km NW Rural) *42.37388, 1.35777* **Camping Castellbò- Buchaca, Ctra Lerida-Puigcerdà 127, 25712 Castellbò (Lleida) [973-35 21 55]** Leave Seo de Urgel on N260/1313 twd Lerida. In approx 3km turn N sp Castellbò. Thro vill & site on L, well sp. Steep, narr, winding rd, partly unfenced - not suitable car+c'van o'fits or lge m'vans. Sm, mkd, pt shd, pt sl, wc; chem disp; shwrs inc; EHU (5A) €5.85; lndry; shop; snacks; playgrnd; pool; dogs €3.60; phone; adv bkg acc; quiet. "CL-type site in beautiful surroundings; friendly recep; basic fac's." 1 May-30 Sep. € 30.00 2013*

SITGES *3C3* (2km SW Urban/Coastal) *41.23351, 1.78111* **Camping Bungalow Park El Garrofer, Ctra C246A, Km 39, 08870 Sitges (Barcelona) [93 894 17 80; fax 93 811 06 23; info@garroferpark.com; www.garroferpark.com]** Exit 26 on the C-32 dir St. Pere de Ribes, at 1st rndabt take 1st exit, at 2nd rndabt take 2nd exit, foll rd C-31 to campsite. V lge, mkd, hdstg, hdg, pt shd, serviced pitches; wc (htd); chem disp; mv service pnt; fam bthrm; shwrs inc; EHU (5-10A) €4.10 (poss rev pol); gas; lndry; shop; rest; snacks; bar; playgrnd; pool; beach shgl 900m; entmnt; wifi; TV; 80% statics; dogs €2.65; phone; bus adj; Eng spkn; adv bkg acc; ccard acc; games rm; tennis 800m; car wash; site clsd 19 Dec-27 Jan to tourers; windsurfing; games area; horseriding; bike hire; CKE/CCI. "Great location, conv Barcelona, bus adj; sep area for m'vans; pleasant staff; recep open 0800-2100; gd level site; quiet; gd old & new facs." ♦ 28 Feb-14 Dec. € 49.60 2014*

SPAIN

SORIA *3C1* (2km SW Rural) *41.74588, -2.48456* **Camping Fuente de la Teja, Ctra Madrid-Soria, Km 223, 42004 Soria [975-22 29 67; camping@fuentedelateja.com; www.fuentedelateja.com]** Fr N on N111 (Soria by-pass) 2km S of junc with N122 (500m S of Km 223) take exit for Quintana Redondo, site sp. Fr Soria on NIII dir Madrid sp just past km 223. Turn R into site app rd. Fr S on N111 stake exit for Quintana Redondo & foll site sp. Med, mkd, pt shd, pt sl, wc; chem disp; fam bthrm; shwrs inc; EHU (6A) €3 (poss no earth); gas; lndry; shop nr; rest; snacks; bar; bbq; playgrnd; pool; wifi; TV; 10% statics; dogs; phone; Eng spkn; adv bkg acc; ccard acc; CKE/CCI. "Vg site; excel, gd for NH; vg san facs; interesting town; phone ahead to check site poss open bet Oct & Easter; hypmkt 3km; easy access to site; pitches around 100sqm, suits o'fits upto 10m; quiet; friendly staff; access fr 9am." ♦ 1 Mar-31 Oct. € 23.30 2014*

SUECA *4E2* (5km NE Coastal) *39.30354, -0.29270* **Camping Les Barraquetes, Playa de Sueca, Mareny Barraquetes, 46410 Sueca (València) [961-76 07 23; info@barraquetes.com; www.barraquetes.com]** Exit AP7 junc 58 dir Sueca onto N332. In Sueca take CV500 to Mareny Barraquetes. Or S fr València on CV500 coast rd. Foll sp for Cullera & Sueca. Site on L. Med, mkd, pt shd, wc; chem disp; mv service pnt; fam bthrm; shwrs inc; EHU (10A) €5.88; gas; lndry; shop; bar; bbq; playgrnd; pool; paddling pool; beach sand 350m; twin axles; red long stay; entmnt; wifi; TV; 70% statics; dogs; phone; bus 500m; Eng spkn; ccard acc; games area; waterslide; tennis; windsurfing school; CKE/CCI. "Quiet, family atmosphere; conv touring base & València; quiet beach 8 min walk; helpful staff; gd." ♦ 16 Jan-14 Dec. € 33.00 2017*

TAPIA DE CASARIEGO *1A3* (2km SW Coastal) *43.56394, -6.95247* **Camping Playa de Tapia, La Reburdia, 33740 Tapia de Casariego (Asturias) [985-47 27 21]** Fr Ribadeo pass thro vill of Serentes on N634 past 1st camping sp. Site on L at 546km post, foll sp to site. Med, mkd, hdg, pt shd, pt sl, wc; chem disp; mv service pnt; fam bthrm; shwrs inc; EHU (16A) €4.06; gas; lndry; shop; rest; bar; beach sand 1m; wifi; dogs; phone; bus 800m; Eng spkn; adv bkg acc; quiet; CKE/CCI. "Gd access; busy, well-maintained, friendly site; o'looking coast & harbour; poss ltd hot water; walking dist to delightful town." ♦ Holy Week & 1 Jun-15 Sep. € 23.00 2015*

⊞ **TAPIA DE CASARIEGO** *1A3* (2km W Rural) *43.54870, -6.97436* **Camping El Carbayin, La Penela, 33740 [985-62 37 09; www.campingelcarbayin.com]** Take N634/E70 (old coast rd parallel to A8-E70) to Serantes, foll sp to site bet N634 and sea. Sm, mkd, pt shd, pt sl, wc; chem disp; fam bthrm; shwrs inc; EHU (3A) €3; lndry; shop; rest; bar; playgrnd; beach sand 1km; 10% statics; phone; bus 400m; adv bkg acc; quiet; ccard acc; fishing; watersports; CKE/CCI. "Gd for coastal walks & trips to mountains; gd; new san facs." ♦ € 19.50 2013*

⊞ **TARAZONA** *3B1* (8km SE Rural) *41.81890, -1.69230* **Camping Veruela Moncayo, Ctra Vera-Veruela, 50580 Vera de Moncayo (Zaragoza) [976 64 90 34; antoniogp@able.es]** Fr Zaragoza, take AP68 or N232 twd Tudela/Logroño; after approx. 50km, turn L to join N122 (km stone 75) twd Tarazona; cont 30km & turn L twd Vera de Moncayo; go thro town cent; site on R; well sp. Med, mkd, hdg, pt shd, pt sl, wc; chem disp; shwrs inc; EHU; gas; lndry; shop nr; rest; snacks; bar; playgrnd; wifi; dogs; adv bkg acc; bike hire; CKE/CCI. "Quiet site adj monastery; friendly owner; gd." ♦ € 26.00 2017*

"Satellite navigation makes touring much easier"

Remember most sat navs don't know if you're towing or in a larger vehicle – always use yours alongside maps and site directions.

⊞ **TARIFA** *2H3* (11km W Coastal) *36.07027, -5.69305* **Camping El Jardín de las Dunas, Ctra N340, Km 74, 11380 Punta Paloma (Cádiz) [956-68 91 01; fax 956-69 91 06; info@lasdunascamping.com; www.campingjdunas.com]** W on N340 fr Tarifa, L at sp Punta Paloma. Turn L 300m after Camping Paloma, site in 500m. Lge, hdg, pt shd, serviced pitches; wc; chem disp; fam bthrm; shwrs inc; EHU (6A) €5.16; lndry; shop; rest; snacks; bar; playgrnd; beach 50m; entmnt; TV; phone; ccard acc. "Poss strong winds; unsuitable lge o'fits due tight turns & trees; modern facs; poss muddy when wet." ♦ € 35.00 2014*

⊞ **TARIFA** *2H3* (10km NW Coastal) *36.06908, 5.68036* **Camping Valdevaqueros, Ctra N340 km 75,5 11380 Tarifa [34 956 684 174; fax 34 956 681 898; info@campingvaldevaqueros.com; www.campingvaldevaqueros.com]** Campsite is sp 9km fr Tarifa on the N340 twds Cadiz. Lge, pt shd, pt sl, wc; chem disp; mv service pnt; shwrs; EHU (6A); shop; rest; bar; playgrnd; pool; paddling pool; beach sandy 1km; entmnt; wifi; TV; 50% statics; dogs; phone; Eng spkn; adv bkg rec; games area; bike hire. "Excel site; watersports nrby." € 43.00 2014*

⊞ **TARIFA** *2H3* (3km NW Coastal) *36.04277, -5.62972* **Camping Rió Jara, Ctra N340, km 81, 11380 Tarifa (Andalucia) [956-68 05 70; campingriojara@terra.com]** Site on S of N340 Cádiz-Algeciras rd at km post 81.2; 3km after Tarifa; clearly visible & sp. Med, mkd, pt shd, wc; chem disp; mv service pnt; shwrs inc; EHU (10A) €4; gas; lndry; shop; rest; snacks; bar; playgrnd; beach sand 200m; wifi; dogs €3.50; adv bkg acc; ccard acc; fishing; CKE/CCI. "Gd, clean, well-kept site; friendly recep; long, narr pitches diff for awnings; daily trips to N Africa; gd windsurfing nr; poss strong winds; mosquitoes in summer." ♦ € 45.00 2014*

TARRAGONA *3C3* (1km NE Urban/Coastal) *40.88707, 0.80630* **Camping Nautic, Calle Libertat s/n, 43860 L'Ametlla de Mar Tarragona [34 977 493 031; fax 34 977 456 110; info@campingnautic.com; www.campingnautic.com]** Fr N340 exit at km 1113 sp L'Ametlla de Mar (or A7 exit 39). Over rlwy bdge, foll rd to L. Turn R after park and TO on R, foll signs to campsite. Lge, hdstg, pt shd, terr, wc; chem disp; mv service pnt; fam bthrm; shwrs; EHU; lndry; shop; rest; snacks; bar; playgrnd; pool; paddling pool; beach; wifi; TV; 25% statics; dogs; phone; bus 500m, train 700m; Eng spkn; adv bkg acc; games area; CCI. "Vg site; tennis court; 5 mins to attractive town with rest & sm supmkt; lge Mercadona outsite town; site on different levels." ♦ 15 Mar-15 Oct. € 50.00 2014*

"There aren't many sites open at this time of year"

If you're travelling outside peak season remember to call ahead to check site opening dates – even if the entry says 'open all year'.

TARRAGONA *3C3* (5km NE Coastal) *41.13019, 1.31170* **Camping Las Palmeras, N340, Km 1168, 43080 Tarragona [977-20 80 81; fax 977-20 78 17; laspalmeras@laspalmeras.com; www.laspalmeras.com]** Exit AP7 at junc 32 (sp Altafulla). After about 5km on N340 twd Tarragona take sp L turn at crest of hill. Site sp. V lge, mkd, pt shd, wc; chem disp; fam bthrm; shwrs inc; EHU (6A) inc; gas; lndry; shop; rest; snacks; bar; playgrnd; pool; paddling pool; beach sand adj; red long stay; entmnt; wifi; 10% statics; dogs €5; phone; ccard acc; games rm; tennis; games area; CKE/CCI. "Gd beach; ideal for families; poss mosquito prob; many sporting facs; gd, clean san facs; friendly, helpful staff; naturist beach 1km; supmkt 5km; excel site." ♦ 2 Apr-12 Oct. € 45.00 2012*

TARRAGONA *3C3* (7km NE Coastal) *41.12887, 1.34415* **Camping Torre de la Mora, Ctra N340, Km 1171, 43008 Tarragona-Tamarit [977-65 02 77; fax 977-65 28 58; info@torredelamora.com; www.torredelamora.com]** Fr AP7 exit junc 32 (sp Altafulla), at rndabt take La Mora rd. Then foll site sp. After approx 1km turn R, L at T-junc, site on R. Lge, hdstg, pt shd, terr, wc; chem disp; mv service pnt; fam bthrm; shwrs inc; EHU (6A) €5.20; gas; lndry; shop; rest; snacks; bar; playgrnd; pool; beach sand adj; red long stay; entmnt; wifi; 50% statics; dogs €3.65; bus 200m; Eng spkn; adv bkg acc; ccard acc; tennis; golf 2km; CKE/CCI. "Improved, clean site set in attractive bay with fine beach; excel pool; sports club adj; conv Tarragona & Port Aventura; various pitch sizes, some v sm." ♦ 18 Mar-31 Oct. 2013*

TARRAGONA *3C3* (12km E Coastal) *41.1324, 1.3604* **Camping-Caravaning Tamarit Park, Playa Tamarit, Ctra N340, Km 1172, 43008 Playa Tamarit (Tarragona) [977-65 01 28; fax 977-65 04 51; tamaritpark@tamarit.com; www.tamarit.com]** Fr A7/E15 exit junc 32 sp Altafulla/Torredembarra, at rndabt join N340 by-pass sp Tarragona. At rndabt foll sp Altafulla, turn sharp R to cross rlwy bdge to site in 1.2km, sp. V lge, hdg, hdstg, shd, pt sl, serviced pitches; wc (htd); chem disp; mv service pnt; fam bthrm; shwrs inc; EHU (10A) inc; gas; lndry; shop; rest; snacks; bar; bbq; playgrnd; pool (htd); paddling pool; beach shgl adj; red long stay; entmnt; wifi; TV; 30% statics; dogs €4; phone; Eng spkn; adv bkg rec; ccard acc; games area; watersports; tennis; CKE/CCI. "Well-maintained, secure site with family atmosphere; excel beach; superb pool; private bthrms avail; best site in area but poss noisy at night & w/end; variable pitch prices; beachside pitches avail; cash machine; car wash; take care overhanging trees; Altafulla sh walk along beach worth visit; excel." ♦ 14 Mar-16 Oct. € 65.00 2016*

SPAIN

⊞ **TOLEDO** *1D4* (4km W Rural) *39.86530, -4.04714* **Camping El Greco, Ctra Pueblo Montalban, Km.0.7, 45004 Toledo [925-22 00 90; info@campingelgreco.es; www.campingelgreco.es]** Site on CM4000 between CM40 ring road & Puenta de la Cava bridge. Fr E foll yellow sp. Fr W look for 5 flagpoles on L next to Cirgarral Del Santoangel Custodio. Med, hdg, mkd, hdstg, pt shd, pt sl, wc (htd); chem disp; mv service pnt; shwrs inc; EHU (6-10A) inc (poss rev pol); gas; lndry; shop; rest; snacks; bar; bbq; playgrnd; pool; paddling pool; wifi; dogs; phone; bus to town, train to Madrid fr town; Eng spkn; ccard acc; games area; CKE/CCI. "Clean, tidy, well-maintained; all pitches on gravel; easy parking on o'skts - adj Puerta de San Martín rec - or bus; some pitches poss tight; san facs clean; lovely, scenic situation; excel rest; friendly, helpful owners; vg; dusty; gd rest; site poss neglected during LS; gd pool, closes fr 15 Sept; cheap bar; nice, neat site; mv service pnt basic." ♦ € 31.00 2018*

⊞ **TORDESILLAS** *1C3* (2km SSW Urban) *41.49584, -5.00494* **Kawan Village El Astral, Camino de Pollos 8, 47100 Tordesillas (Valladolid) [983-77 09 53; info@campingelastral.es; www.campingelastral.com]** Fr NE on A62/E80 thro town turn L at rndabt over rv & immed after bdge turn R dir Salamanca & almost immed R again into narr gravel track (bef Parador) & foll rd to site; foll camping sp & Parador. Poorly sp. Fr A6 exit sp Tordesillas & take A62. Cross bdge out of town & foll site sp. Med, hdstg, mkd, hdg, pt shd, wc (htd); chem disp; mv service pnt; fam bthrm; shwrs inc; EHU (5A-10A) €3.60-5 (rev pol); gas; lndry; shop; rest; snacks; bar; playgrnd; pool; wifi; TV; 10% statics; dogs €2.35; phone; Eng spkn; quiet; ccard acc; tennis; bike hire; rv fishing; site open w/end Mar & Oct; CKE/CCI. "V helpful owners & staff; easy walk to interesting town; pleasant site by rv; vg, modern, clean san facs & excel facs; vg rest; popular NH; excel site in every way, facs superb; various size pitches; worth a visit; conv o'night on rte to S of Spain/Portugal; pull thro pitches avail; excel." ♦ € 26.40 2017*

TORLA *3B2* (2km N Rural) *42.63948, -0.10948* **Camping Ordesa, Ctra de Ordesa s/n, 22376 Torla (Huesca) [974-11 77 21; fax 974-48 63 47; camping@campingordesa.es; www.campingordesa.es]** Fr Ainsa on N260 twd Torla. Pass Torla turn R onto A135 (Valle de Ordesa twd Ordesa National Park). Site 2km N of Torla, adj Hotel Ordesa. Med, pt shd, serviced pitches; wc; chem disp; fam bthrm; shwrs; EHU (6A) €5.50; lndry; shop nr; rest; bar; playgrnd; pool; 10% statics; dogs €3; phone; bus 1km; Eng spkn; adv bkg rec; quiet; ccard acc; tennis; CKE/CCI. "V scenic; recep in adj Hotel Ordesa; excel rest; helpful staff; facs poss stretched w/end; long, narr pitches & lge trees on access rd poss diff lge o'fits; ltd facs LS; no access to National Park by car Jul/Aug, shuttlebus fr Torla." 28 Mar-30 Sep. € 22.60 2016*

TORLA *3B2* (1km NE Rural) *42.63181, -0.10685* **Camping Rió Ara, Ctra Ordesa s/n, 22376 Torla (Huesca) [974-48 62 48; campingrioara@ordesa.net; www.campingrioara.com]** Leave N260/A135 on bend approx 2km N of Broto sp Torla & Ordesa National Park. Drive thro Torla; as leaving vill turn R sp Rió Ara. Steep, narr rd down to & across narr bdge (worth it). Sm, pt shd, pt sl, wc; chem disp; mv service pnt; fam bthrm; shwrs inc; EHU (6A) €4.25; lndry; shop; rest nr; snacks; bar; bbq; wifi; TV; dogs; bus 500m; adv bkg acc; CKE/CCI. "Attractive, well-kept, family-run site; mainly tents; conv for Torla; bus to Ordesa National Park (high ssn); not rec for lge o'fits due to steep app; gd walking & birdwatching; excel; fantastic views; wonderful area.""Facs dated but functional and clean; 20m walk to Toria." ♦ 1 Apr-30 Sep. € 24.00 2015*

⊞ **TORRE DEL MAR** *2H4* (2km S Coastal) *36.7342, -4.1003* **Camping Torre del Mar, Paseo Maritimo s/n, 29740 Torre del Mar (Málaga) [952-54 02 24; fax 952-54 04 31; info@campingtorredelmar.com; www.campingtorredelmar.com]** Fr N340 coast rd, at rndabt at W end of town with 'correos' on corner turn twds sea sp Faro, Torre del Mar. At rndabt with lighthouse adj turn R, then 2nd R, site adj big hotel, ent bet lge stone pillars (no name sp). Lge, hdg, hdstg, mkd, shd, serviced pitches; wc; chem disp; mv service pnt; shwrs inc; EHU (16A) €4.40 (long lead req); gas; lndry; shop; rest nr; bar nr; playgrnd; pool; paddling pool; beach shgl 50m; red long stay; TV (pitch); 39% statics; phone; quiet; tennis; CKE/CCI. "Tidy, clean, friendly, well-run site; some sm pitches; site rds tight; gd, clean san facs; popular LS; constant hot water; poorly laid out; poss flooding in parts of site; conv for town." ♦ € 39.00 2016*

⊞ **TORRE DEL MAR** *2H4* (1km W Coastal) *36.72976, -4.10285* **Camping Laguna Playa, Prolongación Paseo Maritimo s/n, 29740 Torre del Mar (Málaga) [952-54 06 31; fax 952-54 04 84; info@lagunaplaya.com; www.lagunaplaya.com]** Fr N340 coast rd, at rndabt at W end of town with 'correos' on corner turn twds sea sp Faro, Torre del Mar. At rndabt with lighthouse adj turn R, then 2nd R, site sp in 400m. Med, pt shd, wc; chem disp; mv service pnt; fam bthrm; shwrs inc; EHU (10A) inc; gas; lndry; shop; rest; snacks; bar; bbq; playgrnd; pool; beach sand 1km; wifi; 80% statics; dogs; Eng spkn; adv bkg acc; quiet; fishing; games rm. "Popular LS; sm pitches; excel, clean san facs; gd location, easy walk to town; NH only; well shaded; pleasant little seaside town." ♦ € 33.00 2013*

TORRE DEL MAR *2H4* (3km W Coastal) *36.72526, -4.13532* **Camping Almayate Costa, Ctra N340, Km 267, 29749 Almayate Bajo (Málaga) [952-55 62 89; fax 952-55 63 10; almayatecosta@campings.net; www.campings.net/almayatecosta]** E fr Málaga on N340/E15 coast rd. Exit junc 258 dir Almería, site on R 3km bef Torre del Mar. Easy access. Lge, mkd, hdstg, shd, wc; chem disp; mv service pnt; shwrs inc; EHU (10A) €5.10; gas; lndry; shop; bar; bbq; playgrnd; pool; beach sand 50m; red long stay; phone; Eng spkn; adv bkg acc; quiet; ccard acc; golf 7km; car wash; games rm; CKE/CCI. "Helpful manager; pitches nr beach tight for lge o'fits & access rds poss diff; vg resort; excel." ♦ 15 Mar-30 Sep. € 43.50 2013*

🅃 **TORRE DEL MAR** *2H4* (7km W Coastal) *36.71967, -4.16471* **Camping Valle Niza Playa, Ctra N340, km 264, 1 29792 Valle Niza/Malaga [952-51 31 81; fax 952-51 34 76; info@campingvalleniza.es; www.campingvalleniza.es]** A7 Malaga-Motril exit 265 Cajiz Costa; at T-junc L onto N340 Coast rd twd Torre Del Mar. Site in 1km. Med, mkd, pt shd, wc; chem disp; fam bthrm; shwrs inc; EHU (6-10A); lndry; shop; rest; snacks; bar; bbq; pool; beach shgl; twin axles; wifi; TV; 50% statics; dogs; bus 0.3km; adv bkg acc; ccard acc; games area. "Gdn ctr adj; poor beach across rd; gd." € 33.00 2017*

🅃 **TORREVIEJA** *4F2* (7km SW Rural) *37.97500, -0.75111* **Camping Florantilles, Ctra San Miguel de Salinas-Torrevieja, 03193 San Miguel de Salinas (Alicante) [965-72 04 56; fax 966-72 32 50; camping@campingflorantilles.com; www.campingflorantilles.com]** Exit AP7 junc 758 onto CV95, sp Orihuela, Torrevieja Sud. Turn R at rndabt & after 300m turn R again, site immed on L. Or if travelling on N332 S past Alicante airport twd Torrevieja. Leave Torrevieja by-pass sp Torrevieja, San Miguel. Turn R onto CV95 & foll for 3km thro urbanisation 'Los Balcones', then cont for 500m, under by-pass, round rndabt & up hill, site sp on R. Lge, mkd, hdstg, hdg, pt shd, terr, wc; chem disp; mv service pnt; shwrs inc; EHU (10A) inc; gas; lndry; shop; snacks; bar; bbq; playgrnd; pool; paddling pool; beach sand 5km; TV; 20% statics; adv bkg acc; ccard acc; golf nr; horseriding 10km; games rm; CKE/CCI. "Popular, British owned site; fitness studio/keep fit classes; workshops: calligraphy, card making, drawing/painting, reiki, sound therapy etc; basic Spanish classes; 3 golf courses nrby; no o'fits over 10m; recep clsd 1330-1630; walking club; friendly staff; many long-stay visitors & all year c'vans; suitable mature couples; own transport ess; gd cyling, both flat & hilly; conv hot spa baths at Fortuna & salt lakes." ♦ € 30.00 2017*

TORROELLA DE MONTGRI *3B3* (6km SE Coastal) *42.01111, 3.18833* **Camping El Delfin Verde, Ctra Torroella de Montgrí-Palafrugell, Km 4, 17257 Torroella de Montgrí (Gerona) [972-75 84 54; fax 972-76 00 70; info@eldelfinverde.com; www.eldelfinverde.com]** Fr N leave A7 at junc 5 dir L'Escala. At Viladamat turn R onto C31 sp La Bisbal. After a few km turn L twd Torroella de Montgrí. At rndabt foll sp for Pals (also sp El Delfin Verde). At the flags turn L sp Els Mas Pinell. Foll site sp for 5km. V lge, mkd, pt shd, pt sl, wc; chem disp; mv service pnt; fam bthrm; shwrs inc; EHU (6A) inc; lndry; shop; rest; snacks; bar; bbq; playgrnd; pool; beach sand adj; entmnt; wifi; TV; 40% statics; dogs (except 18/7-21/8, at LS €4); quiet; ccard acc; games rm; horseriding 4km; windsurfing; fishing; tennis; bike hire; CKE/CCI. "Superb, gd value site; winter storage; excel pool; wide range of facs; sportsgrnd; hairdresser; disco; no o'fits over 8m high ssn; money exchange; clean, modern san facs; all water de-salinated fr fresh water production plant; bottled water rec for drinking & cooking; mkt Mon." ♦ 17 May-20 Sep. € 58.00 2017*

🅃 **TORROX COSTA** *2H4* (2km NNW Urban) *36.73944, -3.94972* **Camping El Pino, Urbanización Torrox Park s/n, 29793 Torrox Costa (Málaga) [952-53 00 06; fax 952-53 25 78; info@campingelpino.com; www.campingelpino.com]** Exit A7 at km 285, turn S at 1st rndabt, turn L at 2nd rndabt & foll sp Torrox Costa N340; in 1.5km at rndabt turn R to Torrox Costa, then L onto rndabt sp Nerja, site well sp in 4km. App rd steep with S bends. Fr N340 fr Torrox Costa foll sp Torrox Park, site sp. Lge, mkd, shd, terr, wc; chem disp; shwrs inc; EHU €3.80 (long lead req); gas; lndry; shop; rest nr; bar; bbq; playgrnd; pool; beach sand 800m; red long stay; wifi; 35% statics; dogs €2.50; phone; Eng spkn; games area; golf 8km; car wash; CKE/CCI. "Gd size pitches but high kerbs; narr ent/exit; gd hill walks; conv Malaga; Nerja caves, Ronda; gd touring base; noise fr rd and bar; san facs adequate." ♦ € 18.00 2013*

See advertisement

TOSSA DE MAR *3B3* (3km SW Coastal) *41.71509, 2.90672* **Camping Cala Llevado, Ctra Tossa-Lloret, Km 3, 17320 Tossa de Mar (Gerona) [972-34 03 14; fax 972-34 11 87; info@calallevado.com; www.calallevado.com]** Exit AP7 junc 9 dir Lloret. In Lloret take GI 682 dir Tossa de Mar. Site well sp. V lge, mkd, shd, terr, wc; chem disp; mv service pnt; fam bthrm; shwrs inc; EHU (5-10A) €3.50; gas; lndry; shop; rest; snacks; bar; playgrnd; pool; paddling pool; beach shgl adj; entmnt; wifi; TV; 10% statics; phone; Eng spkn; adv bkg acc; quiet; waterskiing; tennis; windsurfing; games area; boat trips; sports facs; fishing; CKE/CCI ♦ Holy Week & 1 May-30 Sep. € 29.00 2016*

UNQUERA *1A4* (5km W Rural) *43.3750, -4.56416* **Camping Colombres (formerly El Mirador de Llavandes), Vegas Grandes, 33590 Colombres (Asturias) [985-41 22 44; info@campingcolombres.com; www.campingcolombres.com]** Fr N634 12km W of San Vicente de la Barquera turn at km 283/284 dir Noriega, site in 1.3km. Med, mkd, pt shd, terr, wc; fam bthrm; EHU (6A); gas; lndry; shop; rest; bar; bbq; playgrnd; pool; paddling pool; beach sand 2km; twin axles; wifi; TV; dogs; Eng spkn; adv bkg rec; quiet; ccard acc; games area. "Peaceful setting; excel for touring Picos; new owners, new amenities, new pool(2015); vg site; mkd cycle rtes; patrolled grnds." ♦ 23 Mar-20 Sep. € 30.00 2015*

⊞ **VALENCIA** *4E2* (10km S Urban/Coastal) *39.38880, -0.33196* **Camping Park El Saler, Ctra del Riu 548, 46012 El Saler (València) [961-83 02 44; info@campingparkelsaler.com; www.campingparkelsaler.com]** Fr València foll coast rd or V15 to El Saler. Site adj rndabt just N of El Saler. Med, mkd, hdstg, hdg, pt shd, wc; chem disp; shwrs inc; EHU (6A) inc; lndry; shop; rest; bar; pool; beach sand 300m; 50% statics; dogs; phone; bus at gate; Eng spkn; CKE/CCI. "Very conv València - hourly bus; tow pin adaptor needed for conn cable; narr pitches & ent." € 25.00 2014*

⊞ **VALENCIA** *4E2* (16km S Rural) *39.32302, -0.30940* **Camping Devesa Gardens, Ctra El Saler, Km 13, 46012 València [961-61 11 36; fax 961-61 11 05; contacto@devesagardens.com; www.devesagardens.com]** S fr València on CV500, site well sp on R 4km S of El Saler. Med, mkd, hdstg, pt shd, wc (htd); chem disp; mv service pnt; fam bthrm; EHU (7-15A) €5; gas; lndry; shop; rest; bar; bbq; playgrnd; pool; beach 700m; 70% statics; phone; bus to València; adv bkg acc; quiet; ccard acc; horseriding; tennis. "Friendly, helpful staff; site has own zoo (clsd LS); lake canoeing; excel; san facs being refurb (2017); lovely pool area; easy access to tourist areas." ♦ € 35.00 2017*

See advertisement

"That's changed – Should I let the Club know?"

If you find something on site that's different from the site entry, fill in a report and let us know. See camc.com/europereport.

VALENCIA *4E2* (9km S Coastal) *39.39638, -0.33250* **Camping Coll Vert, Ctra Nazaret-Oliva, Km 7.5, 46012 Playa de Pinedo (València) [961-83 00 36; fax 961-83 00 40; info@collvertcamping.com; www.collvertcamping.com]** Fr S on V31 turn R onto V30 sp Pinedo. After approx 1km turn R onto V15/CV500 sp El Salar to exit El Salar Platjes. Turn L at rndabt, site on L in 1km. Fr N bypass València on A7/E15 & after junc 524 turn N twd València onto V31, then as above. Turn L at rndabt, site in 1km on L. Med, hdg, mkd, shd, wc; shwrs inc; EHU €4.81 6A; gas; lndry; shop; bar; bbq; playgrnd; pool; paddling pool; beach sand 500m; red long stay; entmnt; 20% statics; dogs €4.28; phone; bus to city & marine park; Eng spkn; adv bkg acc; quiet; ccard acc; car wash; games area. "Hourly bus service fr outside site to cent of València & marine park; helpful, friendly staff; san facs need update sm pitches; conv for F1." ♦ 16 Feb-14 Dec. € 22.00 2014*

⊞ **VALLE DE CABUERNIGA** *1A4* (1km E Rural) *43.22800, -4.28900* **Camping El Molino de Cabuérniga, Sopeña, 39510 Cabuérniga (Cantabria) [942-70 62 59; fax 942-70 62 78; cmcabuerniga@campingcabuerniga.com; www.campingcabuerniga.com]** Sopeña is 55 km. SW of Santander. Fr A8 (Santander - Oviedo) take 249 exit and join N634 to Cabezón de la Sal. Turn SW on CA180 twds Reinosa for 11 km. to Sopeña (site sp to L). Turn into vill (car req - low bldgs), cont bearing R foll sp to site. Med, shd, wc; shwrs inc; EHU (6A) check earth; gas; lndry; shop; snacks; bar; playgrnd; wifi; dogs €1.50; phone; bus 500m; adv bkg acc; quiet; ccard acc; tennis; fishing; CKE/CCI. "Excel site & facs on edge of vill; no shops in vicinity, but gd location, rds to site narr in places; lovely." ♦ € 24.00 2016*

VECILLA, LA *1B3* (1km N Rural) *42.85806, -5.41155* **Camping La Cota, Ctra Valdelugueros, LE321, Km 19, 24840 La Vecilla (León) [987-74 10 91; lacota@campinglacota.com; www.campinglacota.com]** Fr N630 turn E onto CL626 at La Robla, 17km to La Vecilla. Site sp in vill. Med, mkd, shd, wc; chem disp; fam bthrm; shwrs inc; EHU €3.50; lndry; snacks; bar; wifi; 50% statics; train nr; quiet; ccard acc; games area. "Pleasant site under poplar trees - poss diff to manoeuvre lge o'fits; open w/ends out of ssn; gd walking, climbing nr; interesting mountain area; NH only." 23 Mar-30 Sep. € 21.00 2013*

⊞ **VEJER DE LA FRONTERA** *2H3* (10km S Coastal) *36.20084, -6.03506* **Camping Pinar San José, Ctra de Vejer-Caños de Meca, Km 10.2, Zahora 17, 11159 Barbate (Cadiz) [956-43 70 30; fax 956-43 71 74; info@campingpinarsanjose.com; www.campingpinarsanjose.com]** Fr A48/N340 exit junc 36 onto A314 to Barbate, then foll dir Los Caños de Meca. Turn R at seashore rd dir Zahora. Site on L, 2km beyond town. Med, mkd, shd, wc; chem disp; mv service pnt; fam bthrm; shwrs; EHU inc; lndry; shop; rest; playgrnd; pool; paddling pool; beach sand 700m; wifi; TV (pitch); 10% statics; dogs €2 (LS only); adv bkg acc; quiet; games area; tennis. "Excel, modern facs." ♦ € 64.00 2013*

See advertisement

VELEZ MALAGA *2G4* (16km NNW Rural) *36.87383, -4.18527* **Camping Rural Presa la Vinuela, Carretera A-356, Km 30, 29712 La Viñuela Málaga [952-55 45 62; fax 952-55 45 70; campingpresalavinuela@hotmail.com; www.campinglavinuela.es]** Site is on A356 N of Velez Malaga adjoining the W shore of la Vinuela lake. Fr junc with A402, foll sp to Colmenar/Los Romanes. Stay on A356(don't turn off into Los Romanes). Site is on R approx 2.5km after turn for Los Romanes, next to rest El Pantano. Sm, hdstg, hdg, mkd, pt shd, terr, wc; chem disp; shwrs; EHU (5A); lndry; shop; rest; snacks; bar; pool; wifi; TV; 20% statics; dogs €1.10; Eng spkn; adv bkg acc; games area; games rm. "Excel site." 1 Jan-30 Sep. € 21.00 2014*

"I like to fill in the reports as I travel from site to site"

You'll find report forms at the back of this guide, or you can fill them in online at camc.com/europereport.

VIELHA *3B2* (6km SE Rural) *42.70005, 0.87060* **Camping Era Yerla D'Arties, Ctra C142, Vielha-Baquiera s/n, 25599 Arties (Lleida) [973-64 16 02; fax 973-64 30 53; yerla@coac.net; www.aranweb.com/yerla]** Fr Vielha take C28 dir Baquiera, site sp. Turn R at rndabt into Arties, site in 30m on R. Med, shd, wc (htd); chem disp; fam bthrm; shwrs inc; EHU (4-10A) €4.25-5.15; gas; lndry; shop nr; rest nr; snacks; bar; pool; 10% statics; phone; bus 200m; quiet; ccard acc; CKE/CCI. "Pleasant site & vill; skiing nr; ideal for ski resort; san facs gd; gd walking; OK for sh stay." 1 Dec-14 Sep. € 23.70 2012*

⊞ **VILANOVA I LA GELTRU** *3C3* (3km NW Urban) *41.23190, 1.69075* **Camping Vilanova Park, Ctra Arboç, Km 2.5, 08800 Vilanova i la Geltru (Barcelona) [938-93 34 02; fax 938-93 55 28; info@vilanovapark.com or reservas@vilanovapark.com; www.vilanovapark.com]** Fr N on AP7 exit junc 29 onto C15 dir Vilanova; then take C31 dir Cubelles. Leave at 153km exit dir Vilanova Oeste/L'Arboç to site. Fr W on C32/A16 take Vilanova-Sant Pere de Ribes exit. Take C31 & at 153km exit take BV2115 dir L'Arboc to site. Fr AP7 W leave at exit 31 onto the C32 (A16); take exit 16 (Vilanova-L'Arboc exit) onto BV2115 to site. Parked cars may block loop & obscure site sp. V lge, mkd, hdstg, hdg, pt shd, terr, serviced pitches; wc (htd); chem disp; mv service pnt; fam bthrm; shwrs inc; EHU (10A) inc (poss rev pol); gas; lndry; shop; rest; snacks; bar; bbq (elec, gas); playgrnd; pool (covrd, htd); paddling pool; beach sand 3km; sw nr; red long stay; entmnt; wifi; TV; 50% statics; dogs €12.50; phone; bus directly fr campsite to Barcelona; Eng spkn; adv bkg req; ccard acc; horseriding 500m; games rm; bike hire; sauna; fishing; tennis; golf 1km; CKE/CCI. "Gd for children; excel san facs; gd rest & bar; gd winter facs; jacuzzi; spa; fitness cent; helpful staff; gd security; some sm pitches with diff access due trees or ramps; conv bus/train Barcelona, Tarragona, Port Aventura & coast; mkt Sat; excel site; superb." ♦ € 61.00 2013*

See advertisement

⊞ **VILLAFRANCA** *3B1* (2km S Rural) *42.26333, -1.73861* **Camping Bardenas, Ctra NA-660 PK 13.4, 31330 Villafranca [34 94 88 46 191; info@campingbardenas.com; www.campingbardenas.com]** Fr N leave AP15 at Junc 29 onto NA660 sp Villafranca. Site on R 1.5km S of town. Med, hdstg, unshd, wc (htd); chem disp; fam bthrm; shwrs; lndry; shop; rest; bar; bbq; pool; 30% statics; ccard acc; games rm; CCI. "Gd site for winter stopover en rte to S Spain; facs ltd in severe weather; excel rest." € 36.50 2014*

VILLAFRANCA DE CORDOBA *2F4* (1km W Rural) *37.95333, -4.54710* **Camping La Albolafia, Camino de la Vega s/n, 14420 Villafranca de Córdoba (Córdoba) [957-19 08 35; informacion@campingalbolafia.com; www.campingalbolafia.com]** Exit A4/E5 junc 377, cross rv & at rndabt turn L & foll sp to site in 2km. Beware humps in app rd. Med, hdg, hdstg, mkd, pt shd, wc; chem disp; mv service pnt; shwrs inc; EHU (10A) inc (long lead poss req); lndry; shop; rest; snacks; bar; bbq; playgrnd; pool; twin axles; wifi; TV; 10% statics; dogs €2.80; phone; bus to Córdoba 500m; Eng spkn; quiet; CKE/CCI. "V pleasant, well-run, friendly, clean site; watersports park nrby; bar and rest clsd end May; ok for stopover." ♦ 15 Feb-9 Dec. € 27.00 2017*

VILLAMANAN *1B3* (6km SE Rural) *42.29527, -5.53777* **Camping Pico Verde, Ctra Mayorga-Astorga, Km 27.6, 24200 Valencia de Don Juan (León) [987-75 05 25; campingpicoverd@terra.es]** Fr N630 S, turn E at km 32.2 onto C621 sp Valencia de Don Juan. Site in 4km on R. Med, mkd, wc; shwrs inc; EHU (6A) inc; lndry; shop; rest; playgrnd; pool (covrd); paddling pool; 25% statics; dogs free; quiet; tennis; CKE/CCI. "Friendly, helpful staff; conv León; picturesque vill; sw caps to be worn in pool; phone ahead to check site open if travelling close to opening/closing dates." ♦ 15 Jun-8 Sep. € 20.00 2014*

VILLANANE *1B4* (1km N Rural) *42.84221, -3.06867* **Camping Angosto, Ctra Villanañe-Angosto 2, 01425 Villanañe (Gipuzkoa) [945-35 32 71; fax 945-35 30 41; info@camping-angosto.com; www.camping-angosto.com]** S fr Bilbao on AP68 exit at vill of Pobes & take rd to W sp Espejo. Turn L 2.4km N of Espejo dir Villanañe, lane to site 400m on R. Med, mkd, pt shd, wc (htd); chem disp; fam bthrm; shwrs inc; EHU €4.15 (long cable poss req - supplied by site); lndry; shop; rest; snacks; bar; bbq; playgrnd; pool (covrd, htd); entmnt; TV; 50% statics; dogs; phone; Eng spkn; quiet. "Beautiful area; friendly, helpful staff; open site - mkd pitches rec if poss; gd rest; conv NH fr Bilbao; wonderful site; in LS site seems close but call nbr on gate; site long way fr main rd; poss narr vill app rds." ♦ 15 Feb-30 Nov. € 27.00 2014*

"We must tell the Club about that great site we found"

Get your site reports in by mid-August and we'll do our best to get your updates into the next edition.

⊞ **VILLARGORDO DEL CABRIEL** *4E1* (3km NW Rural) *39.5525, -1.47444* **Kiko Park Rural, Ctra Embalse de Contreras, Km 3, 46317 Villargordo del Cabriel (València) [962-13 90 82; fax 962-13 93 37; kikoparkrural@kikopark.com; www.kikopark.com/rural]** A3/E901 València-Madrid, exit junc 255 to Villargordo del Cabriel, foll sp to site. Med, mkd, hdstg, pt shd, terr, serviced pitches; wc; chem disp; mv service pnt; shwrs inc; EHU (6A) €3.70; gas; lndry; shop; rest; snacks; bar; pool; sw nr; red long stay; TV; 10% statics; dogs €0.80; Eng spkn; adv bkg rec; ccard acc; canoeing; horseriding; white water rafting; fishing; bike hire; watersports; CKE/CCI. "Beautiful location; superb, well-run, peaceful site; lge pitches; gd walking; vg rest; many activities; helpful, v friendly family run site; gd hdstg; excel." ♦ € 33.00 2015*

⊞ **VINAROS** *3D2* (5km N Coastal) *40.49363, 0.48504* **Camping Vinarós, Ctra N340, Km 1054, 12500 Vinarós (Castellón) [964-40 24 24; fax 964-45 53 28; info@campingvinaros.com; www.campingvinaros.com]** Fr N exit AP7 junc 42 onto N238 dir Vinarós. Straight on at first 2 rndabt, at 3rd rndabt turn L (3rd exit) dir Tarragona. Site on R at Km1054. Lge, mkd, hdstg, hdg, pt shd, serviced pitches; wc (htd); chem disp; mv service pnt; fam bthrm; shwrs inc; EHU (6A) inc; gas; lndry; shop nr; rest nr; snacks; bar; playgrnd; pool; beach shgl 1km; red long stay; wifi; 15% statics; dogs €3; phone; bus adj; Eng spkn; adv bkg rec; quiet; ccard acc; CKE/CCI. "Excel gd value, busy, well-run site; many long-stay winter residents; spacious pitches; vg clean, modern san facs; elec volts poss v low in evening; gd rest; friendly, helpful staff; rec use bottled water; currency exchange; Peñíscola Castle & Morello worth a visit; easy cycle to town; ok stopover." ♦ € 13.00 2017*

VINUELA *2G4* (8km NW Rural) *36.87383, -4.18527* **Camping Presa La Viñuela, Ctra A356, km 30 29712 Viñuela [952-55 45 62; fax 952-55 45 70; campingpresalavinuela@hotmail.com; www.campinglavinuela.es]** Site on A356 N of Velez Malaga adjoining the W shore of la Vinuela lake. Fr junc with A402, foll sp to Colmenar & Los Romanes. Stay on A356 (don't turn off into Los Romanes). Site is on R approx 2.5km after the turn for Los Romanes, next to El Pantano rest. Sm, mkd, hdstg, pt shd, terr, wc; chem disp; shwrs; EHU (5A); lndry; shop; rest; snacks; bar; pool; wifi; TV; 20% statics; dogs €1.10; Eng spkn; adv bkg acc; games rm; games area. "Excel site." 1 Jan-30 Sep. € 21.00 2014*

VINUESA *3B1* (2km N Rural) *41.92647, -2.76285* **Camping Cobijo, Ctra Laguna Negra, Km 2, 42150 Vinuesa (Soria) [975-37 83 31; recepcion@campingcobijo.com; www.campingcobijo.com]** Travelling W fr Soria to Burgos, at Abejar R on SO840. by-pass Abejar cont to Vinuesa. Well sp fr there. Lge, pt shd, pt sl, wc; chem disp; fam bthrm; shwrs inc; EHU (3-6A) €4-5.70 (long lead poss req); gas; lndry; shop; rest; snacks; bar; bbq; playgrnd; pool; wifi; 10% statics; dogs; phone; Eng spkn; quiet; ccard acc; bike hire; CKE/CCI. "Friendly staff; clean, attractive site; some pitches in wooded area poss diff lge o'fits; special elec connector supplied (deposit); ltd bar & rest LS, excel rests in town; gd walks." ♦ 1 Apr-1 Nov. € 28.00 2017*

⊞ **VITORIA/GASTEIZ** *3B1* (6km SW Rural) *42.83114, -2.72248* **Camping Ibaya, Nacional 102, Km 346.5, Zuazo de Vitoria 01195 Vitoria/Gasteiz (Alava) [945-14 76 20; fax 627-07 43 99; info@campingibaia.com; www.campingibaia.com]** Fr A1 take exit 343 sp N102/A3302. At rndabt foll sp N102 Vitoria/Gasteiz. At next rndabt take 3rd exit & immed turn L twd filling stn. Site ent on R in 100m, sp. Sm, mkd, hdstg, pt shd, pt sl, wc; chem disp; shwrs; EHU (5A) €5.40; gas; lndry; shop; rest nr; snacks; bar; bbq; playgrnd; wifi; 10% statics; dogs; phone; Eng spkn; CKE/CCI. "NH only; gd, modern san facs; phone ahead to check open LS; fair site." € 29.50 2016*

VIVER *3D2* (4km W Rural) *39.90944, -0.61833* **Camping Villa de Viver, Camino Benaval s/n, 12460 Viver (Castellón) [964-14 13 34; info@campingviver.com; www.campingviver.com]** Fr Sagunto on A23 dir Terual, approx 10km fr Segorbe turn L sp Jérica, Viver. Thro vill dir Teresa, site sp W of Viver at end of single track lane in approx 2.8km (yellow sp) - poss diff for car+c'van, OK m'vans. Med, hdg, pt shd, terr, serviced pitches; wc (htd); chem disp; mv service pnt; shwrs inc; EHU (6A) €3.80; lndry; shop nr; rest; snacks; bar; playgrnd; pool; red long stay; TV; 10% statics; dogs €3.10; phone; Eng spkn; adv bkg acc; quiet; ccard acc; CKE/CCI. "Improved site; lovely situation - worth the effort." ♦ 1 Mar-1 Nov. € 19.00 2013*

⊞ **ZARAGOZA** *3C1* (6km W Urban) *41.63766, -0.94227* **Camping Ciudad de Zaragoza, Calle San Juan Bautista de la Salle s/n, 50012 Zaragoza [876-24 14 95; fax 876-24 12 86; info@campingzaragoza.com; www.campingzaragoza.com]** Fr S on A23 foll Adva Gómez Laguna, turn L at 2nd rndabt, in 500m bear R in order to turn L at rndabt, site on R in 750m, sp. Fr all other dirs take Z40 ring rd dir Teruel, then Adva Gómez Laguna twd city, then as above. Site well sp. Lge, mkd, hdstg, pt shd, pt sl, wc (htd); chem disp; mv service pnt; fam bthrm; shwrs inc; EHU (10A) €5.75; lndry; shop; rest; snacks; bar; bbq; playgrnd; pool; twin axles; wifi; TV; 50% statics; dogs €3.75; Eng spkn; adv bkg acc; games area; tennis. "Modern san facs but poss unclean LS & pt htd; poss travellers; unattractive, but conv site in suburbs; gd sh stay; poss noisy (campers & daytime aircraft); gd food at bar; helpful staff; site scruffy; gd NH from Bilbao ferry." ♦ € 26.50 2018*

⊞ **ZARAUTZ** *3A1* (3km NE Coastal) *43.28958, -2.14603* **Gran Camping Zarautz, Monte Talaimendi s/n, 20800 Zarautz (Guipúzkoa) [943-83 12 38; fax 943-13 24 86; info@grancampingzarautz.com; www.grancampingzarautz.com]** Exit A8 junc 11 Zarautz, strt on at 1st & 2nd rndabt after toll & foll site sp. On N634 fr San Sebastián to Zarautz, R at rndabt. On N634 fr Bilbao to Zarautz L at rndabt. Lge, mkd, hdstg, hdg, pt shd, pt sl, terr, wc (htd); chem disp; mv service pnt; fam bthrm; shwrs inc; EHU (6-10A) inc; gas; lndry; shop; rest; bar; bbq; playgrnd; beach 1km; wifi; TV; 50% statics; dogs; phone; train/bus to Bilbao & San Sebastian; Eng spkn; adv bkg acc; ccard acc; golf 1km; games rm; CKE/CCI. "Site on cliff o'looking bay; excel beach, gd base for coast & mountains; helpful, friendly staff; some pitches sm with steep access & o'looked fr terr above; sans facs upgraded but poor standard and insufficient when cr; excel rest; pitches poss muddy; NH for Bilbao ferry; rec arr early to secure pitch; v steep walk to beach (part naturist); gd for NH; excel train service to San Sebastian; gd shop on site; ACSI discount." ♦ € 31.00 2017*

ZARAUTZ *3A1* (8km E Coastal) *43.27777, -2.12305* **Camping Orio Kanpina, 20810 Orio (Guipúzkoa) [943-83 48 01; fax 943-13 34 33; info@oriokanpina.com; www.oriokanpina.com]** Fr E on A8 exit junc 33 & at rndabt foll sp Orio, Kanpin & Playa. Site on R. Or to avoid town cent (rec) cross bdge & foll N634 for 1km, turn L at sp Orio & camping, turn R at rndabt to site. Lge, mkd, pt shd, pt sl, wc; chem disp; mv service pnt; fam bthrm; shwrs inc; EHU (5A) inc; gas; lndry; shop; rest nr; snacks; bar; playgrnd; pool; paddling pool; beach sand adj; wifi; 50% statics; phone; Eng spkn; adv bkg acc; quiet; ccard acc; car wash; tennis; CKE/CCI. "Busy, well maintained site; flats now built bet site & beach & new marina adj - now no sea views; walks; gd facs; friendly staff; useful NH bef leaving Spain; interesting sm town." ♦ 1 Mar-12 Nov. € 36.00 2017*

"I need an on-site restaurant"

We do our best to make sure site information is correct, but it is always best to check any must-have facilities are still available or will be open during your visit.

ZEANURI *3A1* (3km SE Rural) *43.08444, -2.72333* **Camping Zubizabala, Otxandio, 48144 Zeanuri [944-47 92 06 or 660-42 30 17 (mob); zubizabala@gmail.com; www.zubizabala.com]** E fr Bilbao on A8, exit at Galdakao junc 19 onto N240 dir Vitoria/Gasteiz. 2.5km S of Barazar Pass, turn L on minor rd B3542 to Otxandio. Site 300m on R, sp. Sm, mkd, unshd, wc; chem disp; mv service pnt; fam bthrm; shwrs inc; EHU €3.55; lndry; shop; bar; playgrnd; phone; bus at site ent; Eng spkn; adv bkg acc; quiet; games area; CKE/CCI. "V pleasant, tranquil site in woods; superb countryside; pool 4km; conv Bilbao & Vitoria/Gasteiz." ♦ 15 Jun-15 Sep. € 25.00 2012*

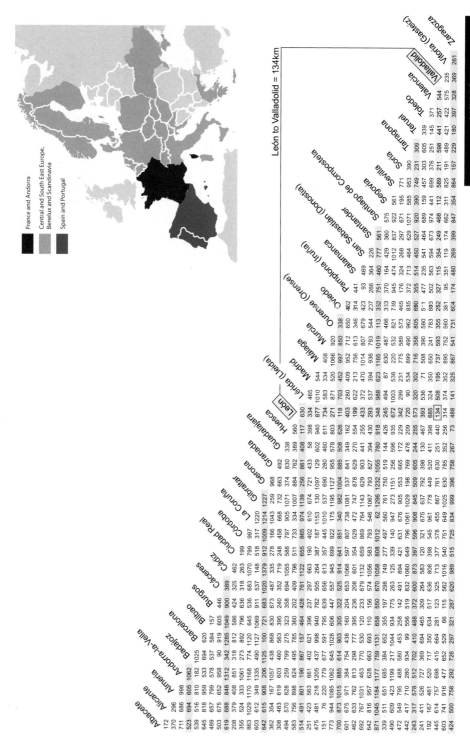

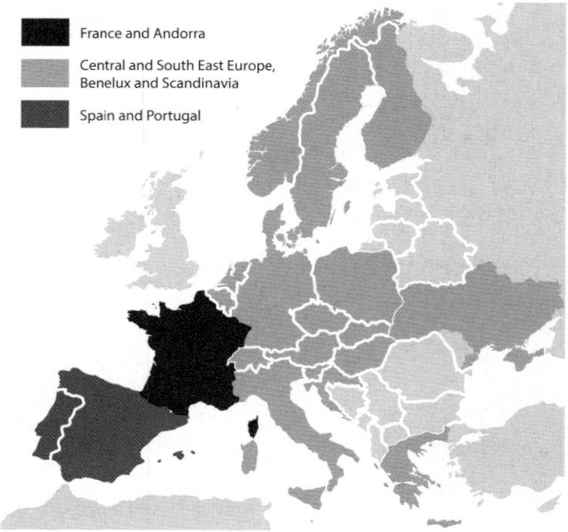

León to Valladolid = 134km

	Albacete	Alicante	Almeria	Andorra-la-Vella	Badajoz	Barcelona	Bilbao	Burgos	Cáceres	Cádiz	Ciudad Real	Córdoba	La Coruña	Gibraltar	Gerona	Granada	Guadalajara	Huesca	León	Lérida (Lleida)	Madrid	Málaga	Murcia	Ourense (Orense)	Oviedo	Pamplona (Iruña)	Salamanca	San Sebastián (Donostia)	Santander	Santiago de Compostela	Segovia	Sevilla	Soria	Tarragona	Teruel	Toledo	Valencia	Valladolid	Vitoria (Gasteiz)
Alicante	172																																						
Almeria	370	296																																					
Andorra-la-Vella	711	686	998																																				
Badajoz	523	694	605	1062																																			
Barcelona	539	516	810	182	1025																																		
Bilbao	645	818	959	533	694	620																																	
Burgos	489	657	799	578	537	584	157																																
Cáceres	503	675	652	958	90	919	605	446																															
Cádiz	619	688	848	1312	342	1285	1049	900	389																														
Ciudad Real	208	379	408	851	318	812	586	424	325	462																													
Córdoba	355	524	333	1061	273	892	796	636	318	263	199																												
La Coruña	863	1029	1170	1168	774	1120	645	536	683	1070	799	997																											
Gibraltar	603	612	349	1335	490	1127	1069	911	537	148	518	317	1220																										
Gerona	642	615	908	206	1125	100	721	683	1020	1379	912	1009	1214	1227																									
Granada	362	354	167	1057	438	868	830	673	487	335	278	166	1043	259	968																								
Guadalajara	308	483	619	603	460	563	395	240	352	719	248	458	668	732	663	492																							
Huesca	494	570	828	259	799	275	323	358	694	1055	588	797	905	1071	374	830	338																						
León	583	756	898	824	495	785	360	202	409	796	511	733	334	1007	884	762	389	560																					
Lérida (Lleida)	514	491	801	196	867	157	464	428	761	1122	655	865	974	1139	256	861	408	117	630																				
Madrid	251	423	563	661	402	621	396	237	297	663	190	402	610	674	721	433	58	398	334	465																			
Málaga	475	481	218	1205	437	998	940	782	505	264	387	187	1153	130	1097	129	602	940	877	1010	544																		
Murcia	151	76	220	779	677	591	795	639	656	613	357	445	1010	537	690	280	460	611	734	583	334	408																	
Ourense (Orense)	773	944	1085	1062	645	1028	606	447	557	945	699	922	175	1195	1127	955	578	803	271	871	520	1066	920																
Oviedo	700	873	1015	885	614	903	305	322	525	914	641	851	340	982	1004	885	508	626	118	703	452	997	850	338															
Pamplona (Iruña)	601	675	971	384	754	438	160	204	653	1068	597	807	738	1081	537	841	349	162	403	280	409	952	712	650	462														
Salamanca	462	633	762	813	298	777	395	236	208	601	354	529	472	747	878	629	270	554	199	622	213	756	613	346	314	441													
San Sebastián (Donostia)	692	767	1031	463	770	530	120	233	679	1132	659	869	764	1143	629	903	441	255	433	372	470	1014	807	679	423	93	469												
Santander	642	816	957	628	663	693	110	156	574	1056	583	793	546	1067	793	827	394	430	293	537	394	936	793	544	227	266	364	226											
Santiago de Compostela	871	1045	1184	1177	759	1131	658	550	670	1058	808	1012	62	1296	1232	1055	780	918	348	988	623	1165	1019	113	352	751	460	777	561										
Segovia	339	511	651	685	384	652	355	197	298	749	277	497	560	761	750	519	144	426	245	494	87	630	487	466	333	370	164	429	360	575									
Sevilla	490	609	423	1199	217	1044	934	775	263	125	339	140	947	273	1151	256	596	935	672	1003	538	220	532	821	739	945	474	1012	837	922	561								
Soria	472	549	795	488	580	453	258	142	491	894	421	631	676	905	553	665	172	229	342	299	231	775	589	573	465	176	324	268	297	671	195	771							
Tarragona	442	417	710	280	934	99	556	519	832	1060	649	796	1061	1029	198	769	476	209	720	90	534	899	490	962	835	372	713	464	629	1071	585	953	390						
Teruel	243	317	578	512	702	410	488	372	600	873	397	596	908	845	509	605	244	255	573	320	302	716	358	805	690	355	514	450	527	920	390	749	231	309					
Toledo	241	411	526	727	369	694	465	309	264	583	120	321	675	637	792	398	130	467	393	536	71	508	390	580	511	477	235	541	464	689	159	457	303	605	339				
Valencia	192	167	461	520	717	350	634	517	636	808	398	545	961	778	449	520	411	398	685	324	350	650	241	783	803	502	563	594	673	974	441	699	376	251	145	371			
Valladolid	445	614	757	698	415	664	281	123	325	713	377	578	455	867	761	630	251	440	134	508	195	737	593	355	252	327	115	354	249	468	112	589	211	598	441	257	544		
Vitoria (Gasteiz)	603	741	916	477	652	529	66	115	562	1016	540	751	649	1025	630	785	352	256	314	374	352	895	752	560	381	95	351	119	174	662	311	825	191	489	421	422	575	235	
Zaragoza	424	500	758	292	726	297	321	287	620	989	515	725	834	999	396	758	267	73	488	141	325	867	541	731	604	174	480	269	399	847	354	864	157	229	180	397	328	369	261

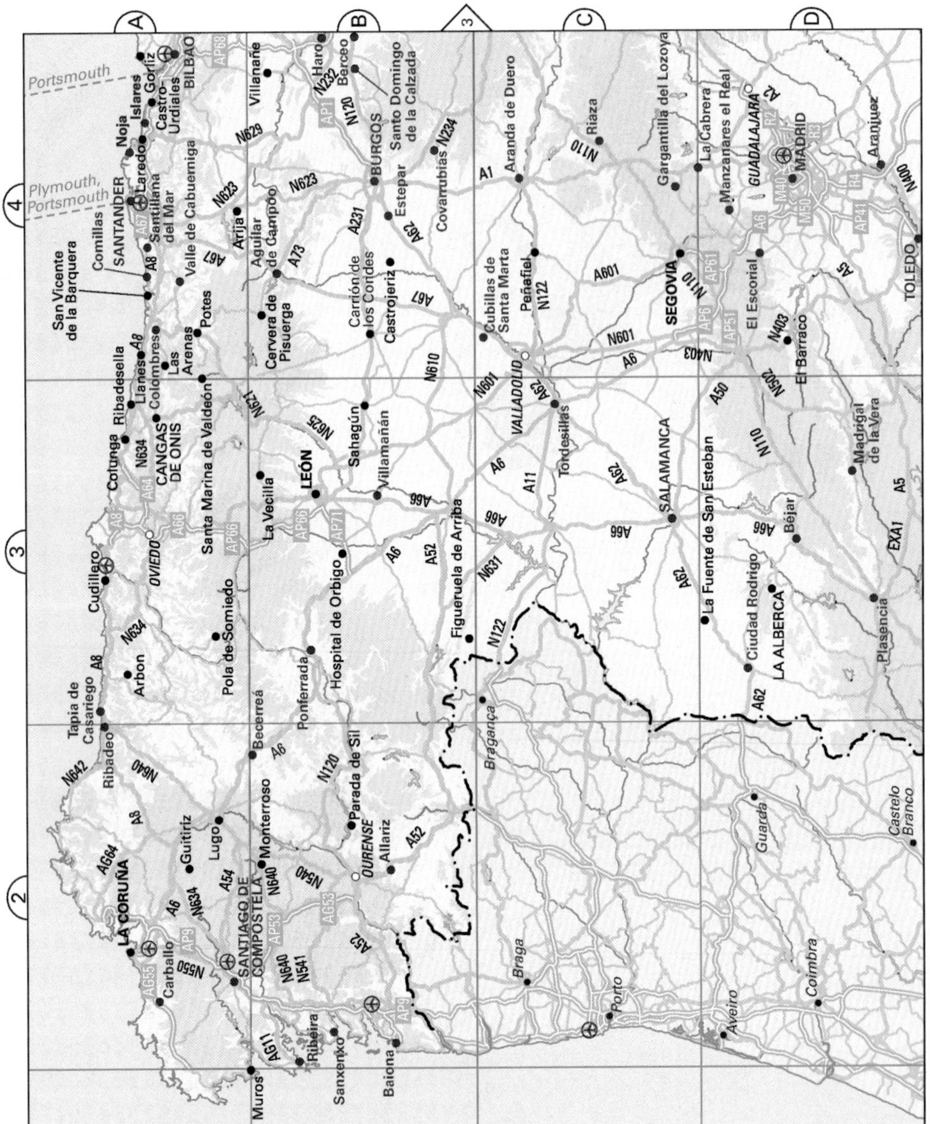

A
B
C
D
1
2
3
4
N
E
S
W
Portsmouth
Plymouth, Portsmouth
BILBAO
Gorliz
Castro-Urdiales
Islares
Noja
Laredo
SANTANDER
Santillana del Mar
Comillas
San Vicente de la Barquera
Valle de Cabuérniga
Arija
Villanañe
Haro
Berceo
Santo Domingo de la Calzada
BURGOS
Estepar
Covarrubias
Aguilar de Campoo
Cervera de Pisuerga
Carrión de los Condes
Castrojeriz
Potes
Las Arenas
Colombres
Llanes
Ribadesella
Colunga
CANGAS DE ONIS
Santa Marina de Valdeón
Sahagún
Villamañán
LEÓN
La Vecilla
OVIEDO
Cudillero
Pola de Somiedo
Arbon
Tapia de Casariego
Ribadeo
Becerreá
Ponferrada
Hospital de Orbigo
Figueruela de Arriba
Bragança
Lugo
Guitiriz
Monterroso
Parada de Sil
OURENSE
Allariz
LA CORUÑA
SANTIAGO DE COMPOSTELA
Carballo
Muros
Ribeira
Sanxenxo
Baiona
Braga
Porto
Aveiro
Coimbra
Guarda
Castelo Branco
Aranda de Duero
Riaza
Cubillas de Santa Marta
Peñafiel
VALLADOLID
Tordesillas
SALAMANCA
SEGOVIA
Gargantilla del Lozoya
La Cabrera
Manzanares el Real
El Escorial
El Barraco
GUADALAJARA
MADRID
Aranjuez
TOLEDO
La Fuente de San Esteban
Ciudad Rodrigo
LA ALBERCA
Béjar
Madrigal de la Vera
Plasencia
A1
A5
A6
A8
A11
A50
A52
A62
A66
A67
A73
A231
A601
AG64
AP9
AP66
AP71
EXA1
N110
N120
N122
N232
N234
N403
N502
N540
N550
N601
N610
N623
N625
N629
N631
N634
N640
N642

Map 1

PORTUGAL
MOROCCO
LISBOA
Portalegre
Elvas
Évora
Beja
Portimão
Faro
Isla Cristina
Mazagón
El Rocío
CÁCERES
Benquerencia
Miajadas
BADAJOZ
Mérida
Horcajo de los Montes
CIUDAD REAL
Santa Elena
Fuenteheridos
CÓRDOBA
Villafranca de Córdoba
SANTAELLA
SEVILLA
Dos Hermanas
Fuente de Piedra
Humilladero
Algámitas
Olvera
Beas de Granada
GRANADA
El Puerto de Santa María
CÁDIZ
Conil de la Frontera
Vejer de la Frontera
Tarifa
San Roque
Alcalá de los Gazules
La Linea de la Concepción
GIBRALTAR (U.K.)
Ronda
Alhaurín de la Torre
Marbella
MÁLAGA
Benajarafe
Torre del Mar
Torrox Costa
Nerja
MOTRIL
Balerma
E
F
G
H
1
2
3
4
All year site(s)
Seasonal site(s)
No sites listed
200m +
0–200m
0 30 60 90 120 150 km
0 20 40 60 80 100 miles
Motorways
Major roads
Main Roads
© Collins Bartholomew Ltd 2019

Map 2

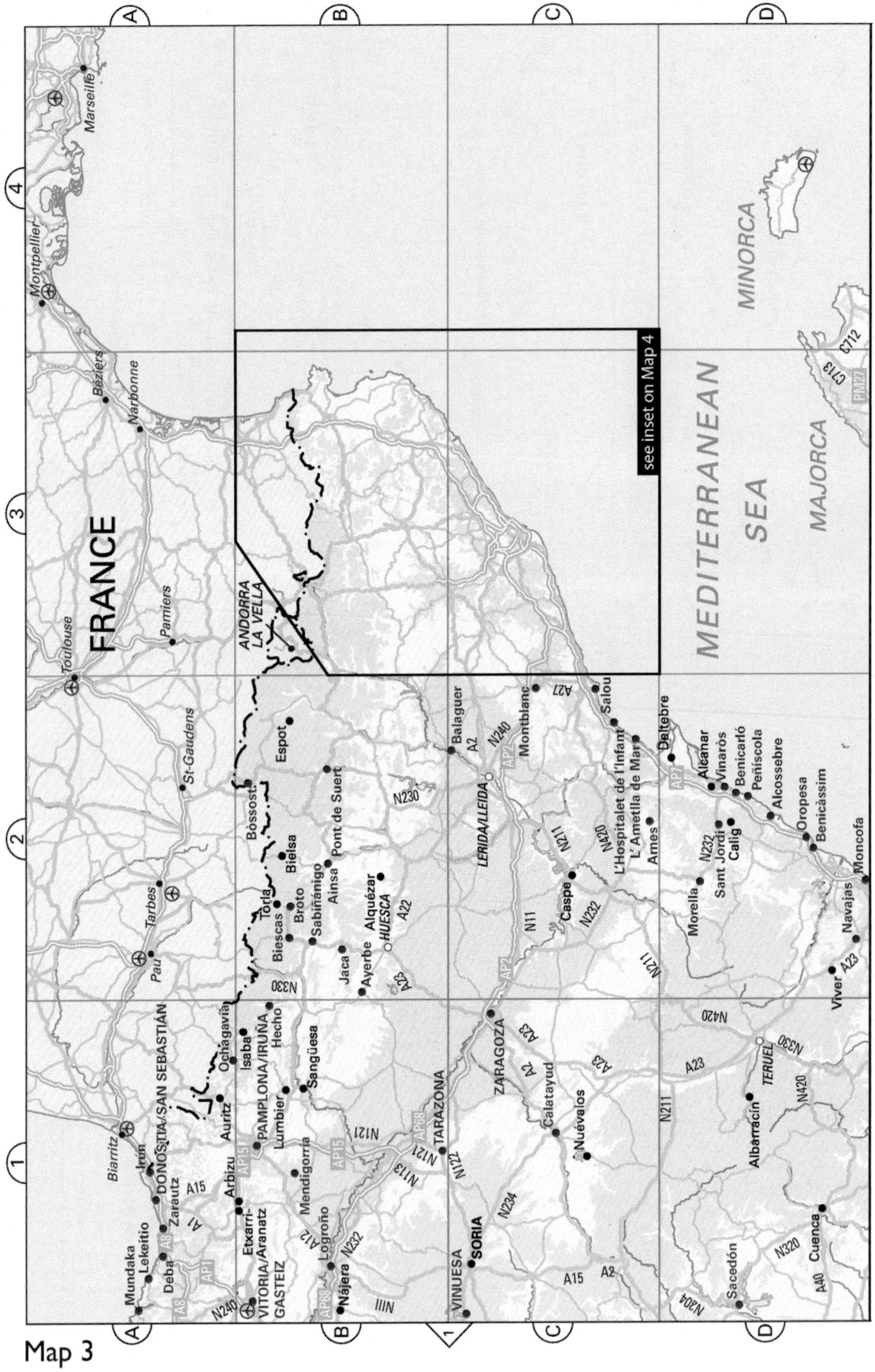
FRANCE
Toulouse
Pamiers
St-Gaudens
Tarbes
Pau
Biarritz
Béziers
Narbonne
Montpellier
Marseille
ANDORRA LA VELLA
see inset on Map 4
MEDITERRANEAN SEA
MINORCA
MAJORCA
Mundaka
Lekeitio
Deba
Zarautz
Irún
DONOSTIA/SAN SEBASTIÁN
Etxarri-Aranatz
Arbizu
Auritz
Ochagavía
Isaba
Hecho
PAMPLONA/IRUÑA
Lumbier
Sangüesa
Mendigorría
VITORIA/GASTEIZ
Logroño
Nájera
TARAZONA
VINUESA
SORIA
ZARAGOZA
Calatayud
Nuévalos
Jaca
Ayerbe
HUESCA
Alquézar
Biescas
Torla
Broto
Sabiñánigo
Ainsa
Bielsa
Pont de Suert
Bòssost
Espot
Balaguer
LERIDA/LLEIDA
Montblanc
Salou
Caspe
L'Hospitalet de l'Infant
L'Ametlla de Mar
Arnes
Deltebre
Alcanar
Vinaròs
Benicarló
Peñíscola
Alcossebre
Oropesa
Benicàssim
Moncofa
Morella
Sant Jordi
Càlig
Navajas
Viver
TERUEL
Albarracín
Cuenca
Sacedón
A8
AP1
N240
A1
A15
AP15
AP68
A12
N232
N111
N121
N113
N122
N234
N330
A23
A22
AP2
A2
N230
N240
N211
N420
N232
N11
A27
AP7
N320
A40
N204
C713
C712
PM27
A
B
C
D
1
2
3
4

Map 3

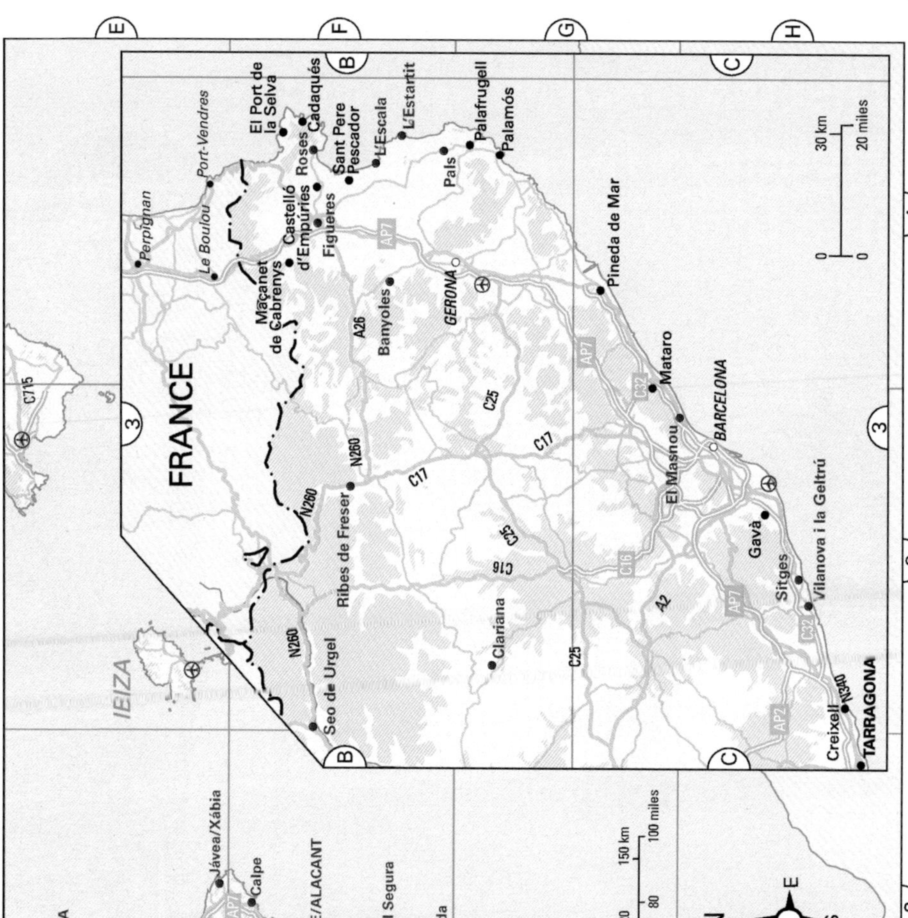
FRANCE
IEIZA
Perpignan
Le Boulou
Port-Vendres
El Port de la Selva
Cadaqués
Roses
Sant Pere Pescador
L'Escala
L'Estartit
Palafrugell
Palamós
Pals
Castelló d'Empúries
Figueres
Maçanet de Cabrenys
Banyoles
GERONA
Pineda de Mar
Mataro
El Masnou
BARCELONA
Gavà
Sitges
Vilanova i la Geltrú
Ribes de Freser
Seo de Urgel
Clariana
Creixell
TARRAGONA
AP7
A26
N260
C17
C25
C16
C32
A2
AP2
N340
C715
30 km
20 miles
VALENCIA
SUECA
Gandia
Jávea/Xàbia
Calpe
Benidorm
ALICANTE/ALACANT
Santa Pola
La Marina
Guardamar del Segura
Torrevieja
Pilar de la Horadada
La Manga del Mar Menor
Bocairent
Sax
Crevillent
San Fulgencio
Fortuna
Puerto de Mazarrón
Águilas
Lorca
Mojácar
Los Gallardos
María
ALMERÍA
Villargordo del Cabriel
Motilla del Palancar
ALBACETE
Alcaraz
Hornos
Cotorios
N320
N322
N330
N332
N344
A3
A7
A30
A31
A32
A33
A35
A43
A91
A92
A92N
AP36
N420
150 km
100 miles
N
E
S
W
Motorways
Major roads
Main Roads
All year site(s)
Seasonal site(s)
No sites listed
200m +
0–200m
© Collins Bartholomew Ltd 2019

Map 4

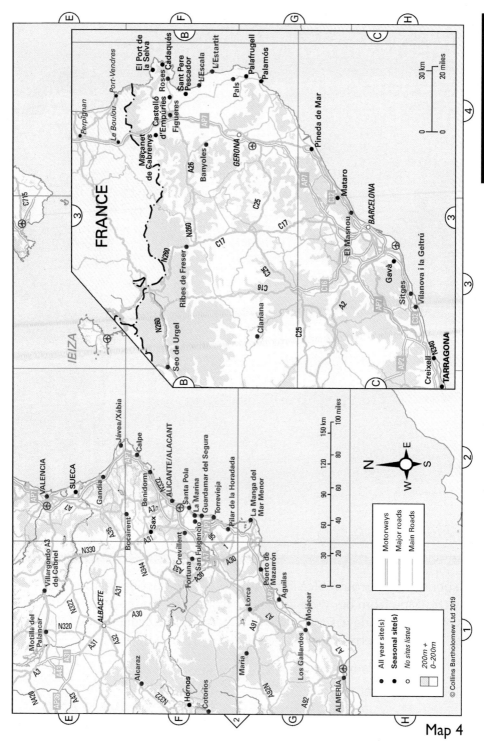

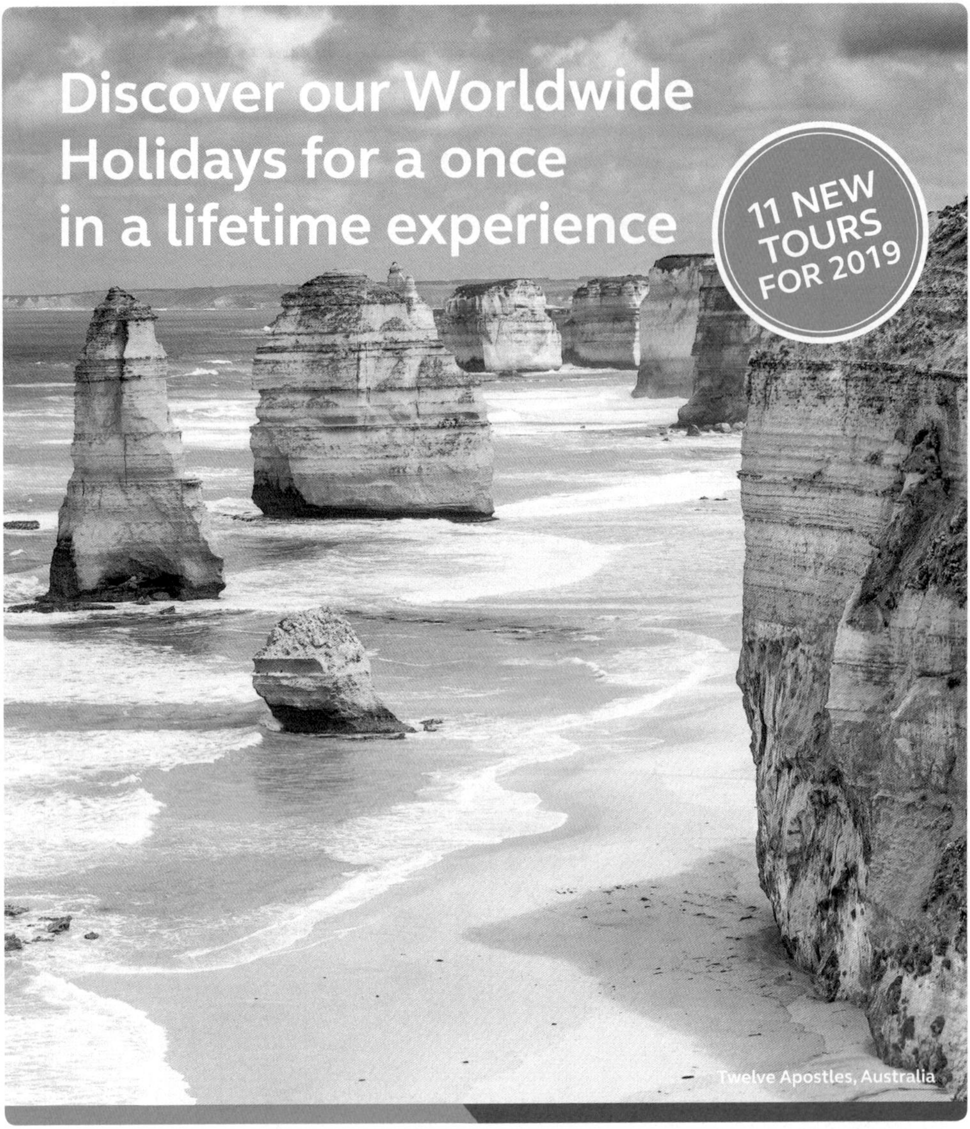

Discover our Worldwide Holidays for a once in a lifetime experience
11 NEW TOURS FOR 2019
Twelve Apostles, Australia

Experience unforgettable sights and create everlasting memories by choosing from our wide range of escorted and independent Worldwide Holidays. Covering USA, Canada, New Zealand and new for 2019 Southern Africa and Australia.

Visit camc.com/worldwide for more information

Club members call: 01342 779 349

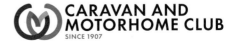

CARAVAN AND MOTORHOME CLUB
SINCE 1907

Site Report Form

If campsite is already listed, complete only those sections of the form where changes apply or alternatively use the Abbreviated Site Report form on the following pages.

Sites not reported on for 5 years may be deleted from the guide

Year of guide used	20..........	Is site listed?	Listed on page no............	Unlisted	Date of visit	......../........./.........

A – CAMPSITE NAME AND LOCATION

Country		Name of town/village site listed under *(see Sites Location Maps)*				
Distance & direction from centre of town site is listed under *(in a straight line)*	...................km	eg N, NE, S, SW	Urban	Rural	Coastal	
Site open all year?	Y / N	Period site is open *(if not all year)*	................../.................. to ..../.................			
Site name			Naturist site	Y / N		
Site address						
Telephone		Fax				
E-mail		Website				

B – CAMPSITE CHARGES

Charge for outfit + 2 adults in local currency	PRICE		

C – DIRECTIONS

Brief, specific directions to site (in km) *To convert miles to kilometres multiply by 8 and divide by 5 or use Conversion Table in guide*	
GPS	Latitude..*(eg 12.34567)* Longitude...*(eg 1.23456 or -1.23456)*

D – SITE INFORMATION

Dogs allowed	DOGS	Y / N	Price per night................................. *(if allowed)*	
Facilities for disabled	♦			
Public Transport within 5km	BUS / TRAM / TRAIN	Adj	Nearby	
Reduction long stay	RED LONG STAY	Credit Card accepted	CCARD ACC	
Advance bookings accepted/recommended/required		ADV BKG ACC / REC / REQ		
Camping Key Europe or Camping Card International accepted in lieu of passport			CKE/CCI	

E – SITE DESCRIPTION

SITE size ie number of pitches	Small Max 50	SM	Medium 51-150	MED	Large 151-500	LGE	Very large 500+	V LGE	Unchanged
Pitch features if **NOT** open-plan/grassy		HDG PITCH	Hedged	HDG PITCH	Marked or numbered	MKD PITCH	Hardstanding or gravel	HDSTG	Unchanged
If site is **NOT** level, is it		PT SL	Part sloping	PT SL	Sloping	SL	Terraced	TERR	Unchanged
Is site shaded?		SHD	Shaded	SHD	Part shaded	PT SHD	Unshaded	UNSHD	Unchanged
ELECTRIC HOOK UP *if not included in price above*		EL PNTS			Price...			Amps......................	
% Static caravans / mobile homes / chalets / cottages / fixed tents on site								................% STATICS	
Serviced Pitched		Y / N			Twin axles caravans allowed?			TWIN AXLES Y / N	

CUT ALONG DOTTED LINE

You can also complete forms online: camc.com/europereport

E – SITE DESCRIPTION CONTINUED…

Phone on site	PHONE		Wifi Internet	WIFI
Television	TV RM	TV CAB / SAT	Playground	PLAYGRND
Entertainment in high season	ENTMNT		English spoken	ENG SPKN
Motorhome Service Point	Y / N			

F – CATERING

Bar	BAR	On site	or	Within 2km	
Restaurant	REST	On site	or	Within 2km	
Shop(s)	SHOP(S)	On site	or	Within 2km	
Snack bar / take-away	SNACKS	On site	Y / N		
Cooking facilities	COOKING FACS	On site	Y / N		
Supplies of bottled gas on site	GAS	Y / N			
Barbecue allowed	BBQ	Charcoal	Gas	Elec	Sep area

G – SANITARY FACILITIES

WC	Heated	HTD WC	Continental	CONT	Own San recommended	OWN SAN REC
Chemical disposal point		CHEM DISP				
Hot shower(s)	SHWR(S)	Inc in site fee?		Y / N		
Child / baby facilities *(bathroom)*		FAM BTHRM		Launderette / Washing Machine		LNDRY

H – OTHER INFORMATION

Swimming pool	POOL	HEATED	COVERED	INDOOR	PADDLING POOL
Beach	BEACH	Adj	or ....km	Sand	Shingle
Alternative swimming *(lake)*	SW	Adj	or ....km	Sand	Shingle
Games /sports area / Games room	GAMES AREA		GAMES ROOM		

I – ADDITIONAL REMARKS AND/OR ITEMS OF INTEREST

Tourist attractions, unusual features or other facilities, eg waterslide, tennis, cycle hire, watersports, horseriding, separate car park, walking distance to shops etc	YOUR OPINION OF THE SITE:	
	EXCEL	
	VERY GOOD	
	GOOD	
	FAIR	POOR
	NIGHT HALT ONLY	
Your comments & opinions may be used in future editions of the guide, if you do not wish them to be used please tick		

J – MEMBER DETAILS

ARE YOU A:	Caravanner		Motorhomer		Trailer-tenter?	
NAME:		MEMBERSHIP NO:				
		POST CODE:				
DO YOU NEED MORE BLANK SITE REPORT FORMS?		YES			NO	

Please use a separate form for each campsite and do not send receipts. Owing to the large number of site reports received, it is not possible to enter into correspondence. Please return completed form to:

The Editor, Overseas Touring Guides, East Grinstead House
East Grinstead, West Sussex RH19 1UA

Please note that due to changes in the rules regarding freepost we are no longer able to provide a freepost address for the return of Site Report Forms. You can still supply your site reports free online by visiting camc.com/europereport. We apologise for any inconvenience this may cause.

Site Report Form

If campsite is already listed, complete only those sections of the form where changes apply or alternatively use the Abbreviated Site Report form on the following pages.

Sites not reported on for 5 years may be deleted from the guide

Year of guide used	20..........	Is site listed?	Listed on page no............	Unlisted	Date of visit	......../........./.........

A – CAMPSITE NAME AND LOCATION

Country			Name of town/village site listed under *(see Sites Location Maps)*			
Distance & direction from centre of town site is listed under *(in a straight line)*		....................km	eg N, NE, S, SW	Urban	Rural	Coastal
Site open all year?	Y / N	Period site is open *(if not all year)*	................../.................. to ..../..................			
Site name					Naturist site	Y / N
Site address						
Telephone			Fax			
E-mail			Website			

B – CAMPSITE CHARGES

Charge for outfit + 2 adults in local currency	PRICE		

C – DIRECTIONS

Brief, specific directions to site (in km) *To convert miles to kilometres multiply by 8 and divide by 5 or use Conversion Table in guide*	
GPS	Latitude..*(eg 12.34567)* Longitude...*(eg 1.23456 or -1.23456)*

D – SITE INFORMATION

Dogs allowed	DOGS	Y / N	Price per night .... *(if allowed)*	
Facilities for disabled	♦			
Public Transport within 5km	BUS / TRAM / TRAIN	Adj	Nearby	
Reduction long stay	RED LONG STAY	Credit Card accepted		CCARD ACC
Advance bookings accepted/recommended/required		ADV BKG ACC / REC / REQ		
Camping Key Europe or Camping Card International accepted in lieu of passport				CKE/CCI

E – SITE DESCRIPTION

SITE size ie number of pitches	Small Max 50	SM	Medium 51-150	MED	Large 151-500	LGE	Very large 500+	V LGE	Unchanged
Pitch features if **NOT** open-plan/grassy		HDG PITCH	Hedged	HDG PITCH	Marked or numbered	MKD PITCH	Hardstanding or gravel	HDSTG	Unchanged
If site is **NOT** level, is it		PT SL	Part sloping	PT SL	Sloping	SL	Terraced	TERR	Unchanged
Is site shaded?		SHD	Shaded	SHD	Part shaded	PT SHD	Unshaded	UNSHD	Unchanged
ELECTRIC HOOK UP *if not included in price above*		EL PNTS			Price...			Amps......................	
% Static caravans / mobile homes / chalets / cottages / fixed tents on site								................% STATICS	
Serviced Pitched		Y / N			Twin axles caravans allowed?			TWIN AXLES Y / N	

You can also complete forms online: camc.com/europereport

CUT ALONG DOTTED LINE

E – SITE DESCRIPTION CONTINUED...

Phone on site	PHONE		Wifi Internet	WIFI
Television	TV RM	TV CAB / SAT	Playground	PLAYGRND
Entertainment in high season	ENTMNT		English spoken	ENG SPKN
Motorhome Service Point	Y / N			

F – CATERING

Bar	BAR	On site	or	Within 2km	
Restaurant	REST	On site	or	Within 2km	
Shop(s)	SHOP(S)	On site	or	Within 2km	
Snack bar / take-away	SNACKS	On site	Y / N		
Cooking facilities	COOKING FACS	On site	Y / N		
Supplies of bottled gas on site	GAS	Y / N			
Barbecue allowed	BBQ	Charcoal	Gas	Elec	Sep area

G – SANITARY FACILITIES

WC	Heated	HTD WC	Continental	CONT	Own San recommended	OWN SAN REC
Chemical disposal point		CHEM DISP				
Hot shower(s)	SHWR(S)	Inc in site fee?		Y / N		
Child / baby facilities *(bathroom)*		FAM BTHRM		Launderette / Washing Machine		LNDRY

H – OTHER INFORMATION

Swimming pool	POOL	HEATED	COVERED	INDOOR	PADDLING POOL
Beach	BEACH	Adj	or ....km	Sand	Shingle
Alternative swimming *(lake)*	SW	Adj	or ....km	Sand	Shingle
Games /sports area / Games room	GAMES AREA		GAMES ROOM		

I – ADDITIONAL REMARKS AND/OR ITEMS OF INTEREST

Tourist attractions, unusual features or other facilities, eg waterslide, tennis, cycle hire, watersports, horseriding, separate car park, walking distance to shops etc	YOUR OPINION OF THE SITE:	
	EXCEL	
	VERY GOOD	
	GOOD	
	FAIR	POOR
	NIGHT HALT ONLY	
Your comments & opinions may be used in future editions of the guide, if you do not wish them to be used please tick		

J – MEMBER DETAILS

ARE YOU A:	Caravanner		Motorhomer		Trailer-tenter?	
NAME:			MEMBERSHIP NO:			
			POST CODE:			
DO YOU NEED MORE BLANK SITE REPORT FORMS?			YES		NO	

Please use a separate form for each campsite and do not send receipts. Owing to the large number of site reports received, it is not possible to enter into correspondence. Please return completed form to:

The Editor, Overseas Touring Guides, East Grinstead House
East Grinstead, West Sussex RH19 1UA

Please note that due to changes in the rules regarding freepost we are no longer able to provide a freepost address for the return of Site Report Forms. You can still supply your site reports free online by visiting camc.com/europereport. We apologise for any inconvenience this may cause.

Site Report Form

If campsite is already listed, complete only those sections of the form where changes apply or alternatively use the Abbreviated Site Report form on the following pages.

Sites not reported on for 5 years may be deleted from the guide

Year of guide used	20..........	Is site listed?	Listed on page no............	Unlisted	Date of visit	......../........./.........

A – CAMPSITE NAME AND LOCATION

Country			Name of town/village site listed under *(see Sites Location Maps)*			
Distance & direction from centre of town site is listed under *(in a straight line)*		...................km	eg N, NE, S, SW	Urban	Rural	Coastal
Site open all year?	Y / N	Period site is open *(if not all year)*	................./................. to ..../.................			
Site name					Naturist site	Y / N
Site address						
Telephone			Fax			
E-mail			Website			

B – CAMPSITE CHARGES

Charge for outfit + 2 adults in local currency	PRICE		

C – DIRECTIONS

Brief, specific directions to site (in km) *To convert miles to kilometres multiply by 8 and divide by 5 or use Conversion Table in guide*	
GPS	Latitude..*(eg 12.34567)* Longitude..*(eg 1.23456 or -1.23456)*

D – SITE INFORMATION

Dogs allowed	DOGS	Y / N	Price per night .... *(if allowed)*
Facilities for disabled	♦		
Public Transport within 5km	BUS / TRAM / TRAIN	Adj	Nearby
Reduction long stay	RED LONG STAY	Credit Card accepted	CCARD ACC
Advance bookings accepted/recommended/required		ADV BKG ACC / REC / REQ	
Camping Key Europe or Camping Card International accepted in lieu of passport			CKE/CCI

E – SITE DESCRIPTION

SITE size ie number of pitches	Small Max 50	SM	Medium 51-150	MED	Large 151-500	LGE	Very large 500+	V LGE	Unchanged
Pitch features if **NOT** open-plan/grassy		HDG PITCH	Hedged	HDG PITCH	Marked or numbered	MKD PITCH	Hardstanding or gravel	HDSTG	Unchanged
If site is **NOT** level, is it		PT SL	Part sloping	PT SL	Sloping	SL	Terraced	TERR	Unchanged
Is site shaded?		SHD	Shaded	SHD	Part shaded	PT SHD	Unshaded	UNSHD	Unchanged
ELECTRIC HOOK UP *if not included in price above*		EL PNTS			Price..			Amps......................	
% Static caravans / mobile homes / chalets / cottages / fixed tents on site								................% STATICS	
Serviced Pitched		Y / N			Twin axles caravans allowed?			TWIN AXLES Y / N	

CUT ALONG DOTTED LINE

You can also complete forms online: camc.com/europereport

E – SITE DESCRIPTION CONTINUED...

Phone on site	PHONE		Wifi Internet	WIFI
Television	TV RM	TV CAB / SAT	Playground	PLAYGRND
Entertainment in high season	ENTMNT		English spoken	ENG SPKN
Motorhome Service Point	Y / N			

F – CATERING

Bar	BAR	On site	or	Within 2km	
Restaurant	REST	On site	or	Within 2km	
Shop(s)	SHOP(S)	On site	or	Within 2km	
Snack bar / take-away	SNACKS	On site	Y / N		
Cooking facilities	COOKING FACS	On site	Y / N		
Supplies of bottled gas on site	GAS	Y / N			
Barbecue allowed	BBQ	Charcoal	Gas	Elec	Sep area

G – SANITARY FACILITIES

WC	Heated	HTD WC	Continental	CONT	Own San recommended	OWN SAN REC
Chemical disposal point		CHEM DISP				
Hot shower(s)	SHWR(S)	Inc in site fee?		Y / N		
Child / baby facilities *(bathroom)*		FAM BTHRM		Launderette / Washing Machine		LNDRY

H – OTHER INFORMATION

Swimming pool	POOL	HEATED	COVERED	INDOOR	PADDLING POOL
Beach	BEACH	Adj	or ....km	Sand	Shingle
Alternative swimming *(lake)*	SW	Adj	or ....km	Sand	Shingle
Games /sports area / Games room	GAMES AREA		GAMES ROOM		

I – ADDITIONAL REMARKS AND/OR ITEMS OF INTEREST

Tourist attractions, unusual features or other facilities, eg waterslide, tennis, cycle hire, watersports, horseriding, separate car park, walking distance to shops etc	YOUR OPINION OF THE SITE:	
	EXCEL	
	VERY GOOD	
	GOOD	
	FAIR	POOR
	NIGHT HALT ONLY	
Your comments & opinions may be used in future editions of the guide, if you do not wish them to be used please tick		

J – MEMBER DETAILS

ARE YOU A:	Caravanner		Motorhomer		Trailer-tenter?	
NAME:			MEMBERSHIP NO:			
			POST CODE:			
DO YOU NEED MORE BLANK SITE REPORT FORMS?			YES		NO	

Please use a separate form for each campsite and do not send receipts. Owing to the large number of site reports received, it is not possible to enter into correspondence. Please return completed form to:

The Editor, Overseas Touring Guides, East Grinstead House
East Grinstead, West Sussex RH19 1UA

Please note that due to changes in the rules regarding freepost we are no longer able to provide a freepost address for the return of Site Report Forms. You can still supply your site reports free online by visiting camc.com/europereport. We apologise for any inconvenience this may cause.

Site Report Form

If campsite is already listed, complete only those sections of the form where changes apply or alternatively use the Abbreviated Site Report form on the following pages.

Sites not reported on for 5 years may be deleted from the guide

Year of guide used	20..........	Is site listed?	Listed on page no............	Unlisted	Date of visit	......../........./.........

A – CAMPSITE NAME AND LOCATION

Country		Name of town/village site listed under *(see Sites Location Maps)*				
Distance & direction from centre of town site is listed under *(in a straight line)*	....................km	eg N, NE, S, SW	Urban	Rural	Coastal	
Site open all year?	Y / N	Period site is open *(if not all year)*	................../.................. to ..../..................			
Site name			Naturist site	Y / N		
Site address						
Telephone		Fax				
E-mail		Website				

B – CAMPSITE CHARGES

Charge for outfit + 2 adults in local currency	PRICE		

C – DIRECTIONS

Brief, specific directions to site (in km) *To convert miles to kilometres multiply by 8 and divide by 5 or use Conversion Table in guide*	
GPS	Latitude..*(eg 12.34567)* Longitude..*(eg 1.23456 or -1.23456)*

D – SITE INFORMATION

Dogs allowed	DOGS	Y / N	Price per night................................ *(if allowed)*
Facilities for disabled	♦		
Public Transport within 5km	BUS / TRAM / TRAIN	Adj	Nearby
Reduction long stay	RED LONG STAY	Credit Card accepted	CCARD ACC
Advance bookings accepted/recommended/required		ADV BKG ACC / REC / REQ	
Camping Key Europe or Camping Card International accepted in lieu of passport			CKE/CCI

E – SITE DESCRIPTION

SITE size ie number of pitches	Small Max 50	SM	Medium 51-150	MED	Large 151-500	LGE	Very large 500+	V LGE	Unchanged
Pitch features if **NOT** open-plan/grassy		HDG PITCH	Hedged	HDG PITCH	Marked or numbered	MKD PITCH	Hardstanding or gravel	HDSTG	Unchanged
If site is **NOT** level, is it		PT SL	Part sloping	PT SL	Sloping	SL	Terraced	TERR	Unchanged
Is site shaded?		SHD	Shaded	SHD	Part shaded	PT SHD	Unshaded	UNSHD	Unchanged
ELECTRIC HOOK UP *if not included in price above*		EL PNTS			Price..			Amps......................	
% Static caravans / mobile homes / chalets / cottages / fixed tents on site								................% STATICS	
Serviced Pitched		Y / N			Twin axles caravans allowed?			TWIN AXLES Y / N	

You can also complete forms online: camc.com/europereport

CUT ALONG DOTTED LINE

E – SITE DESCRIPTION CONTINUED...

Phone on site	PHONE		Wifi Internet	WIFI
Television	TV RM	TV CAB / SAT	Playground	PLAYGRND
Entertainment in high season	ENTMNT		English spoken	ENG SPKN
Motorhome Service Point	Y / N			

F – CATERING

Bar	BAR	On site	or	Within 2km	
Restaurant	REST	On site	or	Within 2km	
Shop(s)	SHOP(S)	On site	or	Within 2km	
Snack bar / take-away	SNACKS	On site	Y / N		
Cooking facilities	COOKING FACS	On site	Y / N		
Supplies of bottled gas on site	GAS	Y / N			
Barbecue allowed	BBQ	Charcoal	Gas	Elec	Sep area

G – SANITARY FACILITIES

WC	Heated	HTD WC	Continental	CONT	Own San recommended	OWN SAN REC
Chemical disposal point		CHEM DISP				
Hot shower(s)	SHWR(S)	Inc in site fee?		Y / N		
Child / baby facilities *(bathroom)*		FAM BTHRM		Laundrette / Washing Machine		LNDRY

H – OTHER INFORMATION

Swimming pool	POOL	HEATED	COVERED	INDOOR	PADDLING POOL
Beach	BEACH	Adj	or ....km	Sand	Shingle
Alternative swimming *(lake)*	SW	Adj	or ....km	Sand	Shingle
Games /sports area / Games room	GAMES AREA		GAMES ROOM		

I – ADDITIONAL REMARKS AND/OR ITEMS OF INTEREST

Tourist attractions, unusual features or other facilities, eg waterslide, tennis, cycle hire, watersports, horseriding, separate car park, walking distance to shops etc	YOUR OPINION OF THE SITE:	
	EXCEL	
	VERY GOOD	
	GOOD	
	FAIR	POOR
	NIGHT HALT ONLY	
Your comments & opinions may be used in future editions of the guide, if you do not wish them to be used please tick		

J – MEMBER DETAILS

ARE YOU A:	Caravanner		Motorhomer		Trailer-tenter?	
NAME:			MEMBERSHIP NO:			
			POST CODE:			
DO YOU NEED MORE BLANK SITE REPORT FORMS?			YES		NO	

Please use a separate form for each campsite and do not send receipts. Owing to the large number of site reports received, it is not possible to enter into correspondence. Please return completed form to:

The Editor, Overseas Touring Guides, East Grinstead House
East Grinstead, West Sussex RH19 1UA

Please note that due to changes in the rules regarding freepost we are no longer able to provide a freepost address for the return of Site Report Forms. You can still supply your site reports free online by visiting camc.com/europereport. We apologise for any inconvenience this may cause.

Site Report Form

If campsite is already listed, complete only those sections of the form where changes apply or alternatively use the Abbreviated Site Report form on the following pages.

Sites not reported on for 5 years may be deleted from the guide

Year of guide used	20..........	Is site listed?	Listed on page no...........	Unlisted	Date of visit	......../........./.........

A – CAMPSITE NAME AND LOCATION

Country			Name of town/village site listed under *(see Sites Location Maps)*			
Distance & direction from centre of town site is listed under *(in a straight line)*		..................km	eg N, NE, S, SW	Urban	Rural	Coastal
Site open all year?	Y / N	Period site is open *(if not all year)*	................./................. to ..../.................			
Site name					Naturist site	Y / N
Site address						
Telephone			Fax			
E-mail			Website			

B – CAMPSITE CHARGES

Charge for outfit + 2 adults in local currency	PRICE		

C – DIRECTIONS

Brief, specific directions to site (in km) *To convert miles to kilometres multiply by 8 and divide by 5 or use Conversion Table in guide*	
GPS	Latitude..*(eg 12.34567)* Longitude...*(eg 1.23456 or -1.23456)*

D – SITE INFORMATION

Dogs allowed	DOGS	Y / N	Price per night .... *(if allowed)*
Facilities for disabled	♦		
Public Transport within 5km	BUS / TRAM / TRAIN	Adj	Nearby
Reduction long stay	RED LONG STAY	Credit Card accepted	CCARD ACC
Advance bookings accepted/recommended/required		ADV BKG ACC / REC / REQ	
Camping Key Europe or Camping Card International accepted in lieu of passport			CKE/CCI

E – SITE DESCRIPTION

SITE size ie number of pitches	Small Max 50	SM	Medium 51-150	MED	Large 151-500	LGE	Very large 500+	V LGE	Unchanged
Pitch features if **NOT** open-plan/grassy		HDG PITCH	Hedged	HDG PITCH	Marked or numbered	MKD PITCH	Hardstanding or gravel	HDSTG	Unchanged
If site is **NOT** level, is it		PT SL	Part sloping	PT SL	Sloping	SL	Terraced	TERR	Unchanged
Is site shaded?		SHD	Shaded	SHD	Part shaded	PT SHD	Unshaded	UNSHD	Unchanged
ELECTRIC HOOK UP *if not included in price above*		EL PNTS			Price..			Amps......................	
% Static caravans / mobile homes / chalets / cottages / fixed tents on site								................% STATICS	
Serviced Pitched		Y / N			Twin axles caravans allowed?			TWIN AXLES Y / N	

CUT ALONG DOTTED LINE

You can also complete forms online: camc.com/europereport

E – SITE DESCRIPTION CONTINUED...

Phone on site	PHONE		Wifi Internet	WIFI
Television	TV RM	TV CAB / SAT	Playground	PLAYGRND
Entertainment in high season	ENTMNT		English spoken	ENG SPKN
Motorhome Service Point	Y / N			

F – CATERING

Bar	BAR	On site	or	Within 2km
Restaurant	REST	On site	or	Within 2km
Shop(s)	SHOP(S)	On site	or	Within 2km
Snack bar / take-away	SNACKS	On site	Y / N	
Cooking facilities	COOKING FACS	On site	Y / N	
Supplies of bottled gas on site	GAS	Y / N		
Barbecue allowed	BBQ	Charcoal	Gas	Elec	Sep area

G – SANITARY FACILITIES

WC	Heated	HTD WC	Continental	CONT	Own San recommended	OWN SAN REC
Chemical disposal point		CHEM DISP				
Hot shower(s)	SHWR(S)	Inc in site fee?		Y / N		
Child / baby facilities *(bathroom)*		FAM BTHRM		Launderette / Washing Machine		LNDRY

H – OTHER INFORMATION

Swimming pool	POOL	HEATED	COVERED	INDOOR	PADDLING POOL
Beach	BEACH	Adj	or ....km	Sand	Shingle
Alternative swimming *(lake)*	SW	Adj	or ....km	Sand	Shingle
Games /sports area / Games room	GAMES AREA		GAMES ROOM		

I – ADDITIONAL REMARKS AND/OR ITEMS OF INTEREST

Tourist attractions, unusual features or other facilities, eg waterslide, tennis, cycle hire, watersports, horseriding, separate car park, walking distance to shops etc	YOUR OPINION OF THE SITE:	
	EXCEL	
	VERY GOOD	
	GOOD	
	FAIR	POOR
	NIGHT HALT ONLY	
Your comments & opinions may be used in future editions of the guide, if you do not wish them to be used please tick		

J – MEMBER DETAILS

ARE YOU A:	Caravanner		Motorhomer		Trailer-tenter?	
NAME:			MEMBERSHIP NO:			
			POST CODE:			
DO YOU NEED MORE BLANK SITE REPORT FORMS?			YES		NO	

Please use a separate form for each campsite and do not send receipts. Owing to the large number of site reports received, it is not possible to enter into correspondence. Please return completed form to:

The Editor, Overseas Touring Guides, East Grinstead House
East Grinstead, West Sussex RH19 1UA

Please note that due to changes in the rules regarding freepost we are no longer able to provide a freepost address for the return of Site Report Forms. You can still supply your site reports free online by visiting camc.com/europereport. We apologise for any inconvenience this may cause.

Site Report Form

If campsite is already listed, complete only those sections of the form where changes apply or alternatively use the Abbreviated Site Report form on the following pages.

Sites not reported on for 5 years may be deleted from the guide

Year of guide used	20..........	Is site listed?	Listed on page no...........	Unlisted	Date of visit	......../........./.........

A – CAMPSITE NAME AND LOCATION

Country		Name of town/village site listed under *(see Sites Location Maps)*				
Distance & direction from centre of town site is listed under *(in a straight line)*	...................km	eg N, NE, S, SW	Urban	Rural	Coastal	
Site open all year?	Y / N	Period site is open *(if not all year)*	................../.................. to ..../..................			
Site name				Naturist site	Y / N	
Site address						
Telephone		Fax				
E-mail		Website				

B – CAMPSITE CHARGES

Charge for outfit + 2 adults in local currency	PRICE		

C – DIRECTIONS

Brief, specific directions to site (in km) *To convert miles to kilometres multiply by 8 and divide by 5 or use Conversion Table in guide*	
GPS	Latitude...*(eg 12.34567)* Longitude..*(eg 1.23456 or -1.23456)*

D – SITE INFORMATION

Dogs allowed	DOGS	Y / N	Price per night................................ *(if allowed)*	
Facilities for disabled	♦			
Public Transport within 5km	BUS / TRAM / TRAIN	Adj	Nearby	
Reduction long stay	RED LONG STAY	Credit Card accepted	CCARD ACC	
Advance bookings accepted/recommended/required		ADV BKG ACC / REC / REQ		
Camping Key Europe or Camping Card International accepted in lieu of passport			CKE/CCI	

E – SITE DESCRIPTION

SITE size ie number of pitches	Small Max 50	SM	Medium 51-150	MED	Large 151-500	LGE	Very large 500+	V LGE	Unchanged
Pitch features if **NOT** open-plan/grassy		HDG PITCH	Hedged	HDG PITCH	Marked or numbered	MKD PITCH	Hardstanding or gravel	HDSTG	Unchanged
If site is **NOT** level, is it		PT SL	Part sloping	PT SL	Sloping	SL	Terraced	TERR	Unchanged
Is site shaded?		SHD	Shaded	SHD	Part shaded	PT SHD	Unshaded	UNSHD	Unchanged
ELECTRIC HOOK UP *if not included in price above*		EL PNTS			Price...			Amps......................	
% Static caravans / mobile homes / chalets / cottages / fixed tents on site								................% STATICS	
Serviced Pitched		Y / N			Twin axles caravans allowed?			TWIN AXLES Y / N	

You can also complete forms online: camc.com/europereport

CUT ALONG DOTTED LINE

E – SITE DESCRIPTION CONTINUED...

Phone on site	PHONE		Wifi Internet	WIFI
Television	TV RM	TV CAB / SAT	Playground	PLAYGRND
Entertainment in high season	ENTMNT		English spoken	ENG SPKN
Motorhome Service Point	Y / N			

F – CATERING

Bar	BAR	On site	or	Within 2km
Restaurant	REST	On site	or	Within 2km
Shop(s)	SHOP(S)	On site	or	Within 2km
Snack bar / take-away	SNACKS	On site	Y / N	
Cooking facilities	COOKING FACS	On site	Y / N	
Supplies of bottled gas on site	GAS	Y / N		
Barbecue allowed	BBQ	Charcoal	Gas	Elec	Sep area

G – SANITARY FACILITIES

WC	Heated	HTD WC	Continental	CONT	Own San recommended	OWN SAN REC
Chemical disposal point		CHEM DISP				
Hot shower(s)	SHWR(S)	Inc in site fee?		Y / N		
Child / baby facilities *(bathroom)*		FAM BTHRM		Launderette / Washing Machine		LNDRY

H – OTHER INFORMATION

Swimming pool	POOL	HEATED	COVERED	INDOOR	PADDLING POOL
Beach	BEACH	Adj	or ....km	Sand	Shingle
Alternative swimming *(lake)*	SW	Adj	or ....km	Sand	Shingle
Games /sports area / Games room	GAMES AREA		GAMES ROOM		

I – ADDITIONAL REMARKS AND/OR ITEMS OF INTEREST

Tourist attractions, unusual features or other facilities, eg waterslide, tennis, cycle hire, watersports, horseriding, separate car park, walking distance to shops etc	YOUR OPINION OF THE SITE:	
	EXCEL	
	VERY GOOD	
	GOOD	
	FAIR	POOR
	NIGHT HALT ONLY	
Your comments & opinions may be used in future editions of the guide, if you do not wish them to be used please tick		

J – MEMBER DETAILS

ARE YOU A:	Caravanner		Motorhomer		Trailer-tenter?	
NAME:		MEMBERSHIP NO:			
		POST CODE:			
DO YOU NEED MORE BLANK SITE REPORT FORMS?		YES		NO	

Please use a separate form for each campsite and do not send receipts. Owing to the large number of site reports received, it is not possible to enter into correspondence. Please return completed form to:

The Editor, Overseas Touring Guides, East Grinstead House
East Grinstead, West Sussex RH19 1UA

Please note that due to changes in the rules regarding freepost we are no longer able to provide a freepost address for the return of Site Report Forms. You can still supply your site reports free online by visiting camc.com/europereport. We apologise for any inconvenience this may cause.

Site Report Form

If campsite is already listed, complete only those sections of the form where changes apply or alternatively use the Abbreviated Site Report form on the following pages.

Sites not reported on for 5 years may be deleted from the guide

Year of guide used	20..........	Is site listed?	Listed on page no............	Unlisted	Date of visit	......../........./.........

A – CAMPSITE NAME AND LOCATION

Country			Name of town/village site listed under *(see Sites Location Maps)*				
Distance & direction from centre of town site is listed under *(in a straight line)*			....................km	eg N, NE, S, SW	Urban	Rural	Coastal
Site open all year?	Y / N	Period site is open *(if not all year)*	................../.................. to ..../..................				
Site name					Naturist site	Y / N	
Site address							
Telephone			Fax				
E-mail			Website				

B – CAMPSITE CHARGES

Charge for outfit + 2 adults in local currency	PRICE		

C – DIRECTIONS

Brief, specific directions to site (in km) *To convert miles to kilometres multiply by 8 and divide by 5 or use Conversion Table in guide*	
GPS	Latitude..*(eg 12.34567)* Longitude...*(eg 1.23456 or -1.23456)*

D – SITE INFORMATION

Dogs allowed	DOGS	Y / N	Price per night................................ *(if allowed)*	
Facilities for disabled	♦			
Public Transport within 5km	BUS / TRAM / TRAIN	Adj	Nearby	
Reduction long stay	RED LONG STAY	Credit Card accepted	CCARD ACC	
Advance bookings accepted/recommended/required		ADV BKG ACC / REC / REQ		
Camping Key Europe or Camping Card International accepted in lieu of passport			CKE/CCI	

E – SITE DESCRIPTION

SITE size ie number of pitches	Small Max 50	SM	Medium 51-150	MED	Large 151-500	LGE	Very large 500+	V LGE	Unchanged
Pitch features if **NOT** open-plan/grassy		HDG PITCH	Hedged	HDG PITCH	Marked or numbered	MKD PITCH	Hardstanding or gravel	HDSTG	Unchanged
If site is **NOT** level, is it		PT SL	Part sloping	PT SL	Sloping	SL	Terraced	TERR	Unchanged
Is site shaded?		SHD	Shaded	SHD	Part shaded	PT SHD	Unshaded	UNSHD	Unchanged
ELECTRIC HOOK UP *if not included in price above*		EL PNTS			Price...			Amps......................	
% Static caravans / mobile homes / chalets / cottages / fixed tents on site								................% STATICS	
Serviced Pitched		Y / N			Twin axles caravans allowed?			TWIN AXLES Y / N	

You can also complete forms online: camc.com/europereport

CUT ALONG DOTTED LINE

E – SITE DESCRIPTION CONTINUED...

Phone on site	PHONE		Wifi Internet	WIFI
Television	TV RM	TV CAB / SAT	Playground	PLAYGRND
Entertainment in high season	ENTMNT		English spoken	ENG SPKN
Motorhome Service Point	Y / N			

F – CATERING

Bar	BAR	On site	or	Within 2km	
Restaurant	REST	On site	or	Within 2km	
Shop(s)	SHOP(S)	On site	or	Within 2km	
Snack bar / take-away	SNACKS	On site	Y / N		
Cooking facilities	COOKING FACS	On site	Y / N		
Supplies of bottled gas on site	GAS	Y / N			
Barbecue allowed	BBQ	Charcoal	Gas	Elec	Sep area

G – SANITARY FACILITIES

WC	Heated	HTD WC	Continental	CONT	Own San recommended	OWN SAN REC
Chemical disposal point		CHEM DISP				
Hot shower(s)	SHWR(S)	Inc in site fee?		Y / N		
Child / baby facilities *(bathroom)*		FAM BTHRM		Launderette / Washing Machine		LNDRY

H – OTHER INFORMATION

Swimming pool	POOL	HEATED	COVERED	INDOOR	PADDLING POOL
Beach	BEACH	Adj	or ....km	Sand	Shingle
Alternative swimming *(lake)*	SW	Adj	or ....km	Sand	Shingle
Games /sports area / Games room	GAMES AREA		GAMES ROOM		

I – ADDITIONAL REMARKS AND/OR ITEMS OF INTEREST

Tourist attractions, unusual features or other facilities, eg waterslide, tennis, cycle hire, watersports, horseriding, separate car park, walking distance to shops etc	YOUR OPINION OF THE SITE:	
	EXCEL	
	VERY GOOD	
	GOOD	
	FAIR	POOR
	NIGHT HALT ONLY	
Your comments & opinions may be used in future editions of the guide, if you do not wish them to be used please tick		

J – MEMBER DETAILS

ARE YOU A:	Caravanner		Motorhomer		Trailer-tenter?		
NAME:		MEMBERSHIP NO:					
		POST CODE:					
DO YOU NEED MORE BLANK SITE REPORT FORMS?		YES		NO			

Please use a separate form for each campsite and do not send receipts. Owing to the large number of site reports received, it is not possible to enter into correspondence. Please return completed form to:

The Editor, Overseas Touring Guides, East Grinstead House
East Grinstead, West Sussex RH19 1UA

Please note that due to changes in the rules regarding freepost we are no longer able to provide a freepost address for the return of Site Report Forms. You can still supply your site reports free online by visiting camc.com/europereport. We apologise for any inconvenience this may cause.

Site Report Form

If campsite is already listed, complete only those sections of the form where changes apply or alternatively use the Abbreviated Site Report form on the following pages.

Sites not reported on for 5 years may be deleted from the guide

Year of guide used	20..........	Is site listed?	Listed on page no............	Unlisted	Date of visit	......../........./.........

A – CAMPSITE NAME AND LOCATION

Country		Name of town/village site listed under *(see Sites Location Maps)*				
Distance & direction from centre of town site is listed under *(in a straight line)*		...................km	eg N, NE, S, SW	Urban	Rural	Coastal
Site open all year?	Y / N	Period site is open *(if not all year)*	................./................. to ..../.................			
Site name					Naturist site	Y / N
Site address						
Telephone			Fax			
E-mail			Website			

B – CAMPSITE CHARGES

Charge for outfit + 2 adults in local currency	PRICE		

C – DIRECTIONS

Brief, specific directions to site (in km) *To convert miles to kilometres multiply by 8 and divide by 5 or use Conversion Table in guide*	
GPS	Latitude...*(eg 12.34567)* Longitude...*(eg 1.23456 or -1.23456)*

D – SITE INFORMATION

Dogs allowed	DOGS	Y / N	Price per night .... *(if allowed)*
Facilities for disabled	♦		
Public Transport within 5km	BUS / TRAM / TRAIN	Adj	Nearby
Reduction long stay	RED LONG STAY	Credit Card accepted	CCARD ACC
Advance bookings accepted/recommended/required		ADV BKG ACC / REC / REQ	
Camping Key Europe or Camping Card International accepted in lieu of passport			CKE/CCI

E – SITE DESCRIPTION

SITE size ie number of pitches	Small Max 50	SM	Medium 51-150	MED	Large 151-500	LGE	Very large 500+	V LGE	Unchanged
Pitch features if **NOT** open-plan/grassy		HDG PITCH	Hedged	HDG PITCH	Marked or numbered	MKD PITCH	Hardstanding or gravel	HDSTG	Unchanged
If site is **NOT** level, is it		PT SL	Part sloping	PT SL	Sloping	SL	Terraced	TERR	Unchanged
Is site shaded?		SHD	Shaded	SHD	Part shaded	PT SHD	Unshaded	UNSHD	Unchanged
ELECTRIC HOOK UP *if not included in price above*		EL PNTS			Price..			Amps......................	
% Static caravans / mobile homes / chalets / cottages / fixed tents on site								................% STATICS	
Serviced Pitched		Y / N			Twin axles caravans allowed?			TWIN AXLES Y / N	

You can also complete forms online: camc.com/europereport

CUT ALONG DOTTED LINE

E – SITE DESCRIPTION CONTINUED...

Phone on site	PHONE		Wifi Internet	WIFI
Television	TV RM	TV CAB / SAT	Playground	PLAYGRND
Entertainment in high season	ENTMNT		English spoken	ENG SPKN
Motorhome Service Point	Y / N			

F – CATERING

Bar	BAR	On site	or	Within 2km	
Restaurant	REST	On site	or	Within 2km	
Shop(s)	SHOP(S)	On site	or	Within 2km	
Snack bar / take-away	SNACKS	On site	Y / N		
Cooking facilities	COOKING FACS	On site	Y / N		
Supplies of bottled gas on site	GAS	Y / N			
Barbecue allowed	BBQ	Charcoal	Gas	Elec	Sep area

G – SANITARY FACILITIES

WC	Heated	HTD WC	Continental	CONT	Own San recommended	OWN SAN REC
Chemical disposal point		CHEM DISP				
Hot shower(s)	SHWR(S)	Inc in site fee?		Y / N		
Child / baby facilities *(bathroom)*		FAM BTHRM		Launderette / Washing Machine		LNDRY

H – OTHER INFORMATION

Swimming pool	POOL	HEATED	COVERED	INDOOR	PADDLING POOL
Beach	BEACH	Adj	or ....km	Sand	Shingle
Alternative swimming *(lake)*	SW	Adj	or ....km	Sand	Shingle
Games /sports area / Games room	GAMES AREA		GAMES ROOM		

I – ADDITIONAL REMARKS AND/OR ITEMS OF INTEREST

Tourist attractions, unusual features or other facilities, eg waterslide, tennis, cycle hire, watersports, horseriding, separate car park, walking distance to shops etc	YOUR OPINION OF THE SITE:	
	EXCEL	
	VERY GOOD	
	GOOD	
	FAIR	POOR
	NIGHT HALT ONLY	
Your comments & opinions may be used in future editions of the guide, if you do not wish them to be used please tick		

J – MEMBER DETAILS

ARE YOU A:	Caravanner		Motorhomer		Trailer-tenter?	
NAME:		MEMBERSHIP NO:				
		POST CODE:				
DO YOU NEED MORE BLANK SITE REPORT FORMS?		YES			NO	

Please use a separate form for each campsite and do not send receipts. Owing to the large number of site reports received, it is not possible to enter into correspondence. Please return completed form to:

The Editor, Overseas Touring Guides, East Grinstead House
East Grinstead, West Sussex RH19 1UA

Please note that due to changes in the rules regarding freepost we are no longer able to provide a freepost address for the return of Site Report Forms. You can still supply your site reports free online by visiting camc.com/europereport. We apologise for any inconvenience this may cause.

Abbreviated Site Report Form

Use this abbreviated Site Report Form if you have visited a number of sites and there are no changes (or only small changes) to their entries in the guide. If reporting on a new site, or reporting several changes, please use the full version of the report form. **If advising prices,** these should be for an outfit, and 2 adults for one night's stay. **Please indicate high or low season prices and whether electricity is included.**

Remember, if you don't tell us about sites you have visited, they may eventually be deleted from the guide.

Year of guide used	20..........	Page No.	....	Name of town/village site listed under	
Site Name				Date of visit	......./......./........
GPS	Latitude..(*eg 12.34567*) Longitude..(*eg 1.23456 or -1.23456*)				
Site is in: Andorra / Austria / Belgium / Croatia / Czech Republic / Denmark / Finland / France / Germany / Greece / Hungary / Italy / Luxembourg / Netherlands / Norway / Poland / Portugal / Slovakia / Slovenia / Spain / Sweden / Switzerland					
Comments:					
Charge for outfit + 2 adults in local currency	High Season	Low Season	Elec inc in price?	Y / N	.............amps
			Price of elec (if not inc)		.............amps

Year of guide used	20..........	Page No.	....	Name of town/village site listed under	
Site Name				Date of visit	......./......./........
GPS	Latitude..(*eg 12.34567*) Longitude..(*eg 1.23456 or -1.23456*)				
Site is in: Andorra / Austria / Belgium / Croatia / Czech Republic / Denmark / Finland / France / Germany / Greece / Hungary / Italy / Luxembourg / Netherlands / Norway / Poland / Portugal / Slovakia / Slovenia / Spain / Sweden / Switzerland					
Comments:					
Charge for outfit + 2 adults in local currency	High Season	Low Season	Elec inc in price?	Y / N	.............amps
			Price of elec (if not inc)		.............amps

Year of guide used	20..........	Page No.	....	Name of town/village site listed under	
Site Name				Date of visit	......./......./........
GPS	Latitude..(*eg 12.34567*) Longitude..(*eg 1.23456 or -1.23456*)				
Site is in: Andorra / Austria / Belgium / Croatia / Czech Republic / Denmark / Finland / France / Germany / Greece / Hungary / Italy / Luxembourg / Netherlands / Norway / Poland / Portugal / Slovakia / Slovenia / Spain / Sweden / Switzerland					
Comments:					
Charge for car, caravan & 2 adults in local currency	High Season	Low Season	Elec inc in price?	Y / N	.............amps
			Price of elec (if not inc)		.............amps

Please fill in your details and send to the address on the reverse of this form.

You can also complete forms online: camc.com/europereport

CUT ALONG DOTTED LINE

Year of guide used	20..........	Page No.	....	Name of town/village site listed under	
Site Name				Date of visit	......./......./........
GPS	Latitude..*(eg 12.34567)* Longitude..*(eg 1.23456 or -1.23456)*				
Site is in: Andorra / Austria / Belgium / Croatia / Czech Republic / Denmark / Finland / France / Germany / Greece / Hungary / Italy / Luxembourg / Netherlands / Norway / Poland / Portugal / Slovakia / Slovenia / Spain / Sweden / Switzerland					
Comments:					
Charge for outfit + 2 adults in local currency	High Season	Low Season	Elec inc in price?	Y / N	............amps
			Price of elec (if not inc)		............amps

Year of guide used	20..........	Page No.	....	Name of town/village site listed under	
Site Name				Date of visit	......./......./........
GPS	Latitude..*(eg 12.34567)* Longitude..*(eg 1.23456 or -1.23456)*				
Site is in: Andorra / Austria / Belgium / Croatia / Czech Republic / Denmark / Finland / France / Germany / Greece / Hungary / Italy / Luxembourg / Netherlands / Norway / Poland / Portugal / Slovakia / Slovenia / Spain / Sweden / Switzerland					
Comments:					
Charge for outfit + 2 adults in local currency	High Season	Low Season	Elec inc in price?	Y / N	............amps
			Price of elec (if not inc)		............amps

Year of guide used	20..........	Page No.	....	Name of town/village site listed under	
Site Name				Date of visit	......./......./........
GPS	Latitude..*(eg 12.34567)* Longitude..*(eg 1.23456 or -1.23456)*				
Site is in: Andorra / Austria / Belgium / Croatia / Czech Republic / Denmark / Finland / France / Germany / Greece / Hungary / Italy / Luxembourg / Netherlands / Norway / Poland / Portugal / Slovakia / Slovenia / Spain / Sweden / Switzerland					
Comments:					
Charge for outfit + 2 adults in local currency	High Season	Low Season	Elec inc in price?	Y / N	............amps
			Price of elec (if not inc)		............amps

Your comments & opinions may be used in future editions of the guide, if you do not wish them to be used please tick ....

Name ....

Membership No. ....

Post Code ....

Are you a Caravanner / Motorhomer / Trailer-Tenter?

Do you need more blank Site Report forms? YES / NO

Please return completed forms to:
The Editor – Overseas Touring Guides
East Grinstead House
East Grinstead
West Sussex
RH19 1FH
Please note that due to changes in the rules regarding freepost we are no longer able to provide a freepost address for the return of Site Report Forms. You can still supply your site reports free online by visiting camc.com/europereport.
We apologise for any inconvenience this may cause.

You can also complete forms online: camc.com/europereport

Abbreviated Site Report Form

Use this abbreviated Site Report Form if you have visited a number of sites and there are no changes (or only small changes) to their entries in the guide. If reporting on a new site, or reporting several changes, please use the full version of the report form. **If advising prices,** these should be for an outfit, and 2 adults for one night's stay. **Please indicate high or low season prices and whether electricity is included.**

Remember, if you don't tell us about sites you have visited, they may eventually be deleted from the guide.

Year of guide used	20..........	Page No.	....	Name of town/village site listed under	
Site Name				Date of visit	......./......./........
GPS	Latitude...(eg 12.34567) Longitude...(eg 1.23456 or -1.23456)				
Site is in: Andorra / Austria / Belgium / Croatia / Czech Republic / Denmark / Finland / France / Germany / Greece / Hungary / Italy / Luxembourg / Netherlands / Norway / Poland / Portugal / Slovakia / Slovenia / Spain / Sweden / Switzerland					
Comments:					

Charge for outfit + 2 adults in local currency	High Season	Low Season	Elec inc in price?	Y / N	.............amps
			Price of elec (if not inc)		.............amps

Year of guide used	20..........	Page No.	....	Name of town/village site listed under	
Site Name				Date of visit	......./......./........
GPS	Latitude...(eg 12.34567) Longitude...(eg 1.23456 or -1.23456)				
Site is in: Andorra / Austria / Belgium / Croatia / Czech Republic / Denmark / Finland / France / Germany / Greece / Hungary / Italy / Luxembourg / Netherlands / Norway / Poland / Portugal / Slovakia / Slovenia / Spain / Sweden / Switzerland					
Comments:					

Charge for outfit + 2 adults in local currency	High Season	Low Season	Elec inc in price?	Y / N	.............amps
			Price of elec (if not inc)		.............amps

Year of guide used	20..........	Page No.	....	Name of town/village site listed under	
Site Name				Date of visit	......./......./........
GPS	Latitude...(eg 12.34567) Longitude...(eg 1.23456 or -1.23456)				
Site is in: Andorra / Austria / Belgium / Croatia / Czech Republic / Denmark / Finland / France / Germany / Greece / Hungary / Italy / Luxembourg / Netherlands / Norway / Poland / Portugal / Slovakia / Slovenia / Spain / Sweden / Switzerland					
Comments:					

Charge for car, caravan & 2 adults in local currency	High Season	Low Season	Elec inc in price?	Y / N	.............amps
			Price of elec (if not inc)		.............amps

Please fill in your details and send to the address on the reverse of this form.

You can also complete forms online: camc.com/europereport

CUT ALONG DOTTED LINE

Year of guide used	20..........	Page No.	....	Name of town/village site listed under		
Site Name					Date of visit	......./......./........
GPS	Latitude...*(eg 12.34567)* Longitude...*(eg 1.23456 or -1.23456)*					
Site is in: Andorra / Austria / Belgium / Croatia / Czech Republic / Denmark / Finland / France / Germany / Greece / Hungary / Italy / Luxembourg / Netherlands / Norway / Poland / Portugal / Slovakia / Slovenia / Spain / Sweden / Switzerland						
Comments:						
Charge for outfit + 2 adults in local currency	High Season	Low Season	Elec inc in price?	Y / N	.............amps	
			Price of elec (if not inc)		.............amps	

Year of guide used	20..........	Page No.	....	Name of town/village site listed under		
Site Name					Date of visit	......./......./........
GPS	Latitude...*(eg 12.34567)* Longitude...*(eg 1.23456 or -1.23456)*					
Site is in: Andorra / Austria / Belgium / Croatia / Czech Republic / Denmark / Finland / France / Germany / Greece / Hungary / Italy / Luxembourg / Netherlands / Norway / Poland / Portugal / Slovakia / Slovenia / Spain / Sweden / Switzerland						
Comments:						
Charge for outfit + 2 adults in local currency	High Season	Low Season	Elec inc in price?	Y / N	.............amps	
			Price of elec (if not inc)		.............amps	

Year of guide used	20..........	Page No.	....	Name of town/village site listed under		
Site Name					Date of visit	......./......./........
GPS	Latitude...*(eg 12.34567)* Longitude...*(eg 1.23456 or -1.23456)*					
Site is in: Andorra / Austria / Belgium / Croatia / Czech Republic / Denmark / Finland / France / Germany / Greece / Hungary / Italy / Luxembourg / Netherlands / Norway / Poland / Portugal / Slovakia / Slovenia / Spain / Sweden / Switzerland						
Comments:						
Charge for outfit + 2 adults in local currency	High Season	Low Season	Elec inc in price?	Y / N	.............amps	
			Price of elec (if not inc)		.............amps	

Your comments & opinions may be used in future editions of the guide, if you do not wish them to be used please tick ....

Name ....

Membership No. ....

Post Code ....

Are you a Caravanner / Motorhomer / Trailer-Tenter?

Do you need more blank Site Report forms? YES / NO

Please return completed forms to:
The Editor – Overseas Touring Guides
East Grinstead House
East Grinstead
West Sussex
RH19 1FH
Please note that due to changes in the rules regarding freepost we are no longer able to provide a freepost address for the return of Site Report Forms. You can still supply your site reports free online by visiting camc.com/europereport.
We apologise for any inconvenience this may cause.

You can also complete forms online: camc.com/europereport

Abbreviated Site Report Form

Use this abbreviated Site Report Form if you have visited a number of sites and there are no changes (or only small changes) to their entries in the guide. If reporting on a new site, or reporting several changes, please use the full version of the report form. **If advising prices,** these should be for an outfit, and 2 adults for one night's stay. **Please indicate high or low season prices and whether electricity is included.**

Remember, if you don't tell us about sites you have visited, they may eventually be deleted from the guide.

Year of guide used	20……….	Page No.	……….	Name of town/village site listed under	
Site Name				Date of visit	……./……./……..
GPS	Latitude……………………………………….*(eg 12.34567)*	Longitude……………………………………..*(eg 1.23456 or -1.23456)*			
Site is in: Andorra / Austria / Belgium / Croatia / Czech Republic / Denmark / Finland / France / Germany / Greece / Hungary / Italy / Luxembourg / Netherlands / Norway / Poland / Portugal / Slovakia / Slovenia / Spain / Sweden / Switzerland					
Comments:					
Charge for outfit + 2 adults in local currency	High Season	Low Season	Elec inc in price?	Y / N	………….amps
			Price of elec (if not inc)		………….amps

Year of guide used	20……….	Page No.	……….	Name of town/village site listed under	
Site Name				Date of visit	……./……./……..
GPS	Latitude……………………………………….*(eg 12.34567)*	Longitude……………………………………..*(eg 1.23456 or -1.23456)*			
Site is in: Andorra / Austria / Belgium / Croatia / Czech Republic / Denmark / Finland / France / Germany / Greece / Hungary / Italy / Luxembourg / Netherlands / Norway / Poland / Portugal / Slovakia / Slovenia / Spain / Sweden / Switzerland					
Comments:					
Charge for outfit + 2 adults in local currency	High Season	Low Season	Elec inc in price?	Y / N	………….amps
			Price of elec (if not inc)		………….amps

Year of guide used	20……….	Page No.	……….	Name of town/village site listed under	
Site Name				Date of visit	……./……./……..
GPS	Latitude……………………………………….*(eg 12.34567)*	Longitude……………………………………..*(eg 1.23456 or -1.23456)*			
Site is in: Andorra / Austria / Belgium / Croatia / Czech Republic / Denmark / Finland / France / Germany / Greece / Hungary / Italy / Luxembourg / Netherlands / Norway / Poland / Portugal / Slovakia / Slovenia / Spain / Sweden / Switzerland					
Comments:					
Charge for car, caravan & 2 adults in local currency	High Season	Low Season	Elec inc in price?	Y / N	………….amps
			Price of elec (if not inc)		………….amps

Please fill in your details and send to the address on the reverse of this form.

You can also complete forms online: camc.com/europereport

CUT ALONG DOTTED LINE

Year of guide used	20..........	Page No.	....	Name of town/village site listed under	
Site Name				Date of visit	......./......./........
GPS	Latitude...(eg 12.34567) Longitude...(eg 1.23456 or -1.23456)				

Site is in: Andorra / Austria / Belgium / Croatia / Czech Republic / Denmark / Finland / France / Germany / Greece / Hungary / Italy / Luxembourg / Netherlands / Norway / Poland / Portugal / Slovakia / Slovenia / Spain / Sweden / Switzerland

Comments:

Charge for outfit + 2 adults in local currency	High Season	Low Season	Elec inc in price?	Y / N	.............amps
			Price of elec (if not inc)		.............amps

Year of guide used	20..........	Page No.	....	Name of town/village site listed under	
Site Name				Date of visit	......./......./........
GPS	Latitude...(eg 12.34567) Longitude...(eg 1.23456 or -1.23456)				

Site is in: Andorra / Austria / Belgium / Croatia / Czech Republic / Denmark / Finland / France / Germany / Greece / Hungary / Italy / Luxembourg / Netherlands / Norway / Poland / Portugal / Slovakia / Slovenia / Spain / Sweden / Switzerland

Comments:

Charge for outfit + 2 adults in local currency	High Season	Low Season	Elec inc in price?	Y / N	.............amps
			Price of elec (if not inc)		.............amps

Year of guide used	20..........	Page No.	....	Name of town/village site listed under	
Site Name				Date of visit	......./......./........
GPS	Latitude...(eg 12.34567) Longitude...(eg 1.23456 or -1.23456)				

Site is in: Andorra / Austria / Belgium / Croatia / Czech Republic / Denmark / Finland / France / Germany / Greece / Hungary / Italy / Luxembourg / Netherlands / Norway / Poland / Portugal / Slovakia / Slovenia / Spain / Sweden / Switzerland

Comments:

Charge for outfit + 2 adults in local currency	High Season	Low Season	Elec inc in price?	Y / N	.............amps
			Price of elec (if not inc)		.............amps

Your comments & opinions may be used in future editions of the guide, if you do not wish them to be used please tick ....

Name ....

Membership No. ....

Post Code ....

Are you a Caravanner / Motorhomer / Trailer-Tenter?

Do you need more blank Site Report forms? YES / NO

Please return completed forms to:
The Editor – Overseas Touring Guides
East Grinstead House
East Grinstead
West Sussex
RH19 1FH
Please note that due to changes in the rules regarding freepost we are no longer able to provide a freepost address for the return of Site Report Forms. You can still supply your site reports free online by visiting camc.com/europereport.
We apologise for any inconvenience this may cause.

You can also complete forms online: camc.com/europereport

Index

Index